INTERNATIONAL UNION OF THEORETICAL
AND APPLIED MECHANICS

DEFORMATION AND FLOW

OF SOLIDS

COLLOQUIUM MADRID SEPTEMBER 26-30, 1955

EDITED BY

R. GRAMMEL

WITH 188 FIGURES

SPRINGER-VERLAG BERLIN HEIDELBERG GMBH 1956

INTERNATIONALE UNION FÜR THEORETISCHE
UND ANGEWANDTE MECHANIK

VERFORMUNG UND FLIESSEN
DES FESTKÖRPERS

KOLLOQUIUM MADRID 26. BIS 30. SEPTEMBER 1955

HERAUSGEGEBEN VON

R. GRAMMEL

MIT 188 ABBILDUNGEN

SPRINGER-VERLAG BERLIN HEIDELBERG GMBH 1956

ISBN 978-3-662-38791-7   ISBN 978-3-662-39690-2 (eBook)
DOI 10.1007/978-3-662-39690-2

URSPRÜNGLICH ERSCHIENEN BEI SPRINGER-VERLAG OHG. IN BERLIN · GÖTTINGEN · HEIDELBERG 1956
SOFTCOVER REPRINT OF THE HARDCOVER 1ST EDITION 1956

Vorwort

Auf einer Sitzung des Büros der Internationalen Union für Theoretische und Angewandte Mechanik (IUTAM) in Brüssel am 21. September 1953 wurde beschlossen, ein Kolloquium über Festkörpermechanik (vorläufiger Titel) für die letzte Septemberwoche 1955 zu planen, und zwar in Madrid, entsprechend einer Einladung, welche die Vertreter der zugehörigen spanischen Organisation Instituto Nacional de Técnica Aeronáutica „Esteban Terradas" (INTA), die Herren A. NÚÑEZ und A. PÉREZ-MARÍN persönlich überbrachten.

Das vom Büro der IUTAM hierfür eingesetzte Wissenschaftliche Komitee bestand aus R. GRAMMEL als Vorsitzendem und den Herren TH. VON KÁRMÁN, M. ROY und G. I. TAYLOR, das spanische Organisations-Komitee aus den Herren A. NÚÑEZ und A. PÉREZ-MARÍN samt einem Stab technischer und administrativer Mitarbeiter.

In drei Beratungen des Wissenschaftlichen Komitees (Brüssel, 27. Juli 1954; Farnborough, 10. September 1954; Paris, 23. Mai 1955) wurde das Programm des Kolloquiums mit dem endgültigen Titel „Verformung und Fließen des Festkörpers" festgelegt und dann in den Tagen vom 26. bis 30. September 1955 gemäß dem Plan von Seite IX verwirklicht.

Jedes der drei Themen (Versetzungen und Plastizität; Nichtlineare Elastizität; Viskoelastizität und Relaxation) und insbesondere jede Sitzung wurde durch einen Allgemeinen Vortrag eingeleitet. Die Titel aller Vorträge gibt das Inhaltsverzeichnis Seite XI und XII an.

Die Vorträge wurden am 26., 27. und 29. September im Consejo Superior de Investigaciones Científicas in Madrid gehalten, am 28. September in Torrejón de Ardoz, verbunden mit einer Besichtigung der Laboratorien des INTA daselbst.

Die Liste der Teilnehmer des Kolloquiums aus den einzelnen Ländern, einschließlich der Vortragenden, ist auf Seite X zusammengestellt. Einige weitere Forscher aus anderen Ländern waren vom Wissenschaftlichen Komitee eingeladen worden, konnten aber nicht erscheinen.

Um die Reisekosten für die Teilnehmer des Kolloquiums wenigstens teilweise zu decken, stellte die UNESCO der IUTAM eine Summe von $ 4000 zur Verfügung. Die Auslagen für die technische Durchführung des Kolloquiums hat in dankenswerter Weise das INTA übernommen, dessen Gastfreundschaft alle Teilnehmer auch beim Empfang und Fest-

bankett in Madrid am 26. September, beim Mittagessen in Torrejón am
28. September und bei einem Ausflug zum Escorial mit Besuch des
berühmten Klosters, mit Mittagessen und mit Besichtigung des Valle de
los Caídos am 30. September in reichem Maße erfahren durften.

Das Wissenschaftliche Komitee möchte der UNESCO und dem INTA
seinen aufrichtigen Dank aussprechen für alle Unterstützung bei diesem
Kolloquium, das den – wohl in vollem Umfange erfüllten – Zweck hatte,
zwei verschiedene Forschertypen zusammenzubringen, die sich und ihre
Arbeitsweise gegenseitig bisher nur wenig kannten, deren Zusammen-
wirken aber weiterhin unerläßlich erscheint, nämlich die Festkörper-
Physiker und die Mechanik-Forscher.

Der Herausgeber der Vorträge dieses Kolloquiums drückt seinen
herzlichen Dank für viele Hilfe insbesondere den Herren PÉREZ-MARÍN,
E. PRADO und Dr. ZOLLER aus und nicht zuletzt dem Springer-Verlag
für seine großzügige Bereitwilligkeit, diesen Bericht zu verlegen und
in bewährter Form und in kurzer Frist der Fachwelt zu übergeben.

Stuttgart, Juni 1956
Technische Hochschule

R. Grammel

Preface

In a meeting held in Brussels on September 21, 1953, the Bureau of the International Union of Theoretical and Applied Mechanics (IUTAM) decided to organize a Colloquium on Solid State and Plasticity (provisional title) during the last week of September 1955. Following an invitation delivered personally by Don A. Núñez and Don A. Pérez-Marín, the representatives of the Spanish adhering organisation, Instituto Nacional de Técnica Aeronáutica "Esteban Terradas" (INTA), Madrid was chosen as location of the Colloquium.

The Scientific Committee named by the Bureau of IUTAM to plan this Colloquium consisted of R. Grammel, chairman, and Th. von Kármán, M. Roy and G. I. Taylor, members. The Spanish Organizing Committee was formed by A. Núñez and A. Pérez-Marín with a staff of technical and administrative collaborators.

In three meetings of the Scientific Committee (in Brussels on July 27, 1954, at Farnborough on September 10, 1954, and in Paris on May 23, 1955) the programme of the Colloquium was settled finally, and that its title should be "Deformation and Flow of Solids" and its dates September 26–30, 1955. Page IX shows the arrangement of the sessions.

Each Session devoted to the particular topics (Dislocations and Plasticity; Non-linear Elasticity; Viscoelasticity and Relaxation) was introduced by a General Lecture. The full titles of the papers may be found in the list of contents on page XI and XII.

The sessions of September 26, 27 and 29 were held at the Consejo Superior de Investigaciones Científicas in Madrid, the session of September 28 in Torrejón de Ardoz, connected with a visit to the laboratories of INTA at this place.

Page X shows the list of participants from the various countries including the lecturers. Several other scientists from other countries had been invited by the Scientific Committee, but were unable to take part in the Colloquium.

UNESCO contributed a sum of $ 4000 to the travelling expenses of participants of the Colloquium. In a most commendable manner INTA covered the expenses connected with the local organisation of the Colloquium. The participants also enjoyed the generous hospitality of INTA on the occasions of a reception and dinner on September 26 in Madrid,

of a lunch in Torrejón on September 28, and of a trip to El Escorial
with visit to the famous Monasterio, with lunch and an excursion to the
Valle de los Caídos on September 30. The Scientific Committee wishes
to express its sincere gratitude to UNESCO and to INTA for all their
support.

The Colloquium on Deformation and Flow of Solids has fully fulfilled
its main object, namely to bring together two groups of scientists which
hitherto had but little connection with each other and hardly knew of
the methods used in the other field: solid state physicists and research
workers in the field of mechanics. Their future cooperation appears to be
essential for further progress.

The editor of these proceedings extends his kindest thanks for much
help and assistance especially to Mr. PÉREZ-MARÍN, Mr. E. PRADO and
DR. ZOLLER and last though not least to the Springer-Verlag for being
willing to publish these proceedings very promptly and with an excellent
standard of production.

Stuttgart, June 1956
Technische Hochschule

R. Grammel

Plan des Kolloquiums — Schedule of the Colloquium

Tag Day	Vorsitzender und Stell- vertreter Chairman and Vice- Chairman	Gegenstand Topic	Vortragende Lecturers
26. IX. 1955 a. m.	R. Grammel M. A. Biot	Versetzungen und Plastizität Dislocations and Plasticity	G. I. Taylor (G. L.) B. Jaoul G. Leibfried
p. m.	H. L. Dryden F. Schultz-Grunow		A. H. Cottrell (G. L.) N. F. Mott J. R. Low jr. F. C. Frank H. G. van Bueren
27. IX. a. m.	M. Roy H. G. van Bueren		A. Seeger (A. V.) C. Crussard E. H. Lee F. Teissier du Cros
p. m.	G. I. Taylor G. Millán		P. G. Hodge jr. (G. L.) M. Velasco de Pando H. G. Hopkins H. Aroeste
28. IX. a. m.	A. Pérez-Marín G. Leibfried	Nichtlineare Elastizität und Vermischtes Non-linear Elasticity and Miscellaneous Problems	H. Kauderer (A. V.) M. Reiner
p. m.	R. Calvo Rodéz G. Millán		L. R. G. Treloar (G. L.) M. S. Plesset E. Volterra
29. IX. a. m.	M. Velasco de Pando E. Volterra	Viskoelastizität und Relaxation Viscoelasticity and Relaxation	M. A. Biot (G. L.) H. le Boiteux F. Schultz-Grunow (A.V.)
p. m.	Th. v. Kármán R. Grammel		I. A. Oding J. G. Oldroyd W. P. Mason

G. L. = General Lecture — A. V. = Allgemeiner Vortrag

Verzeichnis der Teilnehmer — List of Participants

Belgien — Belgium:

P. JANSSENS, J. KESTENS.

Deutschland — Germany:

U. DEHLINGER, R. GRAMMEL, H. KAUDERER, G. LEIBFRIED, K. MARGUERRE, F. SCHULTZ-GRUNOW, A. SEEGER, W. ZERNA.

Frankreich — France:

C. CRUSSARD, R. JACQUESSON, B. JAOUL, M. KIEFFER, H. LE BOITEUX, M. ROY, F. TEISSIER DU CROS.

Großbritannien — Great Britain:

A. H. COTTRELL, F. C. FRANK, H. G. HOPKINS, L. HIMMEL, N. F. MOTT, J. G. OLDROYD, G. I. TAYLOR, L. R. G. TRELOAR.

Israel:

M. REINER.

Niederlande — Netherlands:

H. G. VAN BUEREN, C. ZWIKKER.

Saar:

P. LAURENT.

Schweden — Sweden:

S. O. ASPLUND, K. E. C. NIELSEN.

Spanien — Spain:

J. L. AMORÓS, J. APRÁIZ, E. BALTÁ, F. CACHO, R. CALVO RODÉS, J. DÍAZ, M. FONT-ALTABA, E. GARCÍA SARDINERO, A. GARCÍA POGGIO, L. GONZÁLEZ VÁSQUEZ, M. LÓPEZ RODRÍGUEZ, G. MILLÁN, A. MORA, F. MUÑOZ DEL CORRAL, G. NAVACERRADA Y FARIAS, A. NÚÑEZ RODRÍGUEZ, J. PALACIOS, A. PÉREZ-MARÍN, A. PLANA SANCHO, J. SORIANO SÁNCHEZ, J. TERRAZA, M. VELASCO DE PANDO, L. VILLENA.

Union der sozialistischen Sowjetrepubliken — Soviet Russia:

I. A. ODING

Vereinigte Staaten von Amerika — United States of America:

H. AROESTE, M. A. BIOT, H. L. DRYDEN, P. DUWEZ, P. G. HODGE JR., TH. VON KÁRMÁN, H. LAMPERT, R. LATTER, E. H. LEE, J. R. LOW JR., L. MCKENZIE, W. P. MASON, M. S. PLESSET, PH. J. THEODORIDES, E. VOLTERRA, CH. F. YOST.

Inhaltsverzeichnis – Contents

III
Viskoelastizität und Relaxation
Viscoelasticity and Relaxation

Bemerkung: Zahlen in eckigen Klam-   Note: Numbers in brackets refer to the
mern beziehen sich jeweils auf die Lite-   References (Literatur, Bibliographie) at
ratur (References, Bibliographie) am   the end of each paper.
Schluß des entsprechenden Aufsatzes.

Einführung

Von R. Grammel, Stuttgart

Ich möchte dem Programm dieses Kolloquiums einige allgemeine Bemerkungen vorausschicken, in denen ich versuchen will, zu skizzieren,
wo wir heute stehen.

Die Mechanik, nach der unübertrefflichen Definition von KIRCHHOFF
die Wissenschaft von den Kräften und Bewegungen (zu denen man natürlich auch die Spannungen und Deformationen rechnen muß), war stets
bestrebt, eine exakte Wissenschaft zu sein, d. h. sich durch rein logische
Deduktionen aus wenigen Axiomen aufzubauen, die unsere Erfahrungen
kodifizieren sollten.

In ihrer klassischen Zeit kam die Mechanik diesem Ideal einer exakten
Wissenschaft sehr nahe, und zwar hauptsächlich durch bestimmte *Idealisierungen*, unter denen man mehrere Rangstufen ohne weiteres unterscheiden kann, etwa absolute Idealisierungen, wie den starren Körper
und die ideale Flüssigkeit, oder relative Idealisierungen, wie den ideal-
elastischen, den ideal-plastischen, den starr-plastischen Körper, die
NEWTONsche Flüssigkeit usw., — Idealisierungen, die mathematisch teilweise zugleich Linearisierungen waren.

Man darf es nun als ein charakteristisches Entwicklungsmerkmal der
Mechanik in den letzten Jahrzehnten ansehen, daß sie daran ging, diese
physikalischen und mathematischen Idealisierungen Schritt für Schritt
abzubauen und sich also aus einer *Idealmechanik* zu einer *Realmechanik*
weiterzuentwickeln.

Sie begnügt sich auch schon lange nicht mehr damit, von ihren Fun-
damenten aus durch Lösen von immer neuen Einzelproblemen gewisser-
maßen nur nach oben zu bauen, — sie ist vielmehr mit großer Mühe an
die Revision und Vertiefung ihrer Fundamente selbst gegangen.

Sie ist dabei, die Axiomatik ihres Fundamentes, also ihre materiellen
Grundgesetze, tiefer zu begründen oder, wie man es auch ausdrücken
kann, ihre klassische Axiomatik, die sich um das Warum der Grundgesetze kaum kümmerte, durch eine tiefere Axiomatik zu unterbauen,
die auch dem eigentlichen Wesen der Grundgesetze nachspürt.

Sie nimmt nicht mehr etwa die Elastizitäts- und Plastizitätsgesetze
als gegebene und unerklärte Notwendigkeiten hin, sondern sie will wissen,

wie sich solche Gesetze aus der elektromagnetischen Struktur der Materie ergeben.

Sie nimmt den festen Körper oder die Flüssigkeit und ihre Zwischenzustände nicht mehr einfach als mathematische Kontinua, als Homogenitäten hin, sondern will ihr mechanisches Verhalten aus ihrem molekularen und atomistischen Aufbau, von ihrer kristallinen und kristallitischen Struktur her verstehen.

Sie will wissen, *was* Elastizität, Plastizität, Verfestigung, Sprödigkeit, Relaxation, Viskosität, Thixotropie usw. „wirklich" sind, d. h. wie sie mit den uns geläufigen und verständlichen Grundphänomenen der Physik zusammenhängen und sich womöglich aus ihnen berechnen lassen.

Daß bei dieser neuen Grundlegung der Mechanik des festen Körpers, an der wir zur Zeit arbeiten, bei diesem Fortschreiten der Mechanik von dem nur gedachten zum wirklichen Körper ihre Exaktheit als Wissenschaft an manchen Stellen, wenn auch nicht ganz verlorenging, so doch vorübergehend mehr oder weniger eingeschränkt worden ist, das mag den klassisch erzogenen Mechanikforscher heute wohl sehr bestürzen; aber er kann sich vielleicht mit der Tatsache trösten, daß bei dieser Gelegenheit doch auch Risse im klassischen Fundament entdeckt und beseitigt worden sind.

Und auch er muß — trotz der nicht zu leugnenden Unvollkommenheit des heutigen Zustandes der Festkörpermechanik — die Kühnheit der Gedanken anerkennen, mit der unsere Wissenschaft um eine neue und tiefere Exaktheit ringt.

Unser Programm zeigt deutlich, daß die Mechanik, die früher durchaus eine deduktive Wissenschaft war, heute eine Komposition von empirischer Induktion und theoretischer Deduktion geworden ist, eine Komposition, zu der nunmehr zwei bisher fast immer getrennt nachdenkende Forschertypen sich zusammentun müssen.

Auch für den klassisch gebildeten Forscher gibt es bei dieser Komposition Aufgaben in Hülle und Fülle, wie etwa, um nur einige Beispiele herauszugreifen, die noch ungelösten Probleme einer rationellen, das ganze Gebiet des Festkörpers umfassenden Thermodynamik, oder die statistische Theorie des Polykristalls, oder die Vereinigung der beiden nichtlinearen Zweige der Elastomechanik, also der nicht-Hookeschen Elastizitätstheorie und der Theorie der finiten Deformationen.

Daß bei der Festkörpermechanik heute oft der Physiker gleichberechtigt neben dem Mechanikforscher alten Stils zu Wort kommt, muß beide Partner mit dem Bewußtwerden einer fruchtbaren Zusammenarbeit erfüllen; und gerade die gegenseitige Ergänzung in einem solchen Zusammentreffen zu finden, ist das wesentliche Ziel dieses Kolloquiums.

I

Versetzungen und Plastizität

Dislocations and Plasticity

Strains in Crystalline Aggregate

By G. I. Taylor, Cambridge (England)

1. Aluminium. Much study has been devoted in many countries to the mechanism of plastic strain in crystals by the formation and movement of dislocations. There has also been much experimental work on the macroscopic relationships between stress, strain and the crystal axes of single crystals. With this background of knowledge the man who contemplates a crystal aggregate and tries to understand how it behaves when forced to flow, is in much the same position as one who knows about the kinetic theories of gases and liquids, understands how the statistical effects of molecular collisions are averaged to yield the laws of viscosity, is familiar with POISSON's equations for the flow of viscous fluids, and then sets out to study turbulence. Dr. COTTRELL has used this analogy in a similar case. Even with the very simple laws summarised in POISSON's equations it has proved impossible so far to make any complete representation of turbulent fluid flow. Indeed some people working on that subject have felt so discouraged at their lack of success in finding significant results based on POISSON's equations that they have concluded that these equations do not apply to turbulent flow and have tried to start again, as in the kinetic theory of gases, from known properties of molecules. Such people have had no success in these despairing efforts; nor are they likely to succeed, because POISSON's mathematical expression for the stress-rate of strain relations in a fluid represent the simplest possible averaging process for molecular interactions. The sole reason why we have not succeeded in giving a complete description of turbulent flow is not because we don't know enough about fluids, but because we are not clever enough to find the appropriate solutions of POISSON's equations.

Though this analogy between turbulent flow of fluids and the flow of an aggregate is not an exact one, it does illustrate the reason why workers on aggregates are forced to start with idealised macroscopic

1*

stress-strain relationships determined experimentally with single crystals
rather than attempt to start with dislocations. Such studies will not take
account, in the first instance, of the actual physical conditions at the
grain boundaries, regarding them merely as surfaces at which the displacement and three out of the six components of stress are continuous. Ultimately if a self-consistent description of an aggregate of crystals with
idealised properties can be built up, it may be possible to go further and
apply knowledge gained by the molecular theorists and experimenters on
the physical conditions at crystal boundaries, to aggregates but that is
merely a guess at future developments.

Experiments in which single crystals of certain metals, particularly
those with the face-centred cubic arrangement atoms, are subjected to
uniaxial stress, have established that deformation occurs by slipping in
one direction parallel to one crystal plane and that of all crystallographic-
ally similar planes the slipping occurs only on the one for which the shear
stress in the direction of the slip is greatest. The slipping produces a
rotation of the axes of the specimen relative to the axes of the crystal.
This rotation ultimately brings a second possible slip plane into the most
favourable position for slipping. When this occurs the slipping begins on
the second plane also. These observational laws of macroscopic strain in
certain single crystals have been deduced entirely by measurements of
strain without any reference to surface "slip lines", but these "slip lines"
do, in fact, mark out the intersections of the plane of slip, determined
geometrically from measurements on the surface of the specimen.

In attempting to apply these laws of observation for single crystal
grains to deduce the laws of plasticity of a polycrystalline mass, regarded
as a continuum, some workers have simply imagined each crystal grain
as being subjected to sufficient uniaxial stress in the direction of the
macroscopic stress to make it yield. They then imagined the axes of grains
to be orientated at random and each grain to be unaffected by its neighbours. With this model the sum of the forces due to each grain over any
section of the specimen is equal to the load on it and the stress is $m\,T$
where T is the yield stress for single slipping of each crystal. The value
of m so calculated by SACHS [1] (1928) is $m = 2.238$. This model has little
relationship to polycrystals because neither the stress nor the strain is
continuous at the grain boudaries and grains which originally fitted together will not do so after straining. It has however one advantage, it does
definitely give a lower bound to the value of m because the crystal is
essentially a frictional system and it is a general mechanical principle that
removal of constraints anywhere in such a system decreases the work
necessary to make any given strain in it.

To make irregular grains fit together a combination of small elements
of slipping is necessary at all points. If the strain of a grain boundary is

prescribed and corresponds with a surface of particles in a continuum subjected to a uniform strain, then the material inside the boundary can itself be strained uniformly, but the conditions of stress at the boundary will not be satisfied. It is clear therefore that in the actual polycrystal the strain in an irregular grain would not be uniform even if the strain of its boundary were. On the other hand it is possible to maintain macroscopic continuity within the material by appropriate combinations of slipping on crystal planes whatever the distribution of strain may be inside the grains. In general the slip combinations will vary from point to point in the grain and since each such combination will cause the crystal axis to rotate relative to the axes of strain the grains must necessarily cease to be true crystals and exhibit a spread of crystal directions through their volume. The extreme complexity of the problem makes me doubt whether even a valid statistical picture of the strain in polycrystal in which the stress and strain conditions are satisfied both at the boundaries and within the substance, will ever be attained. Certainly none has yet been proposed.

Failing a complete description it seemed to me some years ago (TAYLOR [5], 1938 a) that one could calculate an upper bound to the stress in an aggregate of face-centred cubic crystals using the principle of virtual work. There is an infinity of ways in which combinations of slips on crystal faces can give rise to a given external strain of an aggreate and satisfy the conditions of continuity of displacement at crystal boundaries. If any one of these occurs the energy wasted is $\sum T ds$ where T is the shear stress in the direction of slipping and ds is an element of shear strain. This must be equal to the work done by the external forces. For an aggregate extending in one direction with longitudinal strain $d\varepsilon$ under a longitudinal stress P the work done is $P d\varepsilon$. Hence in the restricted case where T is constant

$$\frac{P}{T} = \frac{\sum |ds|}{d\varepsilon}.$$

It seemed to me (TAYLOR [5, 6] 1938 a and b) that if we could determine the combination of slips which produce the imposed external strain $d\varepsilon$ and had the minimum value of $\sum |ds|$ we should have solved the problem. All other combinations of slips which satisfy the conditions of continuity at the boundaries of grains would require a higher value of P than $(T/d\varepsilon) \cdot$ (minimum value of $\sum |ds|$) to produce deformation. Thus an upper bound to P might be obtained by finding any combination of slips which satisfied the external strain condition and the condition of continuity of displacement at the grain boundaries. I did not consider the question of whether the stress on all active slip planes could be the same because I imagined that the small elements of slip might occur successively rather than simultaneously. As the stress in the polycrystal increases the shear stress on one plane might rise to the value at which a small slip would occur. This would alter the stress distribution in the same crystal and raise the stress on another plane to the necessary value for slipping. Thus I supposed one could discuss the possibility that during an arbitrary strain an appropriate combination of slips might occur without thinking whether a stress system could occur in which the stress was raised to the slipping value on all active slip planes *simultaneously*. It will be seen later that this question has been investigated much more thoroughly (BISHOP and HILL [3] 1951).

The only kind of strain satisfying continuity of displacement at grain boundaries which I saw any hope of discussing was that in which the strain of each grain was identical with that of the aggregate. I therefore attempted to find the minimum value of $\sum |ds|$ for a crystal whose crystal axes relative to the strain axes of the aggregate was given. In a face-centred cubic crystal there are assumed to be 12 possible slips three on each of the (111) planes in the directions of the diagonals of square faces of the fundamental cube.

Taking the four planes as (111), a; ($1\bar{1}\bar{1}$), b; ($\bar{1}1\bar{1}$), c; ($\bar{1}\bar{1}1$), d; and the directions of slip numbered with suffixes 1, 2, 3 according as they lie in the planes (100), (010), (001) the 12 slips are defined as $a_1 a_2 a_3$, $b_1 b_2 b_3$, $c_1 c_2 c_3$, $d_1 d_2 d_3$ and the components of strain relative to the cubic axes are

$$\left.\begin{aligned}
e_{xx} &= \frac{\partial u}{\partial x} = & a_2 - a_3 + b_2 - b_3 + c_2 - c_3 + d_2 - d_3, \\
e_{yy} &= \frac{\partial v}{\partial y} = & - a_1 + a_3 - b_1 + b_3 - c_1 + c_3 - d_1 + d_3, \\
e_{zz} &= \frac{\partial w}{\partial z} = & a_1 - a_2 + b_1 - b_2 + c_1 - c_2 + d_1 - d_2, \\
2e_{yz} &= \frac{\partial w}{\partial y} + \frac{\partial v}{\partial z} = & - a_2 + a_3 - b_2 + b_3 + c_2 - c_3 + d_2 - d_3, \\
2e_{zx} &= \frac{\partial u}{\partial z} + \frac{\partial w}{\partial x} = & a_1 - a_3 - b_1 + b_3 + c_1 - c_3 - d_1 + d_3, \\
2e_{xy} &= \frac{\partial v}{\partial x} + \frac{\partial u}{\partial y} = & - a_1 + a_2 - b_1 - b_2 + c_1 - c_2 - d_1 + d_2.
\end{aligned}\right\} \quad (1)$$

The object of my work was to find the minimum value of the sum of the absolute values of the twelve slips, subject to the condition (1). It was first shown that this minimum must be one of the cases in which only 5 of the 12 are active, but the number of combinations of 12 things taken 5 at a time is 792 so that the task of selecting the correct choice is likely to be very laborious. This number is considerably reduced by various considerations. First since any direction in a plane can be compounded out of two components at 120° apart, only two slip components are needed from any one slip plane, and corresponding with any given magnitude for an arbitrary shear on, say, the plane a either the vector sum of $a_1 + a_2$ or $a_2 + a_3$ or $a_3 + a_1$ can be used, but if the same shear be added to each of a_1, a_2 and a_3 there are three ways in which the arbitrary shear can be composed of two only, and one can chose the pair which is least of $|a_1| + |a_2|$ or $|a_2| + |a_3|$ or $|a_3| + |a_1|$ without affecting any of the choices of the remaining 3 of the combinations of 5 shears. This reduces very greatly the number of choices for which (1) must be solved, in fact there turn out to be 24 independent choices in which the five shears are chosen so that 2, 2, 1, 0 are on each of the four slip planes respectively. If the choices are made so that 2 are on one plane and one on each of the remaining planes there are 108 of them but 36 of these are not capable of combining into an arbitrarily chosen strain tensor, so that 72 is the correct number, making 96 in all. In my papers on the subject (TAYLOR [5, 6] 1938 a and b) I unfortunately made a mistake in signs in equation (1) and this had the effect of making me draw the false conclusion that choices in which two shears occur on only one of the planes are unable to combine to form an arbitrarily chosen strain tensor. It appears however (BISHOP and HILL [3] 1951) that I did in fact in most cases find the least value of $\sum |ds|$. In most orientations of crystal axes I found in fact usually two different combinations giving the least value. Taking crystal axes with approximately uniform distribution of orientations I obtained a number approximately $m = 3.06$ as an upper bound for the ratio of tensile stress of the aggregate to the shear resistance of a single crystal. Surprisingly this upper bound is very close to the ratio of the observed strength of an aggregate of aluminium to that of a single crystal of the same material.

Though the theory predicted the strength of an aggregate it was incapable of predicting anything about the deformation texture (i.e. the way in which the distribution of orientations of crystal axes is altered by straining) because when the same minimum virtual work is found for several different slip combinations each of these would give rise to different rotations of the crystal axes relative to the specimen. Unless some method can be devised for distinguishing between these combinations or unless the directions of rotation of axes due to the different minimum combinations are not very different from one another no predictions could be made.

Later some theories were put forward (CALNAN and CLEWS [4] 1950, KOCHENDÖRFER [2] 1941) to account for the deformation texture of strained aggregates, none of them could be taken seriously by a mathematically minded person because they all involved strains in which the grains would not fit together after the strain had happened. These authors recognised this, but it did not seem important to them. Their arguments roughly were that there are great irregularities near the grain boundaries so one may perhaps assume that some of the material there flows like a fluid so as to permit single or double slipping to occur without loss of continuity. Such theories, to my mind, have little value. They do not give a true lower bound to the value of m as SACHS [1] does. They do not give a true upper bound as my theory and BISHOP and HILL's theory (which ultimately lead to the same result) do.

This was the position till J. F. W. BISHOP and R. HILL [3] (1951) pointed out that the equations which relate the components of shear stress on the crystal planes in the directions of possible slipping to the six components of stress, referred to the cubic axes, are exactly the same as equations (1) when the six components of strain are replaced by those of stress and the elements of slip by the components of shear stress parallel to the slip directions.

Using this theorem and assuming that the shear stress on each of the active planes were the same, they were able to define 56 states of stress in each of which the components of shear stress were the same on 5 slip planes. They used the principle that the one among the 56 which corresponds with the *greatest* virtual work consistent with a given externally applied strain is the one which operates. Taking as an example of their method the case when the applied strain is an extension in one direction and contractions in two directions perpendicular to it, they were able to 'divide the fundamental triangle of the face-centred cube into regions such that when a point representing the direction of extension fell in one region the particular member of the group of 56 was known.

This very interesting method of analysis yields the same physical result as mine. It uses the same physical assumption that the strain of every grain is the same as that of the aggregate. It gives the same upper bound for m, namely 3.06. It would give the same limits of uncertainty in the predicted deformation texture. In fact the deformation texture diagram given by BISHOP [7] (1954) shows a greater amount of dispersion than mine (TAYLOR [5, 6] 1938) but this is due to the fact that I missed some of the possible combinations of 5 slips. The physical picture presentes is clearer than mine because the distribution of stress among the possible slip planes is considered. Continuity of strain at the boundaries of grains is preserved in both, but continuity of stress is not preserved in either.

A further development in BISHOP and HILL's studies was the calculation of the deviation of the plastic properties of a crystal aggregate regarded as a continuum from the simplest law of ideal plastic continua, namely Mises assumption that the stress tensor and the rate of strain tensor are identical. In this they were not quite so successful as they and

I were in predicting the value of m, but they showed that a crystal aggregate should differ from the ideal Mises plastic body in the way that both copper and aluminium have been observed to do (TAYLOR and QUINNEY [9] 1931, LODE [10] 1926).

With regard to deformation texture the theories of BISHOP, HILL and myself are not capable of predicting anything quantitative because they do not contain any clue as to which of many equidalent combinations of 5 slips operates. BISHOP [7] (1954–55) has analysed the effect of supposing that the choice is determinded by differential hardening of active and passive slip planes, but it can hardly be said that anything very decisive has yet come out of this suggestion.

2. Polycrystalline iron. Experiments with single crystals of iron have established the following facts (TAYLOR and ELAM [8] 1926).

1. When a unidirectional tension is applied the strain is due to slipping on one plane.

2. The direction of slip is one of the four [111] axes.

3. The plane of slip, as determined by external measurements, is not a crystallographic plane. It is very nearly that one of all the planes which pass through the crystallographic [111] axes on which the component of shear stress is a maximum.

This last statement has been criticised by people who have observed the slip lines on surfaces of strained single crystals and identified them as being the traces of various crystallographic planes, of type (101), (121), (231) which contain axes normal to the (111) type planes. This may well be the case but it has no relevance to the question, since the statement depends only on the results of careful measurements of external marks made for the purpose of analysing the total macroscopic strain. How this strain is attained by internal slippings, the motion of dislocations, etc. is an interesting physical question but it is unnecessary to have a complete picture of the microscopic internal condition when formulating laws which represent the results of experiments on macroscopic stress and strain relationships. The experimental result (3) must be interpreted in the sense that the shear stress required to operate a shear strain parallel to any arbitrary plane through the direction [111] is very nearly independent of the orientation of that plane. This result is not generally true for all body-centred cubic metals, in particular it is not true for beta-brass, but it is true to the limit of accuracy of my measurements for iron. The result (3) is not inconsistent with the principle that slippings proceed over crystal planes. Slipping over an arbitrary plane passing through a given [111] axis can occur owing to an appropriate combination of the two nearest crystallographic slip planes.

1b

The maximum angle between the possible neighbouring slip planes [101], [121], [123] is $19°6'$ so that if the resistance to slip on each of those planes were the same, the maximum possible variation in the stress required to operate an arbitrary plane passing through the 111 direction is $1 - \cos 19°33' = 1 - 0.986 = 0.014$ or 1.4 per cent. Thus the experimental law (3) is so slightly at variance with the hypothesis that the shear on a arbitrary plane is due to an appropriate combination of two slips on crystal planes that no existing technique would be capable of distinguishing between them. The advantage in analysis of using the experimental law (3) instead of considering all possible slips on crystal planes is very great. C. S. Barrett for instance (1952) remarks that with four [111] slip directions and 12 possible slip planes passing through each of them, the number of independent slip systems is 48 and since the number of combinations of 48 things taken 5 at a time is 1,712,304 he suggests that the equations connecting compatibility of an arbitrary strain with 5 slip systems would have to solved be this number of times. If, however, the experimental law of slip in single crystals is used, only four independent slips, say s_1, s_2, s_3, s_4, on planes of arbitrary orientation need be considered. Thus the 48 variables are reduced to 8, namely s_1, s_2, s_3, s_4 together with four variables $\alpha_1, \alpha_2, \alpha_3, \alpha_4$, specifying the orientations of the slip planes.

The first step in analysing the problem is to set forth the strain equations connecting an arbitrary strain, expressed in terms of the cubic axes, with the four shears s_1, s_2, s_3, s_4 parallel to the cube diagonal axes $[111], [1\bar{1}\bar{1}], [\bar{1}1\bar{1}], [\bar{1}\bar{1}1]$. The equations of planes through these are

$$
\left.
\begin{aligned}
0 &= \quad x + \alpha_1 y - (1 + \alpha_1) z \quad \text{passes through} \quad x = \quad y = \quad z, \\
0 &= \quad x - \alpha_2 y + (1 + \alpha_2) z \quad \text{passes through} \quad x = -\,y = -\,z, \\
0 &= -\,x + \alpha_3 y + (1 + \alpha_3) z \quad \text{passes through} \quad -\,x = \quad y = -\,z, \\
0 &= -\,x - \alpha_4 y - (1 + \alpha_4) z \quad \text{passes through} \quad -\,x = -\,y = \quad z.
\end{aligned}
\right\} \quad (2)
$$

The strain equations become, after some reduction

$$
\left.
\begin{aligned}
e_{xx} &= \frac{\partial u}{\partial x} = S_1 + S_2 + S_3 + S_4, \\[4pt]
e_{yy} &= \frac{\partial v}{\partial y} = \alpha_1 S_2 + \alpha_2 S_2 + \alpha_3 S_3 + \alpha_4 S_4, \\[4pt]
e_{yz} &= \frac{\partial v}{\partial z} + \frac{\partial w}{\partial y} = -S_1 - S_2 + S_3 + S_4, \\[4pt]
e_{zx} &= \frac{\partial w}{\partial x} + \frac{\partial v}{\partial z} = -\alpha_1 S_1 + \alpha_2 S_2 - \alpha_3 S_3 + \alpha_4 S_4, \\[4pt]
e_{xy} &= \frac{\partial u}{\partial y} + \frac{\partial v}{\partial x} = (1 + \alpha_1) S_1 - (1 + \alpha_2) S_2 - (1 + \alpha_3) S_3 + (1 + \alpha_4) S_4,
\end{aligned}
\right\} \quad (3)
$$

where

$$S_n = \frac{s_n}{\sqrt{6\,(1 + \alpha_n + \alpha_n^2)}} \qquad (n = 1, 2, 3, 4)\,. \tag{4}$$

The virtual work is

$$W = T \sum_n |s_n| = T \sum_n |S_n|\, \sqrt{6\,(1 + \alpha_n + \alpha_n^2)} \tag{5}$$

where T is the shear stress necessary to operate the shear on any one of the planes (2).

If each crystal is forced by external constraints to assume the strain of the aggregate the work done by stresses over its surface is fully accounted for by the stresses taken into account in the expression for virtual work which on the aggregate so that, for instance, if the total strain of the aggregate is that due to a small extension ε along one direction and contractions $1/2\varepsilon$ in all perpendicular directions. The stress P which is necessary to extend the specimen is given by

$$P\varepsilon = T \sum |s|\,. \tag{6}$$

For the same reasons that held in the case of the face centred cubic metals the value of P found from (6) is greater than the true value so that an upper bound to P is found by choosing the smallest value of $\sum |s|$ consistent with (3). In the face-centred case this minimum always occurs when the minimum number, 5, of strains consistent with the strain equations (1) is operative. This is a direct consequence of the fact that the strain equations (1) are linear. The five equations (3) are not linear in all the 8 variables, so that the minimum value of $\sum |s|$ does not necessarily occur when, say, one of the four strains s_1, s_2, s_3, s_4 is zero.

On the other hand if 5 variables, say S_1, S_2, S_3, S_4 and $\alpha_4 S_4$ are regarded as independent (3) can be solved as linear equations to obtain, their values in terms of α_1, α_2, α_3 and the given components of strain. If these be substituted in (5) the problem reduces to that of finding the minimum value of an expression involving α_1, α_2 and α_3 as they are varied.

The problem is therefore that of determining a true minimum value as three parameters vary continuously. This is quite different from that of the face-centred cubic metals in which the least value was required for an expression containing 5 variables selected from 12.

References

[1] SACHS, G.: Z. VDI 72, 734 (1928).
[2] KOCHENDÖRFER, A.: Plastische Eigenschaften von Kristallen, S. 22, Berlin 1941.
[3] BISHOP, J. F. W. and R. HILL: Phil. Mag. 42, 414 (1951).

[4] Calman, E. A. and C. T. B. Clews: Phil. Mag. 41, 1085 (1950).

[5] Taylor, G. I.: J. Inst. Met. 62, 307 (1938).

[6] Taylor, G. I.: Timoshenko 60th Anniversary Volume, p. 218. New York 1938.

[7] Bishop, J. F. W.: J. Mechan. Phys. Solids 3, 130, 259 (1954).

[8] Taylor, G. I. and C. F. Elam: Proc. roy. Soc., Lond., Ser. A 112, 337 (1926).

[9] Taylor, G. I. and Y. H. Quinney: Phil. Trans. roy. Soc., Lond., Ser. A 230, 323 (1931).

[10] Lode, E.: Z. Phys. 36, 913 (1926).

Diskussionsbemerkung zum Vortrag von Sir G. I. Taylor

Von U. Dehlinger, Stuttgart

Bei der Mittelbildung, die nötig ist, wenn man die Fließverfestigung des Vielkristalls aus der des Einkristalls berechnen will, wird oft vorausgesetzt, daß der Verfestigungszustand im Einkristall isotrop ist, d. h. daß die Verfestigung der latenten Gleitsysteme gleich der der betätigten ist. Nach früheren Einkristallversuchen über Doppelgleitung ist dies bei Aluminium zu erwarten, während bei Kupfer und noch mehr bei α-Messing die latente Verfestigung größer als die betätigte ist, was inzwischen nach A. Seeger auch atomistisch erklärt werden konnte.

Diese Isotropie läßt sich nun für das vielkristalline Material experimentell nachprüfen, da sie nach R. Hill die direkte Voraussetzung des Fließgesetzes bei mehrachsigem Spannungszustand ist, das sich nach Prager in der Form schreibt:

$$\frac{d\varepsilon_{ik}}{d\sigma_2} = \frac{1}{C(\sigma_2)}\,\sigma_{ik}.$$

Darin ist σ_2 die invariante Wurzel aus der sog. Gestaltänderungsenergie. Diese Beziehung besagt, daß man für alle Spannungswege $\int d\sigma$, auch solche, bei welchen die Achsen des Spannungstensors σ_{ik} gedreht werden, die gleiche Verfestigungskurve erhält, wenn man die Invariante σ_2 gegen das Integral $\int(d\varepsilon)_2$ der Invarianten der Dehnungsinkremente $d\varepsilon_{ik}$ aufträgt. Außerdem fordert sie, daß bei Drehung der Achsenrichtung des Spannungstensors und konstantem σ_2 keine Deformation auftritt.

Alle diese Folgerungen wurden an vielkristallinem Al bei Fließversuchen mit Überlagerung von Torsion und Zug bis zu einer Gesamtdehnung von 10% mit einer Genauigkeit von 3% bestätigt. Dagegen zeigten sich bei α-Messing (33% Zn) charakteristische Abweichungen: Die Verfestigungskurve für reinen Zug ist um etwa 8% höher als die für reine Torsion. Außerdem hat eine Drehung der Achsen des Spannungstensors eine zusätzliche Verfestigung der weiteren Dehnung von etwa 10% zur Folge. Beide Effekte zeigen, daß der Verfestigungszustand nicht mehr ganz isotrop ist, sondern eine durch die anfänglichen Deformationsrichtungen bestimmte eingeschränkte Symmetrie aufweist[1].

---

[1] Vgl. W. Sautter, A. Kochendörfer u. U. Dehlinger: Z. Metallkde. 44, 44, 442 (1953). — U. Dehlinger, J. Diehl u. J. Meissner: Z. Naturforsch. 11a, 37 (1956).

Influence des joints intergranulaires sur l'écrouissage des métaux

Par **B. Jaoul**, Paris

Avec 11 figures

1. Introduction. La détermination de la courbe de traction d'un polycristal en fonction de celle du monocristal, c'est-à-dire l'étude de l'influence des joints sur la consolidation a fait l'objet de diverses théories. Nous rappelerons d'abord les deux principales dues, l'une à TAYLOR [1], l'autre à KOCHENDÖRFER [2]; puis, par une analyse de la forme des courbes de traction, nous chercherons à déterminer comment varie la consolidation intergranulaire et par quel schéma il est possible de la représenter.

D'après TAYLOR, un grain d'un polycristal ne peut pas se déformer comme un cristal isolé par suite de la présence des grains voisins, mais, quelle que soit son orientation, on peut admettre que sa déformation d'ensemble est la même que celle de l'éprouvette; dans ces conditions, il faut faire intervenir simultanément dans un grain cinq systèmes de glissement. TAYLOR a calculé quelle devait être leur répartition dans le cas des métaux cubiques à faces centrées et, des diverses solutions possibles, il a choisi celle qui correspond au moindre glissement total. En supposant que le cisaillement critique soit le même pour tous les systèmes actifs, il a calculé la résistance du polycristal en fonction de celle du monocristal. La courbe obtenue correspond à peu près aux résultats expérimentaux pour les grandes déformations.

Mais cette théorie comporte de nombreuses hypothèses qui ne semblent pas pouvoir être toujours admises. En effet, d'après les observations micrographiques, tous les grains ne se déforment pas de la même manière et les rotations des grains que l'on peut calculer ne correspondent pas à celles qui ont été observées expérimentalement [3]. D'autre part, la constance du cisaillement relatif aux divers glissements n'est pas vérifiée. Enfin, TAYLOR ne fait pas intervenir la résistance des joints eux-mêmes alors qu'elle n'est pas négligeable (fig. 10).

C'est en considérant principalement cette résistance des joints qu'il a appelé la «consolidation de tensions» que KOCHENDÖRFER a établi une autre théorie:

Il suppose que dans le domaine plastique la consolidation du polycristal pour une déformation ε est égale à la valeur moyenne de la consolidation des monocristaux pour une même déformation. Le calcul de cette valeur moyenne a été fait par SACHS [4] dans le cas des métaux cubiques à faces centrées: les courbes de déformation du monocristal (cisaillement τ et glissement γ) et du polycristal (tension σ et allongement ε) peuvent être reliées par les relations

$$\sigma = 2{,}24\,\tau, \quad \gamma = 2{,}24\,\varepsilon.$$

Pratiquement, d'après KOCHENDÖRFER, la courbe corrigée du monocristal est parallèle à celle du polycristal et s'en déduit par une translation égale à la consolidation de tensions. Celle-ci, correspondant aux tensions élastiques se formant au voisinage des joints, reste constante au cours de la déformation et égale à $(\sigma_0 - 2{,}24\,\tau_0)$, l'indice 0 étant relatif aux valeurs à la limite élastique.

Ces deux théories ont été vérifiées pour de grandes déformations. Mais, pour les faibles déformations, on ne peut pas parler de parallèlisme entre les courbes du monocristal et du polycristal et l'hypothèse d'une consolidation intergranulaire constante doit être rejetée, au moins au début de la déformation.

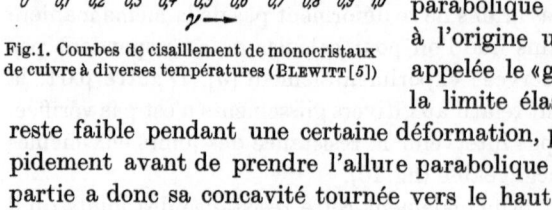

Fig. 1. Courbes de cisaillement de monocristaux de cuivre à diverses températures (BLEWITT [5])

Nous allons maintenant définir la forme exacte des courbes de déformation et essayerons d'expliquer les différences en faisant intervenir, à l'intérieur des grains, un système de tensions complexes dû à la présence des joints et des grains voisins.

2. Forme des courbes de traction des monocristaux. Il a été maintenant observé sur de nombreux métaux, l'aluminium, l'argent, le cuivre, le fer, les laitons, les métaux hexagonaux ... que la courbe de traction du monocristal n'avait pas une allure parabolique simple mais présentait à l'origine une anomalie qui a été appelée le «glissement facile»: après la limite élastique, la consolidation reste faible pendant une certaine déformation, puis se met à croître rapidement avant de prendre l'allure parabolique classique. La première partie a donc sa concavité tournée vers le haut et se termine par une branche asymptotique correspondant à une vitesse d'écrouissage constante. D'après les résultats de BLEWITT [5] sur le cuivre et de STAUB-

WASSER [6] sur l'aluminium, cette première partie ne semble pas dépendre de la température; mais la tension pour laquelle apparait l'écrouissage parabolique croit quand la température décroit (fig. 1). Dans le cas de l'aluminium à température ambiante, cette tension est très basse et la partie à consolidation sensiblement linéaire n'est pas atteinte. Les courbes du cisaillement en fonction du glissement présentent alors un point d'inflexion (fig. 2).

Nous avons étudié [7] les formes des courbes de traction de monocristaux d'aluminium et d'alliages et avons recherché dans quelles conditions le glissement facile pouvait se développer.

La fig. 3 représente la forme des courbes de déformation de monocristaux d'aluminium raffiné à 99,99%, préparés par la méthode de l'écrouissage critique. Ces courbes sont rapportées au glissement et au cisaillement dans le plan et

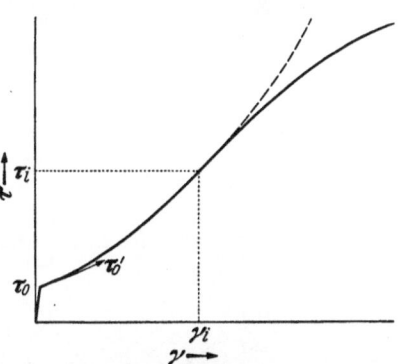

Fig. 2. Courbe de cisaillement d'un monocristal d'aluminium à la température ambiante

la direction de glissement principal. Nous avons défini un premier domaine de déformation comme étant celui qui s'étend jusqu'au point d'inflexion (τ_i, γ_i) de la courbe. On remarque que le développement de celui-ci varie dans de larges limites; de 1,4 à 7,8% selon l'orientation et, d'une manière

générale, croît quand on va de la zône (100)−(111) vers la direction [110]. Divers auteurs ont signalé que pour certaines orientations (celles qui permettent un double glissement) le glissement facile n'existait pas [8, 9]; or, nous l'avons trouvé dans tous nos essais; cependant il est très faible dans certains cas et une simple petite déformation accidentelle peut le faire disparaître.

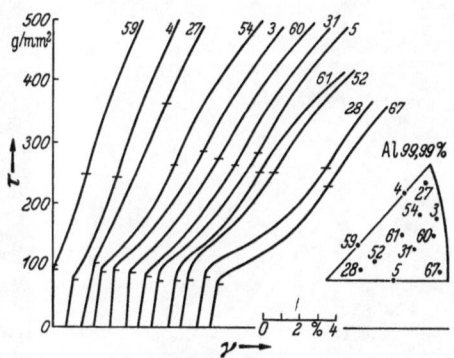

Fig. 3. Courbes de cisaillement de monocristaux de diverses orientations (Al 99,99%)

On remarque, par contre, que le cisaillement au point d'inflexion est indépendant de l'orientation et que la loi de SCHMID [10] relative à la constance du cisaillement critique peut être étendue au cisaillement au point d'inflexion.

Les variations du glissement au point d'inflexion peuvent être liées à la structure de la déformation. Les cristaux peuvent en effet se déformer sous l'action de deux systèmes de glissement (fig. 4), d'un système principal et d'un système secondaire, celui-ci se présentant souvent sous la forme de bandes de glissements secondaires (fig. 5), ou d'un seul système de glissement et de bandes de pliage (fig. 6). Plus le second système de

Fig. 4. Deux systèmes de glissement équivalents (Al 99,99%, $\varepsilon = 5\%$, $G = 150$)

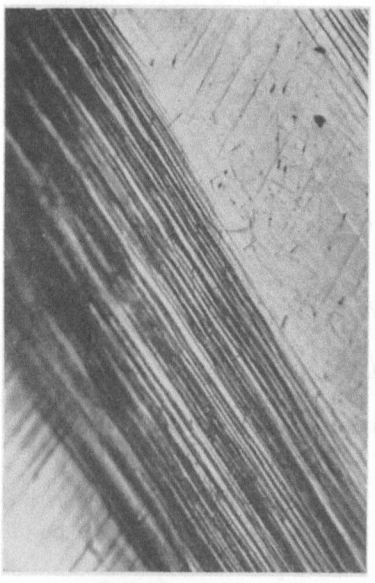

Fig. 5. Bandes de glissements secondaires
(Al 99,99%, $\varepsilon = 5\%$, $G = 150$)

Fig. 6. Bandes de pliage
(Al 99,99%, $\varepsilon = 5\%$, $G = 150$)

glissement est important, plus le glissement facile est court, c'est-à-dire que la pente à l'origine des courbes de cisaillement est forte.

On peut donc relier la consolidation pendant le glissement facile aux intersections de glissements. Celà nous a conduit [11] à une représentation mathématique des courbes vérifiant avec une bonne approximation les résultats expérimentaux:

$$\tau = \tau_0 - \frac{C}{L_n{}^2} L_n \left(1 - \frac{\gamma}{z\,\gamma_i}\right),$$

C étant la consolidation plastique $(\tau_i - \tau_0)$. Cette relation tient compte du fait expérimental que le rapport des pentes au point d'inflexion et à l'origine des courbes est toujours égal à 2 (à 5% près). Mais elle suppose que la vitesse d'écrouissage constante, que l'on obtiendrait à basse température, est infinie; rigoureusement, la relation devrait être écrite en axes obliques.

Après le point d'inflexion, la courbe de déformation prend l'allure parabolique classique et peut être représentée par une relation de la forme:

$$\tau = \tau_0' + B'\,\gamma^{0,3}.$$

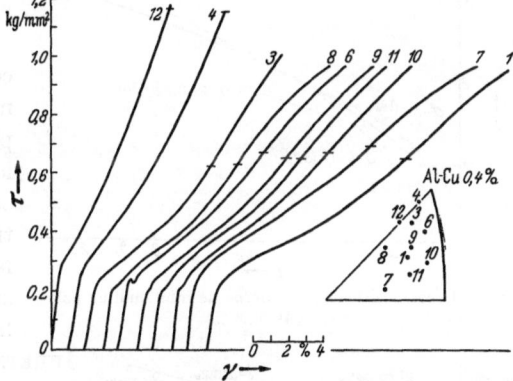

Fig. 7. Courbes de cisaillement de monocristaux de diverses orientations (Al–Cu à 0,4%)

La courbe de déformation des monocristaux se compose donc de deux parties: l'une, indépendante de la température, dans laquelle la consolidation est liée aux barrières de COTTRELL formées aux intersections des glissements; la seconde, qui dépend de la température et qui a la forme parabolique classique. Le point d'inflexion correspond à un net changement de régime de déformation.

On retrouve les mêmes résultats avec les alliages d'aluminium [12], mais l'étendue du glissement facile est augmentée par la présence d'impuretés (fig. 7), ainsi qu'il a été constaté sur d'autres métaux [9, 13]. Celà est dû à ce que la présence d'impuretés limite le développement du second système de glissement [14].

3. Forme des courbes de traction des polycristaux. Cette décomposition des courbes de déformation en deux parties se retrouve avec les polycristaux, mais elle est moins apparente. Cependant, une analyse précise de la forme des courbes de traction des polycristaux [15] montre que la représentation des courbes par une parabole unique n'est qu'approchée et que, pour être rigoureux, il faut représenter les courbes par deux relations du type suivant:

$$\sigma = \sigma_0 + A\,\varepsilon^m \quad \text{jusqu'à} \quad \varepsilon = \varepsilon_p,$$
$$\sigma = \sigma'_0 + B\,\varepsilon^n \quad \text{au delà de } \varepsilon = \varepsilon_p,$$

σ et ε étant la tension et l'allongement.

La fig. 8 représente la décomposition des courbes en deux parties séparées par un point que nous avons appelé le «point de transition» ou la courbe présente un brusque changement de courbure.

Si le grain devient plus gros, la première parabole devient de plus en plus plate et on passe d'une manière continue de la courbe du polycristal à celle du monocristal (fig. 9), le point de transition devenant le point d'inflexion.

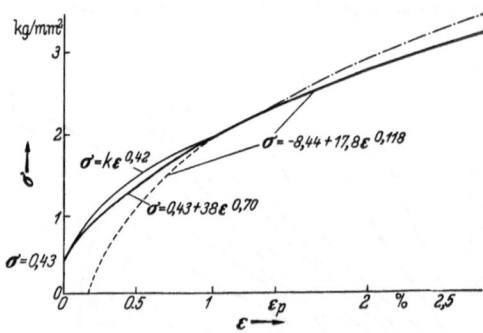

Fig. 8. Décomposition de la courbe de traction d'un polycristal (Al 99,99 %)

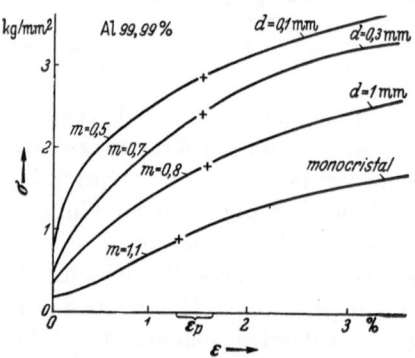

Fig. 9. Influence de la grosseur du grain sur les courbes de traction de polycristaux

Une étude détaillée de ce point de transition [16] nous a permis de le lier à plusieurs phénomènes: l'allongement correspondant est égal à l'écrouissage critique de recristallisation, la tension au point de transition est une limite des fluages rapides (point de Juretzek et Sauerwald [17]). Mais c'est l'étude de la structure par les rayons X qui nous a montré qu'à partir de ε_p les grains commençaient à se fragmenter. Il est à remarquer que, pour une température bien définie, l'allongement ε_p reste constant et peut être considéré comme caractéristique d'un alliage; par contre, il croît quand la température décroît.

Cette décomposition de la courbe de déformation en deux domaines a été observée sur plusieurs métaux.

On retrouve donc sur les courbes de déformation des polycristaux les mêmes particularités que celles observées sur les courbes des monocristaux, mais le changement de régime est masqué par la consolidation due aux joints. Ceux-ci sont en effet responsables d'une structure de déformation très différente de celles observées sur les courbes des monocristaux ; les fig. 10 montrent l'allure des glissements dans un polycristal faiblement déformé. En comparant avec les figures 4, 5 et 6, on voit que beaucoup plus de systèmes de glissement apparaissent et qu'il n'y a pas de bandes de déformation.

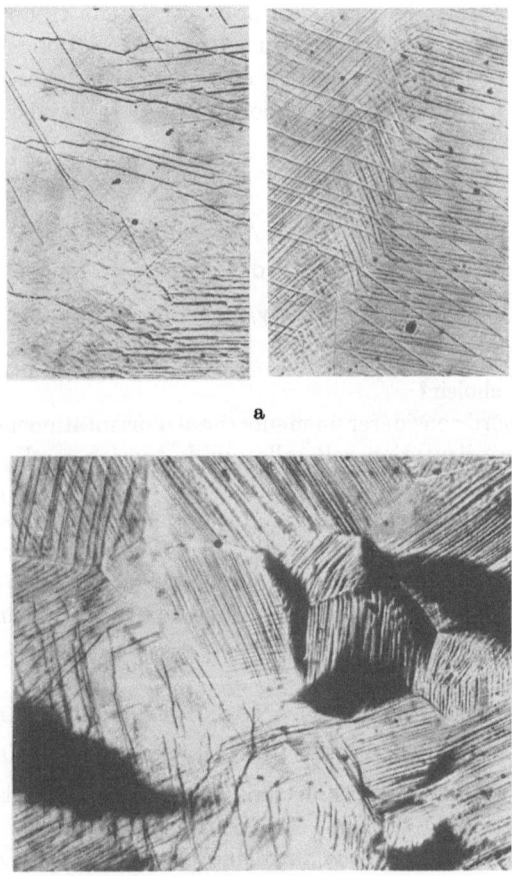

a

b

Fig. 10. Glissements dans les polycristaux a) Al 99,99%, $\varepsilon = 5\%$, $G = 90$;
b) Al–Si 0,2%, $\varepsilon = 6\%$, $G = 90$

La forme des courbes de traction des polycristaux varie en fonction de la grosseur du grain, mais pour des grains de diamètre inférieur à 0,1 mm (pour des éprouvettes de 3 mm de diamètre), les propriétés restent constantes et pour l'aluminium raffiné les équations de la courbe peuvent s'écrire

$$\left. \begin{aligned} \sigma &= 0,4 + 15\,\varepsilon^{0,5} \quad \text{jusqu'à} \quad \varepsilon = \varepsilon_p = 0,015\,, \\ \sigma &= -\,2,3 + 10\,\varepsilon^{0,2} \quad \text{au delà de } \varepsilon = \varepsilon_p\,. \end{aligned} \right\} \text{(en kg/mm}^2\text{)}.$$

4. Comparaison des courbes des monocristaux et des polycristaux.
Nous avons donc à comparer la courbe de déformation du polycristal,
c'est-à-dire les valeurs de la tension normale en fonction de l'allongement,
à une série de courbes de monocristaux qui, elles, sont dans le système
de coordonnées cisaillement-glissement.

Pour transformer la courbe du polycristal et la traduire en cisaillement et glissement, nous pouvons considérer l'orientation moyenne définie
par SACHS, donc diviser les tensions et multiplier les allongements par le
coefficient 2,24. La courbe de cisaillement du polycristal sera alors, dans
le cas de l'aluminium raffiné

$$\tau = \quad 0,18 + 4,5\,\gamma^{0,5} \quad \text{jusqu'à} \quad \gamma = 0,035,$$
$$\tau = -\,1,03 + 4,0\,\gamma^{0,2} \quad \text{au delà de } \gamma = 0,035.$$

Cette courbe est à comparer à une des courbes du monocristal:

$$\tau = 0,085 + 0,24\,L_n(1 - \gamma/2\,\gamma_i) \quad \text{jusqu'à} \quad \gamma = \gamma_i,$$
$$\tau = \tau'_0 + B'\,\gamma^{0,3} \qquad\qquad \text{au delà de } \gamma = \gamma_i,$$

mais, laquelle choisir?

Il faut d'abord considérer un monocristal d'orientation moyenne, c'est-
à-dire que son orientation soit telle que la tension appliquée soit dans
le rapport 2,24 avec le cisaillement effectif; mais, il y en a une infinité
présentant toutes les structures de déformation. Puisque les polycristaux
présentent plusieurs systèmes de glissement, nous choisirons le monocristal dont l'axe a une orientation située sur la zône (100)−(111), c'est-
à-dire une éprouvette telle que la No. 59 de la fig. 3. Pour une telle orientation, le glissement au point d'inflexion est de 2% et les équations de
la courbe sont

$$\tau = \quad 0,085 - 0,24\,L_n(1 - 25\,\gamma) \quad \text{jusqu'à} \quad \gamma = 0,02,$$
$$\tau = -\,0,49 \quad + 2,4\,\gamma^{0,3} \qquad\qquad \text{au delà de } \gamma = 0,02.$$

Ces deux courbes du cisaillement en fonction du glissement sont
représentées sur la fig. 11. Nous avons superposé les points de transition
et d'inflexion car ils sont dus au même phénomène; ils correspondent
cependant à des déformations différentes, mais, dans le polycristal, la
présence des joints gêne le mécanisme de fragmentation et le retarde.

Remarquons d'abord que les cisaillements critiques sont dans le
rapport 2,1. Cette valeur se retrouve avec d'autres métaux; nous avons,
par exemple (en g/mm²)

$$\text{Al–Cu} \quad 0,4\ \% : (\tau_0)_m = \ 210 \qquad (\tau_0)_p = \ 470 \to 2,2$$
$$\text{Al–Cu} \quad 1,0\ \% : (\tau_0)_m = \ 390 \qquad (\tau_0)_p = \ 890 \to 2,3$$
$$\text{Al–Zn} \quad 3,0\ \% : (\tau_0)_m = \ 220 \qquad (\tau_0)_p = \ 535 \to 2,4$$
$$\text{Cuivre } 99,98\% : (\tau_0)_m = \ \ 95\ [9] \qquad (\tau_0)_p = \ 220 \to 2,3$$
$$\text{Fer Armco} \quad : (\tau_0)_m = 3600\ [18] \quad (\tau_0)_p = 8000 \to 2,3$$

c'est-à-dire que la présence des joints augmente la résistance quand il n'y a en principe pas encore de glissements.

Au point de transition, les cisaillements sont, dans le cas de l'aluminium raffiné

$$\text{monocristal: } \tau_i = 250 \text{ gr/mm}^2,$$

$$\text{polycristal: } \tau_p = 1020 \text{ gr/mm}^2.$$

Par suite des glissements qui se sont développés, la tension est montée beaucoup plus rapidement dans le polycristal et elle est devenue quatre fois plus forte que dans le mono-cristal.

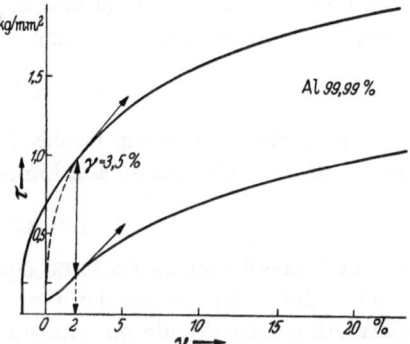

Au point de transition et au point d'inflexion, les pentes des courbes sont les mêmes et égales à 12 kg/mm². Les courbes ayant au delà du point de transition sensiblement la même forme (parabole à faible exposant) sont à peu près parallèles et la consolidation aux joints ne croît plus.

Nous pouvons traduire ces résultats de la manière suivante:

Fig. 11. Courbes de cisaillement du monocristal et du polycristal d'aluminium raffiné

Dans une éprouvette polycristalline, la présence des joints crée à l'intérieur des grains un système de tensions transversales σ_2, σ_3 qui diminue le cisaillement effectif, donc augmente la tension axiale nécessaire pour produire la déformation.

En cherchant les valeurs de ces tensions transversales pour que le cisaillement effectif dans le monocristal et le polycristal soit le même, on obtient

$$\sigma_2 = \sigma_3 = \frac{k-1}{k}\,\sigma_1,$$

k étant le rapport de la résistance du polycristal à celle du monocristal.

k variant de 2,2 à la limite élastique à 4 au point de transition les tensions transversales seront

$$\text{à la limite élastique: } \sigma_2 = \sigma_3 = 0{,}55\,\sigma_1,$$

$$\text{au point de transition: } \sigma_2 = \sigma_3 = 0{,}75\,\sigma_1.$$

Elles resteront constantes au delà du point de transition et, dans le cas de l'aluminium raffiné, seront égales à 1,75 kg/mm², la tension au point de transition étant 2,3 kg/mm².

Nous ferons maintenant une nouvelle hypothèse, dont nous trouverons une vérification avec les éprouvettes tubulaires: Supposons que chacune des tensions, la tension appliquée et les tensions fictives de consolidation, entraine dans sa direction une déformation liée à la tension par la même relation :

$$\sigma = \sigma_0 + A \, \varepsilon^m.$$

Nous aurons au point de transition :

$$\varepsilon_2 = \varepsilon_3 = - \left(\frac{0{,}75\,\sigma_p - \sigma_0}{A} \right)^2,$$

soit $\varepsilon_2 = \varepsilon_3 = -0{,}75\%$ pour $\varepsilon_1 = 1{,}5\%$ en prenant les valeurs obtenues avec l'aluminium raffiné.

C'est la valeur de la déformation transversale correspondant à la condition de conservation du volume

$$\varepsilon_2 = \varepsilon_3 = - \frac{1}{2}\, \varepsilon_1 .$$

Pour des déformations plus faibles, le rapport $|\varepsilon_2/\varepsilon_1|$ décroît car le coefficient k décroît. On sait qu'à la limite élastique on a

$$\varepsilon_2 = \varepsilon_3 = - \nu\, \varepsilon_1 ,$$

ν étant le coefficient de Poisson, égal sensiblement à 1/3.

On a donc, dans le premier domaine de déformation un coefficient de contraction transversale qui passera, assez rapidement de la valeur 1/3 correspondant au domaine élastique, à la valeur 1/2, cette dernière valeur n'étant cependant atteinte qu'au point de transition où la déformation devient complètement plastique ; jusqu'à cette déformation, il reste des zônes sous forte tension élastique.

Nous avons supposé que, dans le cas du monocristal, il n'y avait pas de tensions transversales ; pratiquement, on peut supposer qu'elles existent aussi et qu'elles sont responsables de l'apparition de glissements sur des plans peu favorisés (fig. 5) par suite de la modification du cisaillement maximum effectif dû à la formation de barrières. Mais ces tensions de consolidation intracristalline sont les mêmes dans le monocristal et le polycristal ; en les supposant proportionnelles à la tension appliquée, on peut les négliger.

5. Déformation d'éprouvettes tubulaires. Nous avons eu une confirmation de l'existence des tensions transversales de consolidation en étirant des éprouvettes dans lesquelles les déformations n'étaient plus forcément symétriques.

Des essais sur éprouvettes tubulaires d'aluminium [*16*], à grains fins par rapport à l'épaisseur et sans texture nous ont montré que la contraction du diamètre moyen d'une éprouvette creuse est plus faible que celle d'une éprouvette pleine. Si le tube est très mince, les trois déformations principales sont

$$\varepsilon_1 = \varepsilon , \quad \varepsilon_2 = - \frac{1}{3}\, \varepsilon , \quad \varepsilon_3 = - \frac{2}{3}\, \varepsilon .$$

Supposons, pour simplifier, que la courbe de déformation puisse être représentée par une parabole unique de la forme

$$\sigma = A \, \varepsilon^m$$

avec $m = 0{,}4$ (fig. 8). Notre hypothèse de proportionnalité des tensions aux déformations donne alors le système de tensions

Eprouvette pleine: $\sigma_2 = \sigma_3 = 0{,}76 \sigma_1$,

Eprouvette crouse: $\sigma_2 = 0{,}64$ et $\sigma_3 = 0{,}85 \sigma_1$.

Le développement des glissements à l'intérieur des grains doit avoir lieu pour la même valeur du cisaillement; en appliquant le critère de déformation de von Mises, on trouve que l'éprouvette creuse doit être plus résistante que l'éprouvette pleine dans le rapport 1,33.

Or, c'est ce que l'on obtient expérimentalement: la courbe de traction de l'éprouvette creuse se trouve au-dessus de celle de l'éprouvette pleine dans le rapport 1,4 (la relation $\sigma = A \, \varepsilon^m$ que nous avons utilisée est approchée).

Mais, il est à remarquer que cette différence s'atténue quand la grosseur du grain argumente et un monocristal tubulaire a la même résistance qu'un monocristal plein. Ce phénomène est donc bien dû à la présence des joints qui se traduit par une modification du cisaillement effectif à l'intérieur des grains.

6. Conclusions. La différence de forme entre les courbes de traction d'un monocristal et d'un polycristal ne permet pas de passer de l'une à l'autre par une transformation simple. Mais, il est possible de représenter la consolidation intergranulaire par un système de tensions transversales qui s'oppose au développement des boucles de dislocations; ces tensions, modifiant la valeur du cisaillement effectif en grandeur et en direction sont responsables de l'augmentation de résistance et de l'apparition de nombreux systèmes.

A la limite élastique ces tensions, que l'on suppose égales par symétrie dans une éprouvette ronde, atteignent déjà la valeur de la moitié de la tension appliquée, car la déformation «élastique» comporte des glissements.

Au début de la déformation plastique, elles croissent rapidement pour atteindre une valeur maximum au point de transition; durant ce premier domaine de déformation, la contraction transversale, égale à 1/3 dans le domaine élastique, croit jusqu'à la valeur 1/2 qui n'est atteinte qu'à la fin. Si la forme de l'éprouvette est telle que les déformations transversales ne soient plus forcément égales, les deux tensions transversales, qui varient comme les déformations correspondantes, seront différentes; celà pourra se traduire par une variation de la résistance des éprouvettes en fonction de leur forme.

Pour de plus grandes déformations, le métal est complètement plastique et les tensions de consolidation ne croissent plus; l'augmentation d'écrouissage est alors principalement d'origine intracristalline, par suite de la fragmentation des grains qui apparait à partir du point de transition [7, 16]. La résistance des joints ayant atteint son maximum, les courbes de cisaillement du monocristal et du polycristal sont alors parallèles et l'hypothèse de Kochendörfer est vérifiée dans ce second domaine de déformation plastique.

Bibliographie

[1] Taylor, G. I.: J. Inst. Met. **62**, 307 (1938).
[2] Kochendörfer, A.: Plastische Eigenschaften der Metalle, Berlin 1941.
[3] Barrett, C. S. et L. H. Levenson: Trans. AIME **137**, 112 (1940).
[4] Sachs, G.: Z. VDI **72**, 34 (1929).
[5] Blewitt, T. H., R. R. Coltman et J. K. Redman: Defects in Cristalline Solids, p. 369, Phys. Soc. London 1955.
[6] Staubwasser, W.: Thèse, Université de Göttingen (1954).
[7] Jaoul, B. et I. Bricot: Rev. Métall. **52**, H. 8, 643 (1955).
[8] Lucke, K. et H. Lange: Z. Metallkde. **43**, 55 (1952); **44**, 514 (1953).
[9] Rosi, F. D.: J. Metals **6**, H. 9, 1009 (1954).
[10] Schmid, E. et W. Boas: Kristallplastizität, Berlin 1935.
[11] Jaoul, B.: C.R. Acad. Sci., Paris **241**, 161 (1955).
[12] Jaoul, B.: C.R. Acad. Sci., Paris **240**, 2532 (1955).
[13] Masing, G. et J. Raffelsieper: Z. Metallkde. **41**, 65 (1950).
[14] Murphy, H. M. et E. A. Calnan: Acta Métall. **3**, H. 3, 268 (1955).
[15] Crussard, C. et B. Jaoul: Rev. Métall. **47**, H. 8, 599 (1950).
[16] Jaoul, B.: Publ. sci. techn. Ministère Air., Bull. Serv. techn. No. 290 (1954).
[17] Juretzek et Sauerwald: Z. Phys. **83**, 483 (1933).
[18] Holden, A. N. et F. N. Kunz: J. appl. Physics **23**, H. 7, 799 (1952).

Versetzungen und Gittertheorie

Von **G. Leibfried**, Göttingen

Mit 3 Abbildungen

Bei vielen Problemen, bei denen Versetzungen eine Rolle spielen, geht die Gitterstruktur und damit die Gittertheorie in entscheidender Weise ein. An einigen einfachen Beispielen sollen Zusammenhänge dieser Art qualitativ diskutiert werden.

1. Schubspannung bei endlichen Scherungen für die Alkalien. Die Struktur einer Versetzung kann näherungsweise nach einem von PEIERLS [1] angegebenen Verfahren ermittelt werden. Dazu benötigt man den Zusammenhang zwischen der Schubspannung τ und endlichen Scherungen längs einer möglichen Gleitebene. Diese Verknüpfung wird durch die Gitterstruktur entscheidend beeinflußt. Am einfachsten und übersichtlichsten liegen die Verhältnisse bei den (raumzentrierten) Alkalimetallen. Die Energie bei Scherungen ist praktisch ausschließlich elektrostatischer Natur. Die Wechselwirkung der Ionenrümpfe kann vernachlässigt werden, denn der Durchmesser der Rümpfe ist klein gegen den Abstand nächster Nachbarn im Gitter[1]. Der Anteil der quantentheoretisch zu berechnenden Energie hängt praktisch nur vom Volumen der Elementarzelle ab und spielt daher bei Scherungen keine Rolle. Daher können die Energien und Spannungen bei Scherungen allein aus den elektrostatischen Anteilen der Gitterenergie berechnet werden. Die Ladungsdichte besteht aus Punktladungen in den Ionenrümpfen und einer (als konstant angenommenen) Verteilung der Valenzelektronen[2]. Die Änderung der elektrostatischen Energie bei einer Scherung kann nun mit den üblichen Methoden der Gittertheorie ohne Schwierigkeiten berechnet werden [3].

Abb. 1 zeigt die Scherung eines raumzentrierten Gitters der Gitterkonstanten a. Die Schubspannung $\tau(s)$ bei einer homogenen Verformung muß ersichtlich in der Verschiebung die Periode $2a$ besitzen, da bei einer

---

[1] Bei Na ist z.B. der Abstand nächster Nachbarn 3 AE, der Ionendurchmesser etwa 1,6 AE.

[2] Für eine eingehendere Diskussion vgl. [2].

[3] Auf diese Weise hat K. FUCHS [3] als erster die Schubmoduln (infinitesimale Scherungen!) der Alkalien in sehr guter Übereinstimmung mit experimentellen Daten berechnet.

solchen Verschiebung das unverzerrte Gitter wieder hergestellt wird[1]. Also muß $\tau(s)$ in eine Fourierreihe der Form

$$\tau(s) = \sum_{\nu=1}^{\infty} A_\nu \sin \frac{\nu\pi}{a} s \qquad (1)$$

entwickelbar sein. [Es treten nur sin-Glieder auf, da $\tau(s)$ in s antisymmetrisch sein muß.] Der Peierlssche Ansatz besteht in der Annahme, daß der erste Fourierkoeffizient überwiegt:

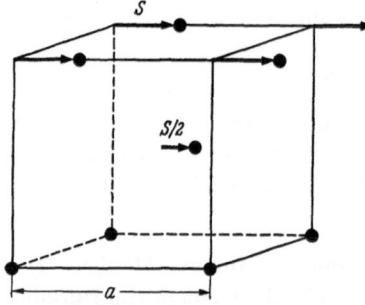

$$\tau(s) = A_1 \sin \frac{\pi}{a} s . \qquad (2)$$

Da für kleine Verformungen (mit dem Schubmodul G)

$$\tau(s) = G \cdot \frac{s}{a} \qquad (3)$$

sein muß, so ist dann A_1 mit dem Schubmodul verknüpft:

$$A_1 \cdot \pi = G . \qquad (4)$$

Abb. 1. Scherung eines raumzentriert-kubischen Gitters der Gitterkonstanten a entlang einer Würfelebene in Richtung einer Würfelkante

Auch bei komplizierteren Scherungen kann man so in $\tau(s)$ die Gitterstruktur verarbeiten, wobei nur die elastischen Daten verwendet werden [4]. Der funktionale Zusammenhang $\tau(s)$ bestimmt dann im wesentlichen die Versetzungsstruktur für die betreffende Ebene.

Abb. 2 zeigt den Verlauf der Schubspannung mit der Peierlsschen Annahme nach (2) und (4), also unter alleiniger Berücksichtigung von A_1. Tatsächlich darf man die höheren Fourierkoeffizienten aber nicht vernachlässigen, wenigstens A_2 sollte noch berücksichtigt werden. Die Rechnung liefert die folgenden Werte (in Einheiten e^2/a^4, wobei e die Elementarladung und a die kubische Gitterkonstante ist):

$$\left. \begin{aligned} A_1 &= 0{,}344 , \\ A_2 &= -0{,}058 , \\ A_3 &= 0{,}003 , \\ A_4 &= -0{,}0003 , \end{aligned} \right\} \quad G \sum_{\nu=1}^{\infty} A_\nu \, \nu \, \pi = 0{,}742 .$$

Abb. 2 zeigt ferner die berechnete Schubspannung unter Berücksichtigung der ersten Fourierkoeffizienten. Die maximale Schubspannung ist verglichen mit dem Peierlsschen Wert G/π um etwa 50% erhöht und das Maximum wird etwas verschoben. Die Rechnungen für die Struktur einer entsprechenden Versetzung sind noch nicht durchgeführt. Der be-

---

[1] Beim kubisch primitiven Gitter wäre die Periode a.

rechnete Verlauf deutet darauf hin[1], daß in diesem Spezialfall die Versetzungsweite gegenüber dem PEIERLSschen Ansatz reduziert und die „PEIERLS-Kraft" entsprechend erhöht wird. In anderen Ebenen erhält man ähnliche Resultate[2].

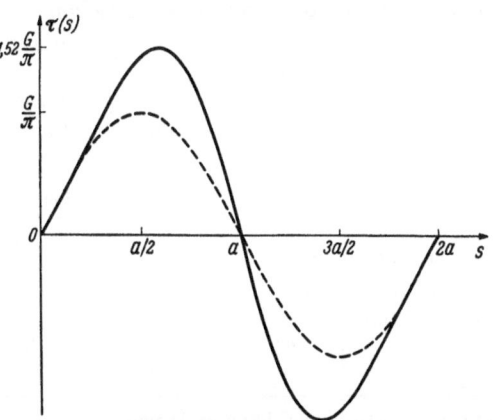

Abb. 2. Schubspannung τ in Abhängigkeit von der Verschiebung s. a) nach PEIERLS (gestrichelt), b) berechnet nach der Gittertheorie

2. Die thermischen Schwankungen der Schubspannung.

Die thermischen Schwankungen der Schubspannungen beeinflussen die Bewegung einer Versetzung in vielfacher Weise. Insbesondere können sie einer Versetzung behilflich sein, Hindernisse atomarer Dimensionen zu überwinden. Im allgemeinen entnimmt man die Schwankungsgröße aus der Thermodynamik, andererseits kann man sie gittertheoretisch zumindest näherungsweise berechnen. Es stellt sich heraus, daß die so berechneten Schwankungen einen qualitativ verschiedenen Temperaturverlauf zeigen. Hier soll erläutert werden, welchen Sinn die makroskopisch-thermodynamisch und die mikroskopisch-gittertheoretisch berechneten Schwankungen haben. Den gitter-theoretisch bestimmten Werten ist im allgemeinen der Vorzug zu geben, da sie die eigentliche physikalische Fragestellung beschreiben.

In vielen Fällen handelt es sich darum, eine Wahrscheinlichkeitsverteilung von Schubspannungen in einem Volumen V anzugeben:

$$w(\tau)\,d\tau = \frac{1}{\sqrt{2\pi\overline{\tau^2}}} \cdot e^{-\frac{\tau^2}{2\overline{\tau^2}}}\,d\tau. \tag{6}$$

Dabei ist $w(\tau)\,d\tau$ die Wahrscheinlichkeit dafür, daß die Schubspannung τ in V zwischen τ und $\tau + d\tau$ liegt.

Aus thermodynamischen Überlegungen[3] erhält man z.B.

$$w(\tau) = C \cdot e^{-\frac{\tau^2 V}{2GkT}} \quad \text{mit } \overline{\tau^2} = \frac{G}{V}\,kT. \tag{6a}$$

---

[1] Wenn man die Ergebnisse von FOREMAN und Mitarbeitern [5] auf den hier behandelten Fall verallgemeinern darf. Danach ist das Produkt aus maximaler Schubspannung und Versetzungsweite annähernd konstant.

[2] Die Rechnungen wurden von C. LEHMANN durchgeführt. Die Resultate im einzelnen werden an anderer Stelle veröffentlicht werden.   [3] Vgl. R. BECKER [6].

In der Tabelle sind daneben Schwankungsquadrate aufgeführt, die aus der Gittertheorie für spezielle Fälle berechnet wurden. Die Ergebnisse sind unter Annahme eines DEBYEschen isotropen Spektrums gewonnen worden. Longitudinale und transversale Anteile sind getrennt behandelt[1].

Tabelle. *Schwankungsquadrate der „thermischen" Schubspannungen im kubisch-flächenzentrierten Gitter*[2]

$$\overline{\tau^2}$$

	Hohe Temperaturen	Absoluter Nullpunkt
Aus Thermodynamik	$\dfrac{G}{V}\,kT$	0
Für ein Atom $V = \dfrac{a^3}{4}$	$0,5\,\dfrac{G}{V}\,kT$	$0,2\,\dfrac{G}{V}\,k\,\Theta$
Für N statistisch unabhängige Atome $V = N\,\dfrac{a^3}{4}$	$0,5\,\dfrac{G}{V}\,kT$	$0,2\,\dfrac{G}{V}\,k\,\Theta$
Für einen Würfel aus N Atomen $V = N\,\dfrac{a^3}{4}$	$0,5\cdot\dfrac{G}{V}\,kT$	$0,16\,\dfrac{G}{V}\,k\,\Theta\cdot\left(\dfrac{4}{N}\right)^{1/3}\,!$
Für N Atome eines Quadrats der 111-Ebene $V = N\,\dfrac{a^3}{4}$	$\dfrac{G}{V}\,kT$	$0,22\,\dfrac{G}{V}\,k\,\Theta$
Für N Atome entlang einer $\overline{1}\,\overline{1}\,2$-Geraden $V = N\cdot\dfrac{a^3}{4}$	$0,5\,\dfrac{G}{V}\,kT$	$0,23\,\dfrac{G}{V}\,k\,\Theta$

a kubische Gitterkonstante, G Schubmodul, Θ DEBYE-Temperatur.

Das Verhältnis der beiden Schallgeschwindigkeiten ist zu 2 angenommen worden, was den elastischen Daten bei Aluminium entspricht. Die DEBYE-Temperatur Θ ist durch die Abschneidefrequenz der transversalen Wellen ω_{tr} definiert ($k\Theta = \hbar\,\omega_{tr}$), was praktisch gleichwertig mit der üblichen Definition ist. Zunächst kann man nach der DEBYEschen Näherung, die ja auf einer Kontinuumsbeschreibung beruht, die Spannungsverteilung für einen Punkt des Kontinuums berechnen (Zeile 2 der Tabelle). Das Ergebnis ist hier durch das Volumen, welches einem Atom des flächenzentrierten Gitters zukommt, ausgedrückt. Jedem Atom (jeder Elemen-

---

[1] Die Frequenzen dieses Ansatzes liegen im allgemeinen zu hoch.
[2] Die Tabelle ist von G. BARSCH berechnet worden (Dissertation Göttingen 1955).

tarzelle) kann eine Verzerrung und damit eine Spannung τ_i zugeordnet werden. Die Angaben der Tabelle beziehen sich immer auf die Verteilung der mittleren Schubspannung

$$\tau = \frac{1}{N} \sum_{i=1}^{N} \tau_i$$

in dem vorgelegten Volumen V mit N Atomen. Zeile 3 gibt das Ergebnis für ein Volumen aus N Atomen unter der Annahme, daß die einzelnen τ_i statistisch unabhängig sind. Tatsächlich sind sie das nicht und die weiteren Resultate enthalten den Einfluß der Korrelationen.

Der auffallendste Unterschied ist wohl der, daß das Schwankungsquadrat der Thermodynamik bei tiefen Temperaturen verschwindet, während die gittertheoretischen Werte auch für $T = 0$ noch Schwankungen aufweisen, die vergleichbar mit denen bei Zimmertemperatur sind, da die DEBYE-Temperaturen der meisten Metalle von der Größenordnung Zimmertemperatur sind[1].

Die Nullpunktsschwingungen des Gitters bewirken diese Schwankungen bei tiefen Temperaturen. Man würde sich nun nicht wundern, daß im allgemeinen die thermodynamischen Formeln von den atomistisch berechneten abweichen. Aber für die makroskopischen Volumina (Zeile 1 und 4 der Tabelle), für die die Thermodynamik anwendbar ist, sollten sie wenigstens übereinstimmen[2]. Das ist nur für hohe Temperaturen der Fall (wenn man von dem Faktor 2 absieht, der möglicherweise dadurch zustande kommt, daß die atomistisch berechneten Größen für die mittlere Schubspannung berechnet werden und die DEBYEsche Näherung nicht besonders gut ist). Es kann kein Zweifel sein, daß die berechneten Spannungs- bzw. die äquivalenten Verzerrungsschwankungen am absoluten Nullpunkt tatsächlich auftreten müssen.

Somit liegt die Frage nahe, welche Bedeutung der thermodynamischen Schwankung zukommt. Das instruktivste Beispiel ist die Volumenschwankung. Wenn man sich ansieht, wie die Verteilungen der statistischen Thermodynamik abgeleitet werden, so stellt man folgendes fest: Bei den Größen wie Energie und Teilchenzahl, die als Variable in HAMILTON-Operator oder HAMILTON-Funktion auftreten, hat man bei der Ableitung der Schwankungsformeln keinerlei Schwierigkeiten. Anders ist es bei den Volumenschwankungen. Das Volumen tritt in der üblichen Beschreibung im HAMILTON-Operator als Parameter und nicht als Variable auf. Will man eine Volumenschwankung beschreiben, so ist man gezwungen, das Volumen als Variable einzuführen. Das Resultat hängt

---

[1] CRUSSARD, C. [7] hat ebenfalls auf diesen Tatbestand hingewiesen. Vgl. auch [8].

[2] Das Schwankungsquadrat der Thermodynamik ist proportional zu V^{-1}, das der Gittertheorie bei tiefen Temperaturen proportional zu $V^{-3/4}$.

nun in höchst charakteristischer Weise davon ab, in welcher Weise diese Variable eingeführt wird.

So erhält man die übliche Wahrscheinlichkeitsverteilung für das Volumen eines Systems bei gegebener Temperatur T und festem Außendruck P (Abb. 3)

$$W(V)\,dV = C \cdot e^{-\dfrac{F(V,\,T) + PV}{kT}}\,dV \tag{7}$$

(F freie Energie, $F + PV$ freie Enthalpie) nur dann, wenn man das Volumen durch die Lage eines Stempels definiert, der als klassische Variable (als makroskopisches System) behandelt werden darf.

Durch Entwicklung um das zu P gehörige mittlere Volumen $\bar V$ erhält man eine Gausssche Verteilung mit dem Schwankungsquadrat $\overline{\varDelta^2 V} = \dfrac{\bar V}{K}\,kT$:

$$W(V)\,dV = C' \cdot e^{-\dfrac{K(V - \bar V)^2}{2\,\bar V\,kT}}\,dV, \tag{7a}$$

wobei K der Kompressionsmodul ist, oder wenn man in der gleichen Näherung mit $p = P - K \cdot \dfrac{V - \bar V}{\bar V}$ auf den Druck p umrechnet

$$W(p)\,dp = C'' \cdot e^{-\dfrac{(p - P)^2\,\bar V}{2\,K\,k\,T}}\,dp. \tag{7b}$$

Diese Formel ist das Analogon zu (6a), nur tritt an Stelle des Schubmoduls der Kompressionsmodul auf.

Die Darstellung (7b) für die Druckverteilung ist so zu verstehen, daß nunmehr das Volumen $\bar V$ fest vorgegeben wird (Kasten mit festen Wän-

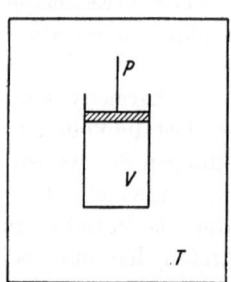

den). P ist dann mit dem mittleren Druck zu identifizieren, ist also die mittlere Kraft auf die Flächeneinheit der Berandung. (7b) kann auch direkt aus den Grundlagen der statistischen Thermodynamik abgeleitet werden, ohne daß das Volumen als Variable eingeführt werden muß. Danach ist völlig allgemein die Druckschwankung

$$\overline{\varDelta^2 p} = -\,kT \cdot \left.\frac{\partial^2 F}{\partial V^2}\right|_{V = \bar V} = \frac{kTK}{\bar V}$$

Abb. 3. Zur Ableitung der Schwankungen des Volumens V eines Systems bei gegebenem Druck P und gegebener Temperatur T

für gegebene Temperatur T und für ein durch feste Wände begrenztes System vom Volumen $\bar V$. Umgekehrt kann man dann (7a) aus (7b) herleiten, wenn man die Beziehung zwischen Druck und Volumen benutzt. Die Relationen zwischen Druck- und Volumenschwankungen sind von gleicher Art wie die zwischen Energie- und Temperaturschwankungen oder zwischen Schwankungen der Teilchenzahl und des chemischen Potentials. Bei den hier interessierenden Schwankungsgrößen handelt es sich aber immer nur um Ausschnitte aus einem Kristall, die nicht durch feste Wände begrenzt sind. Die berechneten Spannungsverteilungen sind zunächst als Verzerrungsverteilungen berechnet. Diese werden dann mit dem HOOKEschen Gesetz auf die entsprechenden Spannungen umgerechnet.

Bei dem einfachsten Modell eines Kristalls, einer linearen Kette aus N Atomen, die mit Federn der Federkonstante f verbunden sind, kann man die Verhältnisse noch einfacher übersehen.

Das Schwankungsquadrat $\overline{\Delta^2 L}$ der Kettenlänge L wäre nach der Thermodynamik

$$\overline{\Delta^2 L} = \frac{N k T}{f}$$

im ganzen Temperaturgebiet. Das bedeutet, daß die Kettenlänge L durch zwei große Massen definiert wird, welche an den Enden der Kette fixiert sind. Die gittertheoretisch berechnete Schwankung stimmt für hohe Temperaturen, wo also die Atome der Kette klassisch behandelt werden dürfen, mit der thermodynamischen Schwankung überein, für $T = 0$ dagegen ergibt sich ein endliches Schwankungsquadrat

$$\overline{\Delta^2 L} = \ln N \cdot \frac{k \Theta}{f},$$

welches nicht proportional zur Kettenlänge ist.

Aus diesen beiden Beispielen wird die Bedeutung der thermodynamischen Verteilung klar. Im allgemeinen kann man also etwa die Verzerrungsschwankungen und die daraus abgeleiteten Spannungsschwankungen nicht aus der Thermodynamik (aus der freien Enthalpie) berechnen, nur bei hohen Temperaturen im Bereich der klassischen statistischen Mechanik ist dies gestattet.

3. Nullpunktsschwingungen und Versetzungen. Hier soll nur kurz diskutiert werden, inwieweit man erwarten kann, daß die Nullpunktsschwankungen, ähnlich wie die thermischen Schwankungen, die Versetzungsbewegung beeinflussen, also ebenfalls zur Überwindung von Hindernissen dienen können. Am besten kann man die Verhältnisse übersehen bei einer einfachen Gitterfehlstelle, einem Loch oder einem besetzten Zwischengitterplatz. Der Zustand „ein Loch in einem Gitterpunkt" ist nicht stabil, es gibt ungeheuer viele andere Zustände gleicher Energie, bei denen sich das Loch an einem anderen Platz befindet. Die Übergangswahrscheinlichkeit für ein Loch zu einer benachbarten Stelle (Tunneleffekt, vgl. die Diskussion zu [7]) hängt entscheidend ab von der Struktur des Grundzustandes des Kristalls, d.h. von seiner Nullpunktsunruhe. Die Diffusion des Lochs wird durch die Nullpunktsunruhe (vor allem also bei kleinen Massen) erleichtert, sie ist auch am absoluten Nullpunkt noch vorhanden.

Die Versetzung ist eine verhältnismäßig stark instabile Gitterstruktur. Die Peierls-Kraft, welche die Versetzung an den (klassisch berechneten) günstigsten Lagen im Kristall festhalten, ist verglichen mit den „thermischen Schubspannungen" außerordentlich klein[1]. Daher sollte man annehmen, daß die „thermischen" Schwankungen auch bei $T = 0$ schon ausreichen, um die Peierls-Kraft unwirksam zu machen. Genauso können die Nullpunktsschwankungen eine Überwindung von atomaren Hindernissen unter äußeren Spannungen ermöglichen. Somit liegt die *Vermutung* nahe, daß die gittertheoretisch berechneten Schwankungen genauso behandelt werden können wie thermische Schwankungen, daß diese Behandlung ein Ersatz für die Berechnung der entsprechenden quantentheoretischen Übergangswahrscheinlichkeiten darstellt und daß dieser Einfluß durchaus vergleichbar mit den Verhältnissen bei Zimmertemperatur ist.

Schließlich kann man die Einwirkung der thermischen Bewegung auch noch in etwas anderer Weise formulieren. Man kann ja eine Versetzung (etwa eine Frank-Read- Quelle) als ein dynamisches System betrachten, kann deren Bewegung näherungsweise in einzelne harmonische Eigenschwingungen zerlegen und diese zunächst klassische Theorie quantenmechanisch behandeln. Die Nullpunktsschwingungen dieses Systems sind dann äquivalent dem Einfluß der Nullpunktsschwankungen der Schubspannung auf die *klassisch zu behandelnde* Versetzung.

Im Grunde hat man also zu erwarten, daß bei tiefen Temperaturen die aus der Thermodynamik abgeleiteten Aktivierungsansätze ungültig werden.

Literatur

[1] Peierls, R.: Proc. phys. Soc., Lond. **52**, 34 (1940). — Für eine zusammenfassende Darstellung vgl. F. R. N. Nabarro: Advances in Physics **1**, 269 (1952).
[2] Leibfried, G.: Thermische und mechanische Eigenschaften von Kristallen. Handbuch der Physik VII/1, Berlin 1955.
[3] Fuchs, K.: Proc. roy. Soc., Lond., Ser. A **157**, 444 (1936).
[4] Leibfried, G. u. H. D. Dietze: Z. Phys. **131**, 113 (1952).
[5] Foreman, A. J., M. A. Jaswon u. J. K. Wood: Proc. phys. Soc., Lond., Ser. A **64**, 156 (1951).
[6] Becker, R.: Phys. Z. **26**, 919 (1925).
[7] Crussard, C.: Solvay-Konferenz Brüssel 1951 (L'Etat Solide).
[8] Leibfried, G.: Z. Phys. **127**, 344 (1950).

---

[1] $\sqrt{\overline{\tau^2}}$ für ein Atom nach Zeile 2 der Tabelle ist von der Größenordnung 100 kp/mm².

Dislocations in Crystals

By A. H. Cottrell, Harwell

With 9 figures

1. Dislocation networks. In this lecture, I shall talk mainly about the mechanical effects of dislocations in crystals. When beginning to think about the mechanical properties of, say, a face-centred cubic crystal we usually start from a picture of the type shown in fig. 1, in which the lines of dislocations are imagined to exist in the crystal in the form of a three-dimensional network. We do not start from a picture of the perfect crystal since X-ray and microscopic investigations [1–5] have shown that most crystals contain dislocations; since the presence of dislocations is an essential condition for crystal growth at practical rates from dilute solutions and vapours [6]; and since most crystals of normally plastic substances begin to deform plastically at stresses far below the theoretical strength for the creation of dislocations in the perfect lattice. Even the most conservative estimates of this theoretical strength lead to values of about $\mu/30$, where μ is the shear modulus; the critical shear stress for plastic flow in a soft metal crystal on the other hand usually lies in the range $10^{-5}\mu$ to $10^{-4}\mu$. A direct demonstration of the extremely high strength of crystals which are too small in diameter to contain dislocations, or to allow them to multiply, is provided by the extremely high strength of tin "whiskers" [7].

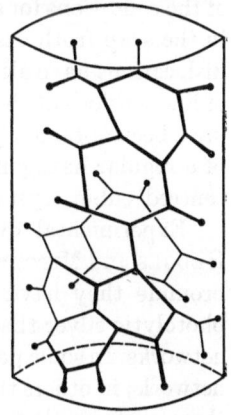

Fig. 1. A network of dislocations in a crystal

These have a diameter of about 10^{-4} cm., and can be bent elastically until the surface strain reaches the range 10^{-2} to 10^{-1}.

Why do we picture the dislocations as forming a network? There are both theoretical arguments and experimental evidence for such a picture. The theoretical arguments are as follows:

1. The elastic energy of a dislocation, obtained by integrating the long-range elastic strain field of a dislocation, is about 5 electron volts per atomic unit of length along the dislocation line. This is far too much to allow dislocations to exist as defects in thermal equilibrium in a crystal

of normal size. In principle, therefore, dislocations ought all to anneal out from crystals. The fact that in practice annealing rarely, if ever, reaches this stage must mean that dislocations in crystals usually settle down into extremely persistent, metastable arrangements. One thinks immediately of stable foam structures as another example of this kind of behaviour.

2. It is geometrically possible to link a number of dislocation lines into a three-dimensional network, each line being joined to its neighbours in a 'node' at each end, provided that the sum of the Burgers vectors[1] of all dislocations that emerge from a node is always zero.

3. The elastic energy of a dislocation provides the dislocation line with a line tension

$$T = \alpha \mu b^2, \tag{1}$$

where b is the length of the Burgers vector, and $\alpha \simeq 1/2$. A line spanning two nodes will thus tend to shorten as much as possible. As in the case of soap froths, certain types of networks can be rendered stable by the tensions of the members linked together in them. The detailed analysis of the conditions for stability of a particular network is more difficult than in the soap froth case, because the long-range elastic interactions of the dislocations have also to be taken into account, as well as the equilibrium of line tensions at the nodes. Solutions for certain simple types of network have been obtained [8]. One important configuration, for example, is that of a regular hexagonal net of dislocation lines on a (111) plane in a face centred cubic crystal.

Experimental evidence for such networks has been provided by Hedges and Mitchell [3]. Using lightly strained single crystals of silver bromide they have observed in the interior of these crystals lines of photolytic silver that show many of the features expected of dislocation networks. In some parts of the crystal these lines form a three-dimensional network; in others they lie in surfaces of "cells" about 10^{-3} cm. across. In these surfaces they sometimes form rows of parallel dislocations which correspond to the well-known picture of a "small-angle tilt boundary" between two crystals slightly rotated with respect to each other about an axis in the boundary [9, 10]. In other cases they form hexagonal nets of the type expected for (111) boundaries between crystals slightly rotated with respect to each other about a [111] axis.

It has also been established in recent years that the points where dislocations emerge on crystal surfaces can be located quite unambigu-

---

[1] The Burgers vector of a dislocation defines the displacement of atoms in one face of a slip plane, relative to those in the other face, which is caused by the passage of the dislocation along the plane; more generally, it is the residue of any closed line integral of displacement taken in a given sense once round a dislocation line.

ously in certain cases by means of etch-pits, which nucleate preferentially at such places. In germanium crystals, for example, rows of etch-pits along small-angle tilt boundaries were obtained; from the known change of crystal orientation across such a boundary, and from the known lattice parameter, the spacing of the dislocations which form the boundary could be calculated, and this agreed very well with the observed spacing of etch-pits [11]. The results are shown in Table 1.

Table 1.

Spacings of dislocations and etch-pits in small-angle boundaries in germanium crystals (after VOGEL, PFANN, COREY and THOMAS [11])

Angle between crystals (seconds)	Calculated spacing of dislocations (cm.)	Observed spacing of etch-pits (cm.)
17.5	$4.7 \cdot 10^{-4}$	$5.3 \cdot 10^{-4}$
65.0	$1.3 \cdot 10^{-4}$	$1.3 \cdot 10^{-4}$
85.0	$0.97 \cdot 10^{-4}$	$0.99 \cdot 10^{-4}$

X-ray reflections give indirect evidence for such networks. The "mosaic" crystal of DARWIN and EWALD was introduced many years ago to account for the large integrated intensity and the breadth of the reflection of X-rays from crystals. In essence this X-ray evidence shows that in many crystals there are small misorientations of the order of 1 degree of arc between pieces of the crystal about 10^{-4} cm. across; the defect structure that will produce such misorientations with least expenditure of energy is a network of dislocations. The densities of dislocations in some crystals have in fact been determined by this method [2].

All these experimental methods lead to the conclusion that the network in a well-annealed metal crystal is usually made up of links each about 10^{-4} to 10^{-3} cm. in length. Similar values for the scale of the network are also suggested by the observed critical shear stresses of annealed crystals. To produce slip in a crystal containing a network it is necessary to expand one of the links into a dislocation ring that sweeps out the slip plane. This involves first bending the dislocation into a semicircular loop of radius $l/2$, where l is the distance between its (fixed) nodes. To pass through this critical stage the force on the dislocation (σb per unit length) from the applied shear stress σ must exceed the force due to the line tension. This gives the minimum stress to expand the dislocation loop into its slip plane as

$$\sigma = \frac{2T}{bl} \simeq \frac{\mu b}{l}. \qquad (2)$$

With $\sigma \simeq 10^{-5}\mu$ to $10^{-4}\mu$ and $b \simeq 2.5 \cdot 10^{-8}$ cm. then $l \simeq 10^{-4}$ to 10^{-3} cm.

This process of producing slip by bending a link of the network into a dislocation loop, which then expands into its slip plane, has an important topological property, discovered by FRANK and READ [12], that makes possible the formation of slip bands in plastically deformed crystals. As fig. 2 shows, when a dislocation loop is expanded round two fixed nodes (fixed, for example, because the other dislocations that join these nodes do not lie in slip planes) the rotating arms of the loop coalesce when they run together behind the nodes and form a complete dislocation ring which can expand into its slip plane and produce a unit of slip, together with another dislocation link between the nodes which is able to bend out again and repeat the process. Thus a single "Frank-Read source", such as this, is able to produce almost unlimited amounts of slip on a single glide plane by repeated rotations about its nodes. The intense localisation of slip on a few active slip planes, which is a notable feature of the plastic deformation of crystals, can thus be explained quite readily from this simple geometrical property of a dislocation network.

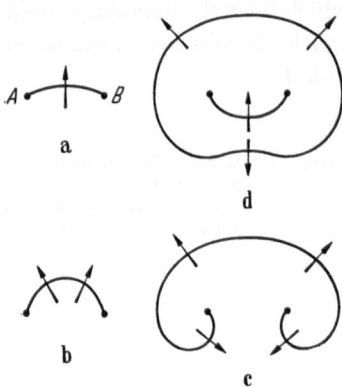

Fig. 2. Stages *a*, *b*, *c*, and *d*, in the formation of a dislocation ring from a Frank-Read source *A B* (by courtesy of Pergamon Press)

2. **Orientations of dislocations in networks.** We have next to consider the orientations of the various dislocation lines in a network. In an elastically isotropic crystal the long-range elastic energy of a straight dislocation line with a given Burgers vector depends on the angle between the line and the Burgers vector but is otherwise in dependent of the orientation of the dislocation in the lattice. A screw dislocation (for which the Burgers vector is parallel to the line) has an energy lower than that of an edge dislocation (Burgers vector perpendicular to the line) by a factor $1 - \nu$, where ν is Poisson's ratio. This leads to a tendency to form screw dislocations, but the need for mechanical stability at the nodes prevents each individual dislocation linked to the network from choosing its own orientation freely. Most of the dislocations must therefore be neither pure screw nor pure edge, but of intermediate orientations.

What crystallographic directions are the dislocation links likely to follow? Any dislocation with an edge component can be associated with a certain crystal plane, the plane defined by its line and its Burgers vector; for geometrical reasons the glide motion of the dislocation can take place only in this "glide plane", although the dislocation can move in other directions by a "climbing" motion in which it absorbs or generate vacancies (vacant atomic sites). In an elastically anisotropic crystal the long-range energy of the dislocation may depend on the orientation of its glide plane. Foreman and Lomer [*13*], for example have shown that the plane of lowest (long-range) energy for dislocations in copper, gold and aluminium is (110). The fact that (111) is the observed slip plane in these metals, and not (110), shows that the choice of the crystallographic plane of slip cannot be explained from the long-range elastic properties of dislocations.

3. The fine structure of dislocations. This difficulty over the crystallographic plane of slip is only one of many problems the solutions of which have had to wait until atomistic theories of the structures at the centres of dislocations could be developed. These theories deal broadly with three main aspects of the fine structure of dislocations: the width of dislocations; jogs; and partial dislocations.

The analysis of PEIERLS and NABARRO [*14, 15*] has shown that the force needed to move a dislocation through an otherwise perfect lattice (called the PEIERLS-NABARRO force) is extremely sensitive to the "width" of the dislocation, i. e. to the extent to which the highly strained region in the core of the dislocation is spread out along the slip direction. Defining w as the width of the region in which the slip displacement of the atoms in the slip plane of the dislocation lies between $^1/_4 b$ and $^3/_4 b$, the stress σ to move the dislocation is given in order of magnitude as

$$\sigma = \mu \exp\left(-\frac{2\pi w}{b}\right), \tag{3}$$

where μ is the shear modulus. For slip to be possible at a stress of $10^{-5}\mu$ the width w has to be greater than $2b$. The value of w depends on the misfit energy in the slip plane at the centre of the dislocation, and thus on the precise form of the laws of interatomic force for the material in question. In cases where hard incompressible atoms are held together by non-directed bonds, as in copper, the dislocation in a close-packed plane can be several atoms wide and extremely mobile; where the atoms are held together by strongly-directed covalent bonds, as in diamond, the dislocation is narrow and hard to move. The theory thus provides a basis for understanding the characteristic malleability of metals and brittleness of non-metals.

It is hardly reasonable to expect that a long dislocation line would remain in one particular atomic plane for its entire length. It must surely step occasionally from one layer of atoms to the next. Each such step is a "jog". These jogs have important properties. On dislocations with screw components they create vacancies or interstitial atoms when they are moved along with the dislocation; the stress to move the dislocation has therefore to be increased to supply the energy necessary to form these point defects [*16, 17*]. Jogs on edge dislocations are centres for the absorption or thermal creation of vacancies, and much of the theory of annealing and high-temperature creep is based on this property [*16*]. On pure edge dislocations jogs do not create point defects when they glide, although they are probably difficult to move (at low temperatures, at least) because the dislocation lines are narrow and have large PEIERLS-NABARRO forces in the vicinity of jogs [18, 19].

The theory of partial dislocations originated in the work of HEIDEN-REICH and SHOCKLEY [20]. Applied in particular to the face-centred cubic lattice the theory shows how a dislocation with a BURGERS vector equal to a unit lattice vector can lower its short-range energy by splitting, in a (111) plane, into two partial dislocations. Each of the latter has a short BURGERS vector, which displaces atoms from normal positions to those of the nearest twinned configuration along the (111) plane, but taken as a pair they still produce one lattice unit of slip. They are joined by a strip of stacking fault in the slip plane, in which the atoms lie in the twinned configuration, and the surface energy of this faulted layer holds the two partials together at a characteristic spacing that is commonly a few interatomic distances.

The potentialities of this theory with regard to the problem of deformation twins are obvious; we expect that under certain states of stress the partials can be pulled apart, so increasing the proportion of twinned material in the crystal [21]. But at the yield stress for slip in soft crystals the partials cannot be separated and must glide in pairs, leaving unfaulted lattice behind them. Their importance for slip lies in the following features:

1. The PEIERLS-NABARRO force of a partial is very small since the BURGERS vector, which determines b in equation (3), is small.

2. Dissociation into partials occurs substantially only on planes where it produces stacking faults of low energy. In face centred cubic metals these are (111), the observed slip planes. In body centred cubic iron there appears to be no plane on which low energy faults can be formed; correspondingly, there is no strongly preferred slip plane in this metal.

3. The energy of a given stacking fault in a lattice depends on the type of interatomic forces present. In copper, where the ion-ion repulsion of hard shells predominates, the fault energy is less than in aluminium, where BRILLOUIN zone effects are important. SEEGER and SCHOECK [22] have deduced from known twin boundary energies the values 40 and 200 erg. cm^{-2} for the stacking fault energies in copper and aluminium, respectively. They conclude that the width of the fault between the partials in copper is about 8 atomic spacings for a screw dislocation and 12 for an edge. In aluminium the width is only about 2 atoms. As a result, only a small activation energy (about 1 eV) is needed to transfer the glide direction of a screw dislocation in aluminium from one slip plane, such as (111), to an intersecting plane such as (11$\bar{1}$). This is held to account for the well-known tendency for slip bands in aluminium to deflect themselves from one such slip plane to an intersecting one with the same slip

direction ("cross-slip"). In copper on the other hand the corresponding
activation energy (about 10 eV) is too large for cross-slip. A similar argument shows that jogs on dislocations can be formed more easily in aluminium than in copper, so that annealing processes that depend on the
"climb" of dislocations can occur more easily in the former metal than
the latter [18].

4. Some types of partial dislocations cannot glide except at stresses
approaching the theoretical strength of the perfect lattice [23]. These
"sessile" dislocations behave in this way because their geometrically
defined glide plane — the plane containing the dislocation line and the
BURGERS vector — is not one in which their stacking fault has low energy.
They are important in theories of work hardening because they form
fixed centres of internal stress.

In studying partial dislocations experimentally it would be an advantage if we could vary the energy of a stacking fault continuously and
observe the effect of this on the plastic properties of the material. An
approach to such an experiment has recently been made by RACHINGER
and myself [24]. We prepared a series of crystals of compounds all of
which had the caesium chloride structure. This consists of a simple cubic
lattice with atoms of type A at
the cube corners and atoms of
type B at the cube centres (see
fig. 3). When there is no difference between A and B, so that
the lattice is in fact body centred cubic, the shortest lattice
vector is that which joins a cube
centre to a neighbouring corner.
This is the BURGERS vector of
the unit dislocation in the body
centred cubic lattice and it generates slip along [111], the observed slip direction for this lattice. But when A and B are different the dislocation is a partial dislocation of the CsCl
structure and is attached to a
stacking fault in which $A-A$

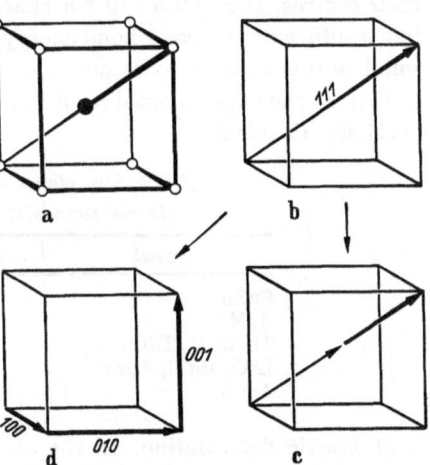

Fig. 3. a) Caesium chloride structure; b) BURGERS
vector of unit [111] dislocation; c) after dissociation
into partial [111] dislocations; d) after dissociation
into unit dislocations of the [100] type

and $B-B$ pairs of nearest neighbours are present. Provided A and B are
similar in size and electrochemical character, however, we expect the
crystal to permit the presence of such partials, in pairs, and to produce
[111] slip by the movement of these pairs. As the difference in A and B
increases, the specific energy of the stacking fault in such a pair also

increases, and to compensate for this the two dislocations take up a closer spacing. When the specific fault energy becomes large they eventually coalesce into a single large dislocation which is unstable and decomposes into unit lattice dislocations that produce slip along [100].

To obtain a working criterion RACHINGER and I argued that the critical surface energy γ to make the pair coalesce is that which exerts a force on the dislocation equal to the theoretical strength of the lattice. According as the fault energy is larger or smaller than this, so the direction of slip should be [100] or [111]. This gives $\gamma = \varepsilon \mu b$, where μ is the shear modulus, b is the length of the BURGERS vector, and $\varepsilon \mu$ is the theoretical shear strength ($\varepsilon \simeq 1/30$). Taking typical values of μ and b, $\gamma \simeq 250$ erg/cm². Expressed in terms of bond energies in a stacking fault on a (110) plane this is about 0.06 electron volts per bond. The observed ordering energy in CuZn corresponds to a bond energy of about 0.015 eV, so that the slip direction in this alloy should be [111], as is observed. However, this bond energy is exceptionally small and in many other alloys the bond energy may be large enough to change the slip direction. As a simple illustration, consider the formula $-z^2 e^2/r$ for the electrostatic energy of two charged ions, $+ze$ and $-ze$, spaced a distance r between their centres. If $r = 2.5 \cdot 10^{-8}$ a charge of only ± 0.1 electrons on each ion is sufficient to give a bond energy of 0.06 eV. An "ionic" character as small as this appears to be possible even in highly "metallic" alloys [25]. In fact, as the experimental results show, [100] slip is more common than [111] slip (Table 2).

Table 2. *Slip planes and slip directions*
in compounds of the CsCl type

Crystal	Slip direction	Slip plane
CuZn	111	110
AgMg	111	321
Tl(Br, I), Tl(Cl, Br) LiTl, MgTl, AuZn AuCd }	100	110

4. Plastic deformation. In the above sections we have sketched the picture of a dislocated crystal usually accepted at present as the starting point for theories of the plastic properties of solids. We shall now consider what happens when such a crystal is strained into the plastic range.

In some respects it is simplest to begin by considering the behaviour of slightly impure crystals (or crystals containing a superabundance of point defects, produced by quenching or by bombardment with fast atomic particles), in which the impurity atoms have reduced their elastic energies by segregating to dislocations. Plastic deformation then begins with a sudden jerk — the yield point phenomenon — as dislocations pull themselves away from these impurity atoms [26]. This break-away process is strongly influenced by thermal fluctuations and the well-known sensitivity of the upper yield point of mild steel to the rate and temperature of straining is thought to originate here [26].

The yield phenomenon has been used recently by ADAMS and myself [27] to establish, in copper crystals, that the first dislocations to become mobile are those that lie near the surface, where the scale of the dislocation network presumably is coarser than in the interior of the crystals. By introducing about 1 percent of zinc into the surface of the specimen, to a depth of a few microns, we were able to produce yield points in copper crystals that normally do not show the yield phenomenon. After the zincified surface had been removed by etching, this yield phenomenon could no longer be produced.

There is also some evidence from the behaviour of pure metals, where the yield point phenomenon is not observed, that the active FRANK-READ sources are longer than average links of the network. The argument is as follows. If the sources were of only average length, slip would begin when the "loop" stress, given by equation (2), was reached. A consequence of this is that the variation of the yield stress σ with the temperature of straining ought to be almost identical with that of the shear modulus, μ. But in practice σ/μ is not constant at low temperatures (e.g. in aluminium it is about 20 percent larger in liquid air than at room temperature), although its variation is much less than that of an impurity yield point. In the case where slip begins at abnormally long sources the yield stress (at low temperatures at least) is not the "loop" stress but the stress needed to make the generated dislocation loops cut through the "forest" of intersecting dislocations provided by the network [16, 17, 26]. This cutting process produces jogs at the points of intersection. If U_0 is the work done at the point of intersection during cutting, l is the distance between points of intersection along the gliding dislocation, σ is the applied stress and d is the displacement of the gliding dislocation during cutting, then the work done by the applied stress at each point of intersection during cutting is $\sigma b l d$, and the remainder,

$$U = U_0 - \sigma b l d, \tag{4}$$

has to be supplied by thermal fluctuations. This leads of course to a decrease of σ/μ with increasing temperature. The values of U_0 and d depend on the orientations and fine structures of the intersecting dislocations. For an edge dislocation cutting a screw in aluminium, $U_0 \simeq 0.5\,\mathrm{eV}$ and $d \simeq b$; in copper, $U_0 \simeq 5\,\mathrm{eV}$ and $d \simeq 10b$ [17, 28]. An additional factor to be taken into account when attempting to interpret observed variations of σ/μ with temperature is that energy has to be supplied to move the jogs formed by the cutting process; the glide of jogs on dis-

3 b

locations is also determined by formulae of the type of equation (4), but with different values of U_0, l and d [28]. As a result the variation of σ/μ with temperature can be rather complicated and may differ quite radically from one metal to another [17]. This is shown by measurements on aluminium [29] and copper [30] crystals (see fig. 4).

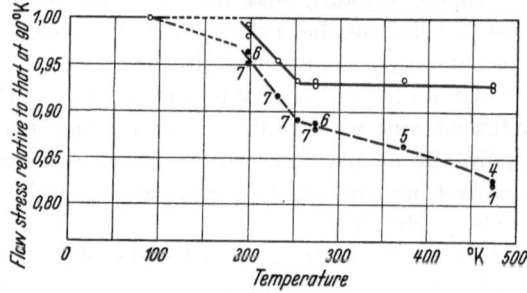

Fig. 4. Variation of flow stress of copper crystals with temperature, before (broken line) and after (full line) correction for the variation of elastic modulus; reference [30] (by courtesy: Philosophical Magazine)

As the gliding dislocations move out into their slip planes they meet other obstacles such as foreign atoms, precipitates, grain boundaries, and surface films. The theory of the behaviour of dislocations under these various circumstances is now fairly well developed [26]. In this lecture, however, I shall consider only the theory of work hardening, which involves interactions of dislocations with one another and which is at present at an interesting stage of development.

5. Work hardening. In a crystal oriented for slip on a single family of planes we picture the glide dislocations as all moving in parallel planes. Two such dislocations on nearby planes may become captured in each other's stress field and so become unable to move independently of each other [31]. We believe that this effect cannot, *by itself*, be the cause of work hardening, for several reasons:

1. The dislocations in such a pair are anchored only to each other, not to the lattice, so that a third dislocation can push them along together.

2. The deformation bands [32] which are observed to form from large groups of such elastically bound dislocations have short-range stress fields which cannot appreciably harden the crystal as a whole [16].

3. If the flow stress σ of the cold-worked crystal were determined entirely by the long-range elastic forces between glide dislocations, then σ/μ ought to be intensitive to temperature. But in practice the sensitivity of σ/μ is hardly altered by cold work [29, 30]. The effect is as if the density of the forest of intersecting dislocations increased in proportion to the flow stress of the material.

4. Face-centred cubic crystals of pure metals work harden only slightly when deformed plastically on a single glide system (see fig. 5); this is the phenomenon of "easy glide" [33—35].

It is usually considered that easy glide ends through the occurrence of small amounts of slip on secondary systems that intersect the primary glide planes. Often this secondary slip occurs almost fortuitously, due to stress concentrations formed where dislocations pile up in deformation bands, or behind precipitates, or beneath surface films. In crystals oriented symmetrically for multiple slip (e. g. axis of tension on or near the [100]–[111] symmetry zone) easy glide does not occur and intense work hardening sets in immediately.

One effect of slip on intersecting systems is to increase the density of the forest. While this undoubtedly contributes

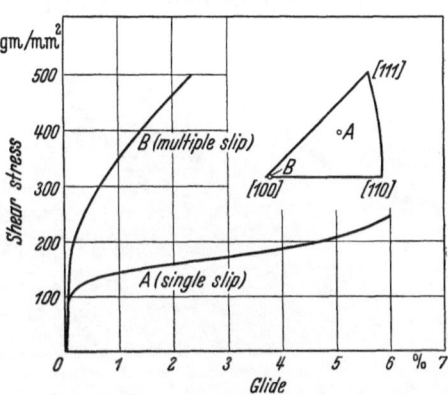

Fig. 5. Stress-strain curves on pure aluminium crystals (after Lücke and Lange); (by courtesy: Oxford Press)

to the hardening of the metal at low temperatures and may explain the sensitivity of σ/μ to temperature in cold worked crystals, it cannot account for the whole of work hardening since, for example, coldworked aluminium crystals retain over three-quarters of their low-temperature strength when heated into the range (above about 200° K) where σ/μ ceases to vary with temperature [29] and where the thermal fluctuations are considered to be strong enough to make the forest practically "transparent" to gliding dislocations [17].

Evidently there must be some additional hardening effect of intersecting slip. We believe that this is provided by the so-called "Lomer-Cottrell barriers", formed along lines of intersection of active slip planes by the coalescence of dislocations [36, 37]. For example, a dislocation in a (111) plane with a Burgers vector along [10$\bar{1}$] is attracted by one in (11$\bar{1}$) with a vector along [011] and they run together to coalesce along the line [1$\bar{1}$0], forming a dislocation with a vector along [110]. This coalesced dislocation is geometrically capable of gliding along (001) but can become completely locked to the lattice by dissociating into a sessile group of three partial dislocations, two of which split off into the planes (111) and (11$\bar{1}$) while the third remains at the line of intersection.

When the stacking fault energy is large, as in aluminium, the separation of these partials is small and a correspondingly small activation energy is needed to close up the group to enable the coalesced dislocation to glide along (001). FRIEDEL [*38*] has used this argument to account for the (100) slip observed in aluminium at high temperatures. In copper, on the other hand, where the stacking fault energy is small this cannot happen and (100) slip is not observed.

As regards evidence for such barriers the photographs obtained by JACQUET [*39*] on lightly-strained and etched brass show numerous examples where dislocations in intersecting glide planes have run together at the lines of intersection and there formed barriers strong enough to withstand the pressure of the dislocations piled up behind them in these glide planes (see fig. 6). Furthermore, as FRIEDEL [*38*] has pointed out, the short, fragmentary slip bands, observed to form in copper [*40*] during the stage of linear work hardening that follows easy glide, are evidence of strong barriers in the primary slip planes at about the spacing (about 10^{-3} cm.) expected for FRANK-READ sources of secondary systems capable of providing the necessary barrier dislocations. Starting from this observation, FRIEDEL has built up a quantitative theory of linear hardening in face-centred cubic metallic crystals [*38*].

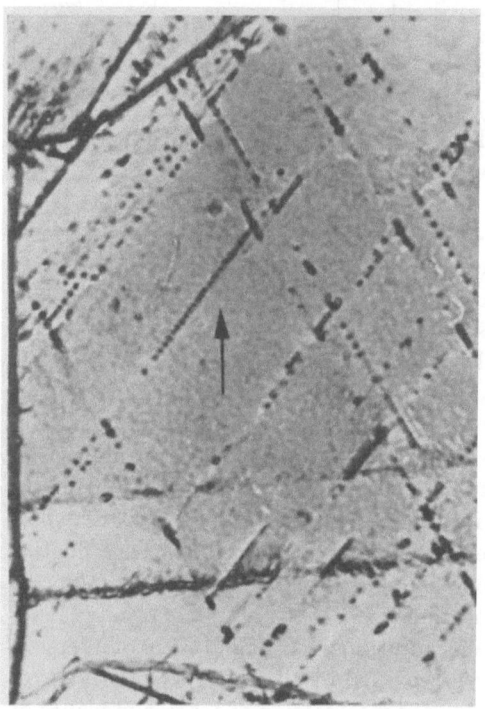

Fig. 6. Etch-pits on slip lines in lightly deformed alpha brass (by courtesy: P. A. JACQUET: Comptes Rendus *257* [1953], p. 1248)

Indirect evidence for such barriers is provided by the phenomenon of "work-softening", which has been observed in crystals of aluminium and copper [*29, 30*]. After being strained heavily in tension at low temperature (e.g. in liquid air), each such crystal showed a large yield phenomenon (e.g. a 10% fall in stress) when strained again at some higher temperature (e.g. room temperature) (see fig. 7 and 8).

The resemblance to an impurity yield point was striking, although subsidiary experiments proved that it was not caused by the migration of impurities or point defects to dislocations. The material in fact appeared to be already fully "strain aged" at the low temperature, although in such a manner that a yield drop could be produced only by increasing the temperature of straining some 100° K or so. Now a barrier dislocation along the line of the leading glide dislocation in an active slip plane is similar in some respects to a line of impurity atoms along the glide dislocation, and we may expect some similarity of behaviour in the two cases. In particular we expect that the barrier may be broken down by the combined action of the applied stress and thermal fluctuations, and that the dislocations freed in this manner can glide forward

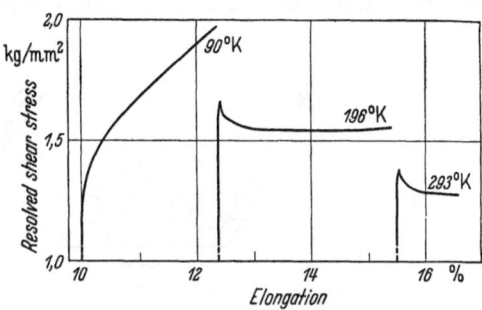

Fig. 7. Yield drops produced in an aluminium crystal by raising the temperature of cold-working: reference [29] (by courtesy: The Royal Society)

and force other barriers to break down. In this way a catastrophic release of dislocations spreads through the crystal and a yield drop is observed.

According to this theory the maximum number of dislocations held up by a single barrier at a certain applied stress must decrease with increasing temperature. This means that the states of cold work developed in crystals at different temperatures (or rates) of straining should differ

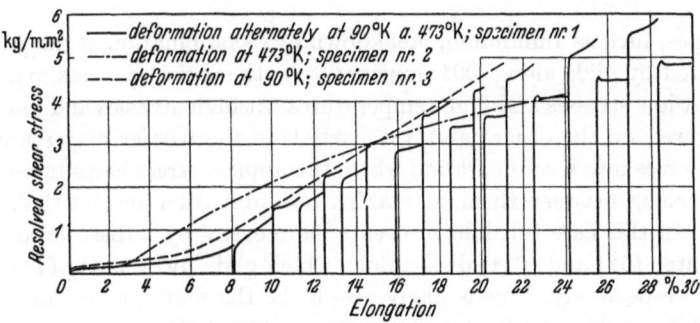

Fig. 8. Effects of temperature on the work hardening behaviour of copper crystals; reference [30] (by courtesy: Philosophical Magazine)

not only in intensity, but also in kind [29]. For example, a crystal cold worked to a given hardness should contain a small number of large piled-up groups if worked at a low temperature, and a large number of small piled-up groups if worked at a high temperature.

The thermal breakdown of LOMER-COTTRELL barriers has also been suggested independently by FRIEDEL [38] to explain the onset of the parabolic stage of hardening, which follows the linear stage and begins

at a stress that decreases with increasing temperature (see fig. 8). The
mechanism of breakdown is at present being studied theoretically by
STROH [*41*] and FRIEDEL [*38*]. There are several ways in which breakdown
can occur and the choice depends on the fine structure of the barrier and
on stress and temperature (see fig. 9). For metals with large stacking fault

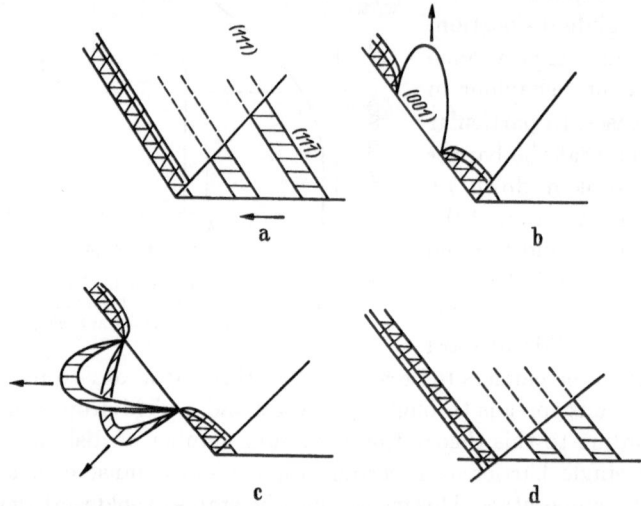

Fig. 9. a) Pile-up of dislocations against a barrier; b) collapse of barrier into dislocation on (001);
c) collapse of barrier into dislocations on (111) and (11$\bar{1}$); d) alternative form of barrier

energies, such as aluminium, breakdown by recombination of the partials
followed by glide along (001) appears to be the easiest process, particularly at low stresses and high temperatures. At high stresses and low temperatures, on the other hand, the partials in the barrier may rearrange
themselves into a configuration where the applied stress helps to increase
the spacing between them, so making recombination and (001) slip difficult. In this case breakdown occurs more easily by a dissociation into
separate [10$\bar{1}$] and [011] dislocations which glide away along (111) and
(11$\bar{1}$), respectively. This is likely also to be the main process in metals
with low stacking fault energies, such as copper [*41*].

   6. Conclusion. The theories of dislocation "forests" and of sessile barriers which I have sketched briefly and qualitatively in the above
sections are still in an active stage of development and it is too early yet
to say that we have a complete theory of work hardening. One question
that has not been convincingly answered so far is why the sensitivity of
the flow stress to temperature is practically constant for a wide variety
of cold-worked states [*29, 30*], which seems to imply that the density of
the dislocation forest manages, in some way that is not entirely clear, to

maintain itself in close proportion to the level of internal stress provided by the sessile barrier hardening. There is also the general question of the relation of the dislocation structures suggested by these theories to the "cellular" structures observed by the X-ray micro-beam method [2].

It would be unwise to judge at this stage whether these problems are likely to upset the present theory of work hardening, or whether they will be resolved by the straight-forward development of all its details; both the theory itself and the experimental observations against which it is being compared are too new. On the other hand, the leading idea in the theory — that the origin of strong work hardening lies in the inter-actions of dislocations on intersecting systems — is hardly likely to be affected by future developments, since it is so firmly supported by the experimental observations of easy glide, of the effects of crystal orien-tation and temperature on work hardening, and of pile-ups in active slip bands.

I am grateful to Drs. J. FRIEDEL, A. N. STROH, and A. SEEGER, for allowing me to refer to unpublished work of theirs on the theory of work hardening.

References

[1] WILLIAMSON, G. K. and W. H. HALL: Acta Metall. 1, 22 (1953).

[2] GAY, P., P. B. HIRSCH and A. KELLY: Acta Metall. 1, 315 (1953).

[3] HEDGES, J. M. and J. W. MITCHELL: Phil. Mag. 44, 223 (1953).

[4] WILSDORF, H. and D. KUHLMANN-WILSDORF: Defects in Crystalline Solids, p. 175. London: Physical Society 1955.

[5] WYON, G. and P. LACOMBE: Defects in Crystalline Solids, p. 187. London: Physical Society 1955.

[6] BURTON, W. K., N. CABRERA and F. C. FRANK: Nature, Lond. 163, 398 (1949).

[7] HERRING, C. and J. K. GALT: Phys. Rev. 85, 1060 (1952).

[8] FRANK, F. C.: Defects in Crystalline Solids, p. 159. London: Physical Society 1955.

[9] BURGERS, J. M.: Proc., nederl. Akad. Wetensch. 42, 293 (1939).

[10] READ, W. T. and W. SHOCKLEY: Phys. Rev. 75, 692 (1949).

[11] VOGEL, F. L., W. G. PFANN, H. E. COREY and E. E. THOMAS: Phys. Rev. 90, 489 (1953).

[12] FRANK, F. C. and W. T. READ: Phys. Rev. 79, 722 (1950).

[13] FOREMAN, A. J. E. and W. M. LOMER: Phil. Mag. 46, 73 (1955).

[14] PEIERLS, R.: Proc. phys. Soc., Lond. 52, 34 (1940).

[15] NABARRO, F. R. N.: Proc. phys. Soc., Lond. 59, 256 (1947).

[16] MOTT, N. F.: Proc. phys. Soc., Lond., Ser. B 64, 729 (1951); Phil. Mag. 43, 1151 (1952); Phil. Mag. 44, 742 (1953).

[17] SEEGER, A.: Defects in Crystalline Solids, p. 328. London: Physical Society 1955.

[18] SEEGER, A.: Defects in Crystalline Solids, p.391. London: Physical Society 1955.

[19] MADDIN, R. and A. H. COTTRELL: Phil. Mag. 46, 735 (1955).

[20] HEIDENREICH, R. D. and W. SHOCKLEY: Report on Strength of Solids, p. 57. London: Physical Society 1948.

[21] COTTRELL, A. H. and B. A. BILBY: Phil. Mag. **42**, 573 (1951). — N. THOMPSON and D. J. MILLARD: Phil. Mag. **43**, 422 (1952).

[22] SEEGER, A. and G. SCHOECK: Acta Metall. **1**, 519 (1953).

[23] FRANK, F. C.: Proc. phys. Soc., Lond., Ser. A **62**, 202 (1949).

[24] RACHINGER, W. and A. H. COTTRELL: Acta Metall., in press (1955).

[25] FRIEDEL, J.: Phil. Mag. **43**, 153 (1952).

[26] COTTRELL, A. H.: Dislocations and Plastic Flow in Crystals. Oxford 1953.

[27] ADAMS, M. A. and A. H. COTTRELL: to be published.

[28] SEEGER, A.: private communication (1955).

[29] COTTRELL, A. H. and R. J. STOKES: Proc. roy. Soc., Lond., in press (1955).

[30] ADAMS, M. A. and A. H. COTTRELL: Phil. Mag., in press (1955).

[31] TAYLOR, G. I.: Proc. roy. Soc., Lond., Ser. A **145**, 362 (1934).

[32] HONEYCOMBE, R. W. K.: J. Inst. Met. **80**, 45, 49 (1951).

[33] ANDRADE, E. N. DA C. and C. HENDERSON: Phil. Trans. roy. Soc., Lond., Ser. A **244**, 177 (1951).

[34] MASING, G. and J. RAFFELSIEPER: Z. Metallkde. **41**, 65 (1950).

[35] LÜCKE, K. and H. LANGE: Z. Metallkde. **43**, 55 (1952).

[36] LOMER, W. M.: Phil. Mag. **42**, 1327 (1951).

[37] COTTRELL, A. H.: Phil. Mag. **43**, 645 (1952).

[38] FRIEDEL, J.: Phil. Mag., in press (1955).

[39] JACQUET, P. A.: Acta Metall. **2**, 752 (1954).

[40] BLEWITT, T. H., R. R. COLTMAN, and J. K. REDMAN: Defects in Crystalline Solids, p. 369. London: Physical Society 1955.

[41] STROH, A. N.: to be published.

Diskussionsbemerkung zum Vortrag von A. H. Cottrell

Von A. Seeger, Stuttgart

Dr. COTTRELL hat im vorangehenden Beitrag gezeigt, wie die Beobachtungen über die Gleitrichtung bei der plastischen Verformung von Verbindungen bzw. geordneten Legierungen mit CsCl-Struktur auf Grund der Annahme interpretiert werden können, daß ein kritischer Wert für die spezifische Energie der Grenzflächen zwischen den geordneten Bereichen über die Auswahl zwischen den Gleitrichtungen ⟨100⟩ und ⟨111⟩ entscheidet.

Hier soll auf eine andere Deutungsmöglichkeit hingewiesen werden, die in den meisten Fällen qualitativ die gleichen Ergebnisse wie die COTTRELLsche vorhersagt, vielleicht aber auf Grund zukünftiger Messungen von ihr experimentell unterschieden werden kann.

Sie geht aus von der naheliegenden Annahme, daß für die BURGERS-Vektoren des größten Teils der in einem Kristall gebildeten Versetzungen die bei der Kristallisation gebildete Struktur maßgebend ist. Dies bedeutet, daß bei einer kubischraumzentrierten Legierung, die bis zum Schmelzpunkt geordnet ist, überwiegend BURGERS-Vektoren ⟨100⟩ vorhanden sein werden, da dies die kürzesten Gittervektoren des betreffenden BRAVAIS-Gitters sind. Bei einer Legierung, die im ungeordneten Zustand erstarrt bzw. rekristallisiert und sich erst bei tieferen Temperaturen in die geordnete Phase umwandelt, sollten demnach die Versetzungen BURGERS-Vektoren $1/2⟨111⟩$ haben. Diese „Halbversetzungen" der geordneten Struktur sind dann durch einen aus einer Bereichsgrenzfläche gebildeten Anordnungsfehler

paarweise miteinander verbunden. Man kann kaum annehmen, daß sich die überwiegende Mehrzahl dieser Halbversetzungen zu „vollständigen" Versetzungen zu-
sammenlagert, da dazu sehr komplizierte Versetzungsbewegungen und Spannungs-
verhältnisse notwendig wären. Man erwartet, wie es auch beobachtet wird, in diesem
Falle die Gleitrichtung $\langle 111 \rangle$.

Im anderen Falle (BURGERS-Vektor $\langle 100 \rangle$) hängt die beobachtete Gleitrichtung
davon ab, ob die äußere Spannung für die Zerlegung der BURGERS-Vektoren nach
dem Schema

$$[100] = \frac{1}{2} [1\bar{1}\bar{1}] + \frac{1}{2} [111] ,$$

die mit Energieaufwand verbunden ist, ausreicht. Man kann also ohne genaue
Kenntnis der Grenzflächenenergie der geordneten Bereiche nicht sagen, welche
Gleitrichtung zu erwarten ist. Eine entsprechende Zerlegung, wie oben angegeben,
scheint auch für das Auftreten der $\langle 211 \rangle$-Gleichrichtung bei sehr zinkreichem
α-Messing maßgebend zu sein.

Zusammenfassend kann man also sagen, daß man auf Grund der hier dargelegten
Auffassung stets die Gleichrichtung $\langle 111 \rangle$ erwartet, wenn die Ordnungs-Unordnungs-
umwandlung unterhalb der Schmelz- bzw. Rekristallisationstemperatur stattfindet,
während es andernfalls von den Einzelheiten abhängt, ob $\langle 111 \rangle$ oder $\langle 100 \rangle$ als
Gleitrichtung beobachtet wird.

Discusión de la Conferencia de A. H. Cottrell

Por J. L. Amorós, Barcelona

Con 4 figuras

Dos cuestiones son de interes en cuanto a la distribución de las dislocaciones en
la superficie del cristal de pirita estudiado por nosotros[1]: relación con las lineas de
deslizamiento del cristal y relación con la poligonización del cristal. En el primer
caso, las dislocaciones se agrupan a lo largo de aquellas lineas y están separadas
a distancias aproximadamente constantes. ESHELBY y FRANK y NABARRO han
elaborado una teoria sobre el equilibrio de una distribución lineal de dislocaciones.
Debido a que las dislocaciones se repelen entre si con intensidad inversa a la distancia
mutua entre ellas, se llega a la conclusión de que deben situarse a distancias aproximadamente iguales entre si. La fig. 1 es un claro ejemplo de esta distribución perio-
dica de las dislocaciones en la pirita, confirmando, de manera indirecta, la teoria.
Según FRANK y READ en una linea de deslizamiento emerge una fila de dislocaciones
helicoidales. Por tanto, mediante una corrosión ligera pueden hacerse visibles, tal
como ha sido demostrado por AMELIOKX en cristales de aluminio convenientemente
tratados. De aquel trabajo se deduce que efectivamente, las dislocaciones se presentan a lo largo de las lineas de deslizamiento. Una confirmación de este hecho lo
tenemos en la fig. 1 antes mencionada, donde la periodicidad es clara y su agrupa-
ción sobre ciertas lineas es patente. En la fig. 2 vemos una exfoliación de corrosión;
las tensiones producidas por esta exfoliación se traducen en una serie de lineas de
deslizamiento, en las cuales se concentran las dislocaciones, primero independientes, (D) después coalescentes (B) y, finalmente, unidas para formar un peldaño
claramente visible (C). Según el peldaño se realiza posteriormente la exfoliación por

---

[1] Esta nota es parte de un trabajo que aparecerá en Publicaciones del Departa-
mento de Cristalografia, Barcelona en colaborociòn con M. Brandoly.

corrosión, típica de los cristales de pirita. Si unimos esto a la observación anterior de que las dislocaciones se producen a intervalos regulares, se ve claramente la

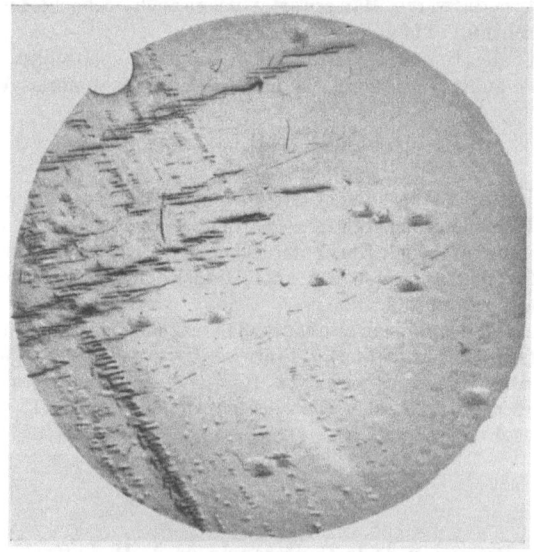

Fig. 1

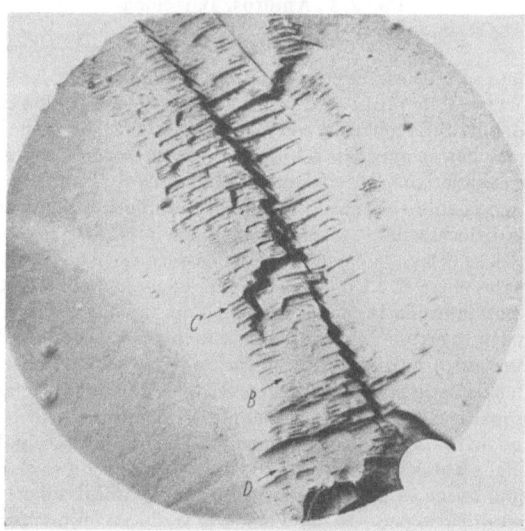

Fig. 2

razón de la formación de bloquecitos en la exfoliación de la fig. 1. Estos bloques son múltiplos enteros de la distancia correspondiente a la separación mútua de las dislocaciones de la fig. 1, como facilmente puede comprobarse.

El proceso antes señalado para la formación de un escalón de deslizamiento confirma la suposición de que las lineas de deslizamiento vengan caracterizadas por la presencia de filas de dislocaciones a lo largo de ellas, dando, por tanto, evidencia

Fig. 3

experimental al mecanismo propuesto por FRANK y READ para este tipo de desliza-mientos. Uno de los hechos más evidentes de la fig. 1 es que las dislocaciones no se presentan distribuidas uniformemente a lo largo de la linea de tensión, sino que se

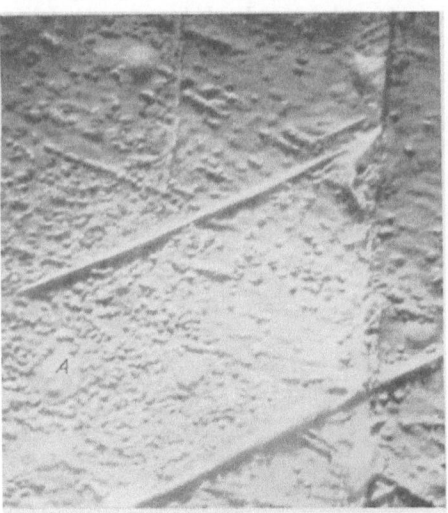

Fig. 4

presentan agrupadas. Este hecho coincide con la previsión teórica de MOTT y la observación de AMELINCKX. Por otra parte, esta distribución es sensiblemente periódica, lo cual puede explicarse sencillamente por ser el cristal un medio periódico.

4*

Siendo el medio cristalino periódico toda acción ejercida será más o menos largo según la resonancia especifica de la acción y red y dependerá, evidentemente, de ambos. Veamos que pasa con una tensión aplicada sobre una linea en el cristal. La manera de disminuir la tensión es la deformación de la red en esa linea, pero la manera estable de la red es el estado sin deformar, luego, el compromiso entre ambas tendencias se logrará, precisamente, la una alternancia de zonas deformadas y zonas sin deformación, tanto más extensas estas últimas cuanto mayor sea la tensión aplicada. La consecuencia de todo esto es la aparición de máximos de deformación a tramos equidistantes de la zona de aplicaciones de la tensión, máximo que serán demostrables por la concentración de ladislocaciones. La fig. 1, anteriormente citada, nos muestra con toda evidencia esta concentración discreta y periódica de las alineaciones de dislocaciones, viéndose que la separación entre las filas de dislocaciones es practicamente constante para cada sistema de planos reticulares. Esta relación la presentamos aquí solo, se forma cualitativa, pues todavía no se conoce la exacta relación entre el número de pozos de corrosión y el número de dislocaciones. Lo que se puede afirmar siempre es que una dislocación puede ser el centro de un pozo de corrosión, pero no que cada dislocación necesariamente origine un pocillo.

En los casos en que no ocurre una distribución de dislocaciones a lo largo de lineas de tensión, es decir, cuando las dislocaciones se distribuyen de manera general en la cara, lo hacen fundamentalmente en los bordes de los granos que constituyen la estructura de poligonización. Puede verse claramente este fenómeno en las fig. 3. Aun en aquellos casos en que una gran masa de dislocaciones ocupan una gran zona, se preservan nitidamente zonas cuyo aspecto es idéntico al citado al describir el poligonización de la superficie cristalina, fig. 4 (A). Luego, podemos inferir que la poligonización es sencillamente, la independencia de determinadas zonas de cristal «bueno», en el sentido de Frank, limitado por estrechas zonas de cristal «malo», es decir, por concentraciones de dislocaciones. Algo semejante a lo que nos estamos refiriendo ha sido descrito en los cristales de bromuro de plata antes citados.

Theories of Fracture in Metals

By **N. F. Mott,** Cambridge

With 3 figures

1. Introduction. One of the objectives of any theory of plastic deformation based on dislocations must be to explain the origin of fracture and its connection with plastic flow. In particular we have to try to understand

a) Brittle fracture,

b) Ductile fracture,

c) Fatigue.

The purpose of this paper is to discuss these three phenomena.

2. Work-hardening. Before attempting to describe possible theories of these processes, it will be useful to outline what is known about work-hardening, particularly of single crystals.

Recent experimental research enables us to distinguish three stages of the work-hardening process, shown as I, II and III in fig. 1. They may be described as follows:

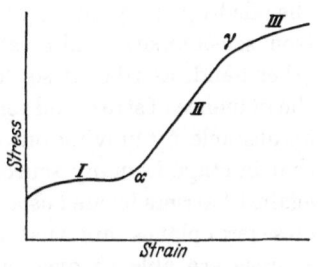

Fig. 1.
Work-hardening of a single crystal

I. *Easy glide.* This is absent in polycrystalline materials and occurs in single crystals only when slip is confined to a single slip system. No detailed theory of work-hardening in this region has been given, but it seems likely that dislocations move a comparatively large distance before being held up, and that such hardening as is observed is due to a gradual accumulation of dislocations in the material.

II. Stage II is certainly associated with slip on more than one set of planes. The point α (fig. 1) at which it begins is markedly dependent on orientation; the hardening in the rapid (linear) region is not strikingly dependent on temperature, orientation or purity. It is widely believed that the rapid hardening is due to the following effect; according to the ideas of COTTRELL (1952) and LOMER (1952) two dislocations on different slip planes and with different BURGERS vectors can join together to form

a "sessile" dislocation, or dislocation which is unable to move. The existence of these barriers, and of other barriers of similar type, greatly limits
the distance which any dislocation can move, so that the "FRANK-READ"
sources must in fact produce many more dislocations to be stored in the
material after a given strain. Exactly how these dislocations produce the
hardening is a matter of dispute; FRIEDEL (1955) has published a detailed
theory in which piled-up groups containing some hundreds of dislocations
form at the ends of the slip lines, against the COTTRELL-LOMER barriers.
These piled-up groups, as first suggested by the present author (MOTT
1952), may have a much stronger influence on the hardness than a random
distribution of dislocations as in TAYLOR's theory of 1933. We shall return
to this theory below.

The point γ on the stress-strain curve at which the rate of hardening
decreases rapidly has the following characteristics:

(i) It moves rapidly with decreasing temperature to higher stresses.

(ii) Only for strains greater than γ do the familiar long slip bands
occur (DIEHL, MADER and SEEGER 1955), which recent research has shown
often to consist of a cluster of elementary lines. For aluminium at room
temperature the points α and γ are hardly separated, but they separate
as the temperature is lowered.

FRIEDEL ascribes the point γ to the breaking down of COTTRELL-
LOMER barriers under the combined influence of stress and temperature,
the piled-up groups greatly accentuating the stress's activity. Dislocations
from selected sites will be able to move a large distance. SEEGER, on the
other hand, ascribes it so "cross slip", a process by which, again under
the influence of stress and temperature, a screw dislocation can pass round
an obstacle by moving out of its slip plane. In any case it seems clear
that in stage II many sources have produced dislocations which pile up
against barriers formed as a consequence of the movement of dislocations
on several planes; and that at the point γ dislocations from a *few* of these
sources are able to overcome the barriers and move right across the
crystal.

As mentioned above, however, the coarse slip bands are observed
under the electron microscope so consist of clusters of elementary lines.
It seems to the writer that one way in which this could be understood is
if we believe thas a series of dislocations, moving across a crystal already
hardened in stage II of the deformation, can *soften* the material through
which they pass, thus allowing further slip to occur preferentially in this
region. If one believes that piled-up groups are formed in stage II it is
not impossible to see how this may occur. In FRIEDEL's model, for
example, in the slip plane of the dislocations that it will generate, a source
is typically surrounded by a ring of COTTRELL-LOMER barriers, against
which piled-up groups of dislocations are formed. This is illustrated in

fig. 2, where the dislocations in the slip plane from the source S are shown by full lines. It is the "back stress" from these dislocations which prevents the source from working. In stage III a dislocation "escapes" from another such arrangement on a different but parallel slip plane; if this slip plane is close to that of S, the dislocation, shown by the dotted line in fig. 2, will be strongly attracted to the piled-up group and form a ring just below or above it. Something of the kind was first proposed by OROWAN (1941), and KUHLMANN-WILSDORF e. a. (1952) has suggested it as a softening mechanism. The soft-

ening will occur because the piled-up group, which initially has a stress field round it which falls off inversely as the distance, will be converted into a dipole, with a field that falls off as $1/r^2$. The stress from it at the source S will be very much decreased, and S will be able to emit more dislocations, which will push through the barriers $ABCD$ either by breaking them down (FRIEDEL 1955), or by cross slip (SEEGER 1955, 1956). In either case a

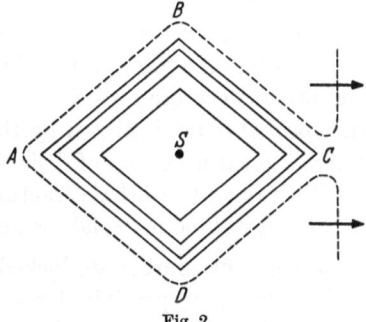

Fig. 2.
Showing dislocation rings on a slip plane

banded structure is obtained, since the process will be repeated whenever a dislocation or group of dislocations passes closely above or below an arrangement of the type shown in fig. 2.

3. Fracture. As mentioned above, we have to discuss brittle fracture, ductile fracture and fatigue. The initiation of the crack in these three drocesses is not necessarily due to the same mechanism. At the time of writing two mechanisms have been put forward to explain brittle and ductile fracture; a mechanism hitherto unpublished will be put forward to explain fatigue.

Brittle fracture. It seems to be widely agreed that the GRIFFITH mechanism — the initiation of fracture by pre-existent surface cracks, though applicable to amorphous materials such as glass, does not correctly describe fracture in metals, particularly the brittle fracture which occurs at low temperatures in ferritic materials. It is widely agreed that fracture in metals and probably most other crystalline materials is preceded by slip, though possibly by slip in one grain only.

Low (1956) in a paper presented to this conference points out that, when a number of screw dislocations from a given source cut a screw dislocation, part of a plane of atoms in the crystal grain is removed. He suggests that, if a tensile stress exists across this plane, a crack may be initiated. In support of this argument HOLLOMON has pointed out to the author that this accounts nicely for the fact that hydrogen embrittles steel much more markedly for *slow* rates of deformation; at slow rates the hydrogen would have time to diffuse into these cracks and force them apart. Against it we have two objections:

a) Although no calculations have been made, it seems to us that (in the absence of hydrogen) the stress that would have to be applied to separate a gap one atom thick would be very large.

b) We no longer believe that the above mechanism *would* remove a plane of atoms, because the jog formed will move by "conservative motion" (i. e. without forming vacancies) sideways along the dislocation.

The present author prefers the model developed by STROH (1954, 1955) and others according to which brittle fracture certainly and ductile fracture probably are initiated by the high stresses at the end of a slip line, where a piled-up group of dislocations is formed.

With regard to brittle fracture, the analysis by STROH suggests that in polycrystalline materials the conditions for fracture should be

a) that the sources are locked, either by impurities (carbon, nitrogen or hydrogen) or possibly by a PEIERLS-NABARRO force of which the magnitude is not known. Under these conditions, once a source begins to act it will at once, without further increase in the applied strain, produce a *large* number of dislocations which pile up against the boundary[1].

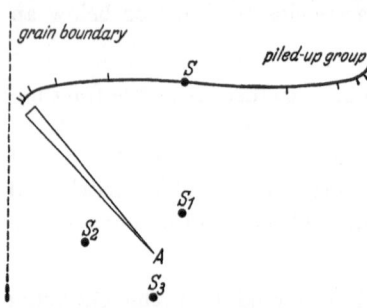

The stress produced by these groups may, according to STROH's analysis, be great enough to initiate a crack.

b) The crack is then pictured as opening as in fig. 3. We then have to ask whether the crack will spread, with a velocity approaching that of sound, through the crystal, or whether it will be damped out by plastic flow round its apex[2]. One assumes that plastic flow will be due to the gene-

Fig. 3. A crack opening at the end of a slip line

ration of dislocations by sources S_1, S_2, S_3 already present at random points in the material. If the dislocations in $S_1, S_2, S_3, \ldots$ are locked, one has to ask whether, in the short time occupied by the passage of the apex A of the crack near to them, these sources will have time to break away from the locking atoms and generate dislocations. According to

---

[1] C. f. the photographs of JACQUET (1955) which show etch pits on the surface of brass due to such piled-up groups.

[2] The effect of hydrogen may perhaps also be explained along these lines. If the deformation is carried out slowly, the hydrogen will have time to diffuse to any newly formed sources and lock them sufficiently for this effect to occur.

STROH's analysis, at low temperatures they will not have time, so that the crack propagates with small loss of energy in deforming the material plastically. At high temperatures, on the other hand, they will be damped out, so that considerable plastic deformation ought to lead to a series of *small* cracks.

All this analysis applies to polycrystals; we are not able to suggest any theory of cleavage in single crystals.

Ductile fracture. Turning now to ductile fracture, this is a phenomenon characteristic of polycrystalline metals; single crystals of face-centred cubic metals at any rate pull down to a point before they break. It seems then very likely that what is happening here is that a number of small cracks (say a micron long) are formed in a heavily cold-worked material by piled-up groups at the ends of the *coarse* slip lines where they meet grain boundaries. The visible fracture in the middle of the test piece should be formed, according to this model, when these cracks coalesce.

We consider that grain boundaries are necessary because piled-up groups at COTTRELL-LOMER barriers will break down the barrier, according to the analysis of STROH, before a crack is formed.

Fatigue. In the last section we have put forward the hypothesis that brittle and ductile fracture have their origin in piled-up groups at the ends of slip lines, and that it is typically here, where the slip line meets a grain boundary, that stresses large enough to initiate fracture occur. We consider that, whatever the ultimate fate of this hypothesis, piled-up groups cannot be responsible for a fatigue crack. Recent experimental work on copper and aluminium (THOMPSON and WADSWORTH 1954, 1956, SMITH 1956) shows that a fatigue crack is normally initiated, in a single crystal or a polycrystalline metal, on the surface and along a slip line that has been formed in the first few thousand cycles. Certain slip lines, in polycrystalline specimens oriented so that the maximum stress is applied in the slip direction, appear to broaden and develop into cracks. The fatigue process can be hindered either by lowering the temperature or by the removal of oxygen, but the work of ROSENBERG (1956) shows that for polycrystalline copper at the temperature of liquid helium the fatigue process occurs as at room temperatures for stresses about 50 per cent. higher.

It seems to us hardly likely that piled-up groups can play a part in the initiation of these cracks. If the hypotheses of the last sections are correct, piled-up groups big enough to initiate cracks can only be formed at the ends of slip lines where they meet the grain boundaries, not all along the slip line. We consider, therefore, that the crack must be formed in some way by the movement backwards and forwards of dislocations in a "coarse" slip line — i. e. a slip line formed by strains in the stage III region of fig. 1. The mechanisms suggested earlier for the formation of coarse slip lines or bands show that these bands will contain dislocations, and it seems reasonable to suppose that some of them can move back-wards and forwards under reversed stressing, though their exact arrange-ment in space and the conditions under which they can move will have to be the subject of future research.

The most promising mechanism by which moving dislocations could produce a crack is by the production of vacancies and interstitial atoms. It is known that these are formed in cold-worked materials. It is known also that under alternating stresses polygonisation of pure metals and precipitation in unstable alloys are accelerated, so the evidence that vacan-cies are formed is strong. The mechanism by which vacancies are formed by moving dislocations is not clear; a full discussion will be given in another paper. Here we shall enquire what the consequences of their production under fatigue conditions are likely to be.

A moving dislocation is thought to produce per unit length a number L/lb of interstitials and vacancies, where L is the distance through which it moves, b the interatomic distance, and l "the distance between jogs", probably of the order of distance between dislocations in the material, say $10^4 b$. Thus if a dislocation moves backwards about 10^4 times, every atom in the slip plane will have been pushed out of position once. Of course they will tend to fall back into position; but the immediate result of a few thousand cycles, at a stress large enough to ensure that there are moving dislocations all along the slip plane, will be the production of a narrow, highly disordered region. This will necessarily expand, the surrounding lattice being deformed by the stresses set up.

When the disorder is great enough, even at low temperatures, this disordered region should recrystallise, taking up the structure and orien-tation of the surrounding crystal grain. At that part of the slip plane where this first occurs a severe tensile stress will be set up, large enough to initiate a crack. We believe that this mechanism is the origin of fatigue, at any rate at low temperatures. At intermediate temperatures (say room temperature in copper), it is believed that the interstitial atoms can diffuse away, leaving an excess of vacancies in the neighbourhood of the slip plane, which in itself could initiate a crack. At higher temperatures still one would expect clusters of vacancies, some of which would not be destroyed by receiving interstitials because many interstitials would attach themselves to dislocations.

If this is a correct model, one has to ask why cracks are observed to form on the surface. One cannot resist the impression that this is due to the presence of gas, and that any gas, even helium at low temperatures, may play a role here. If the surface of a crack is really clean, it is difficult to see why is should not close up due to some further re-arrangement of atoms, when the two sides are allowed to touch; but a gas even weakly adsorbed could play a role in keeping the crack open.

References

COTTRELL, A. H.: Phil. Mag. **43,** 654 (1952).

DIEHL, J., S. MADER, and A. SEEGER: Z. Metallkde. **46,** 650 (1955).

FRIEDEL, J.: Phil. Mag. **46,** 1169 (1955).

JACQUET, P. A.: Acta Metall. **2,** 742 (1954).

KUHLMANN-WILSDORF, D., J. H. VAN DER MERWE, and H. WILSDORF: Phil. Mag. **43,** 632 (1952).

LOMER, W. M.: Phil. Mag. **42,** 1327 (1951).

LOW, G. R.: Madrid Conference on Deformation and Flow of Solids 1956, p. 60.

MOTT, N. F.: Phil. Mag. **43,** 1157 (1952).

OROWAN, E.: Nature, Lond. **147,** 452 (1941).

ROSENBERG, A.: to be published 1956.

SEEGER, A.: Report of the Conference on Defects in Crystalline Solids, p. 328. London: Physical Society 1955.

SEEGER, A.: Madrid Conference on Deformation and Flow of Solids 1956, p. 90.

SMITH, G. G.: J. Inst. Met. (1956) (in press).

STROH, A. N.: Proc. roy. Soc., Lond., Ser. A **223,** 404 (1954).

STROH, A. N.: Proc. roy. Soc., Lond., Ser. A **232,** 548 (1955).

TAYLOR, G. I.: Proc. roy. Soc., Lond., Ser. A **145,** 362 (1933).

THOMPSON, N. and N. J. WADSWORTH: Phil. Mag. **45,** 223 (1954).

THOMPSON, N. and N. J. WADSWORTH: Phil. Mag. **1,** 113 (1956).

Dislocations and Brittle Fracture in Metals

By **John R. Low** jr., Schenectady

With 10 figures

1. Introduction. There now exists a considerable body of evidence that the type of brittle fracture observed in metals at low temperatures is always preceeded by some plastic deformation. Metals are not observed to fracture in the elastic range, and even under severely embrittling conditions, such that the macroscopic stress strain, curve appears to be entirely elastic, evidence can be usually be found for microscopic slip preceeding fracture. For example, the macroscopic ductility of poly-crystalline iron at low temperatures may be varied by varying the grain size; the larger grain sizes in such a series will break brittlely, whereas the finer grain sizes will exhibit considerable ductility. If, in such a series, the *fracture* stresses for the brittle specimens are compared with the yield stresses of the ductile specimens, as in fig. 1 a, it is found that the fracture stresses of the brittle specimens vary in the same way with grain size as do the yield stresses of the ductile specimens. Further, a single curve serves to describe the functional relationship between grain size and fracture stress and grain size and yield stress, i. e., $\sigma = \sigma_0 + k d^{-1/2}$. Where σ is the yield or fracture stress depending upon whether or not the specimen is fine or coarse grained, d is the mean grain diameter, and σ_0 and k are constants depending on the material, the temperature and the strain rate. A similar relationship between grain size and yield stress has been repeatedly found at more elevated temperatures where ductile behaviour is observed [1–3]. A more direct test of the equivalence of the brittle fracture stress and the yield stress may be obtained by comparing the brittle fracture stress in *tension* with the yield stress in *compression* for a series of grain sizes as has been done in fig. 1 b. Within the limits of accuracy of the test methods used, it appears that the stress at which fracture occurs in tension is the same as the stress at which yielding, i. e., slip, occurs in compression.

In view of the evident connection between the beginning of plastic flow and the brittle fracture process, it seems appropriate to investigate the various ways in which the motion of dislocations might lead to the initiation of fracture. In the next section, several dislocation models for crack initiation are discussed and compared with experiment; in the following section the subject of the propagation of a crack, once it has been formed, is discussed, and in particular, the role of dislocations in influencing brittle crack propagation in metals.

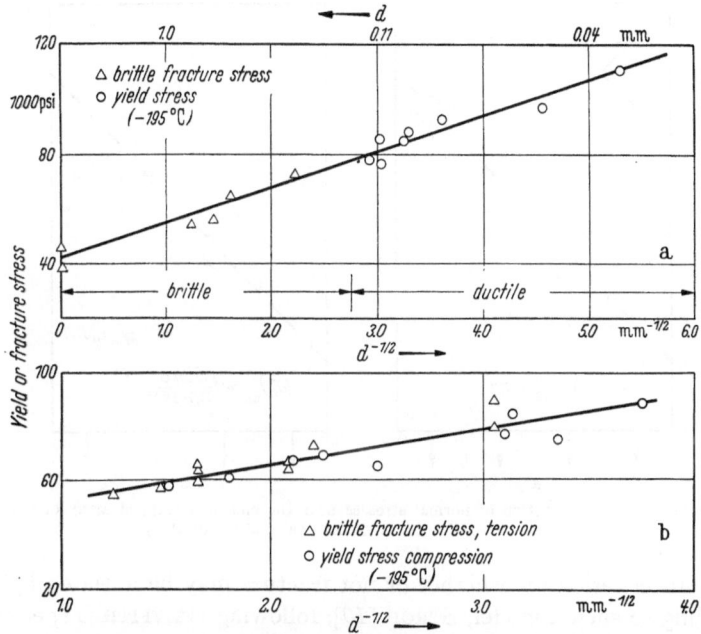

Fig. 1. a) Fracture stresses for coarse grained, brittle, specimens and yield stresses for fine grained, ductile, specimens of 0.03% C iron at — 195° C. Tension tests
b) Comparison of fracture stress in tension with yield stress in compression for a series of grain sizes all of which were brittle in tension at — 195° C. Same material as fig. 1 a

2. Dislocation models for the initiation of fracture. Without specifying any detailed atomic mechanism, ZENER [4] some time ago suggested that the stress concentration at the end of a slip band, stopped by a sufficiently rigid obstacle, might be high enough to nucleate a crack. Later ZENER [5] modified this concept by suggesting that the leading dislocations in a stopped array of dislocations on a slip plane might merge to provide a crack nucleus. This latter suggestion has been investigated in some detail by KOCHENDÖRFER [6] who concludes that such an occurrence is unlikely unless the dislocations which are to merge with a rigidly fixed dislocation are moving with a velocity approaching the velocity of elastic waves in the material.

FRANK [7], and ESHELBY, FRANK and NABARRO [8] have analysed the equilibrium positions of linear arrays of dislocations in a slip plane and the stress fields in the vicinity of such arrays under an external applied shear stress. Using their results, KOEHLER [9] and STROH [10] have in-

vestigated the normal stresses to be expected in the region of around the
end of a stopped slip band. The principal results of these latter investigations are shown schematically in fig. 2a and 2b.

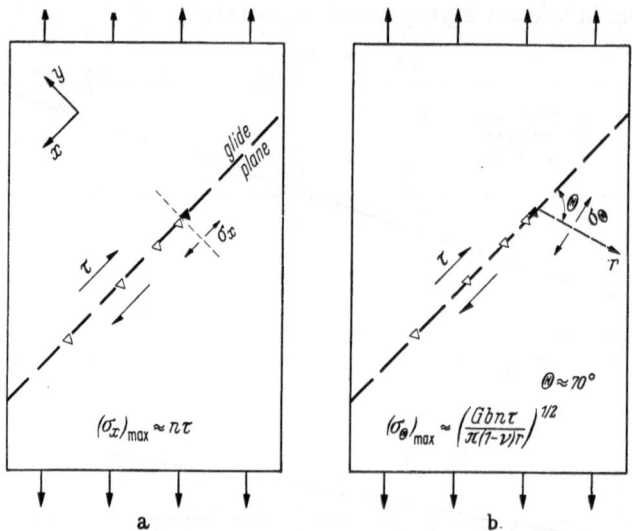

Fig. 2. Schematic representation of normal stresses near the end of a stopped array of edge dislocations according to KOEHLER (2 a) and STROH (2 b)

As his criterion for whether or not fracture may be initiated by slip
according to such a model, STROH [10], following GRIFFITH [11] equates
the surface energy of the two new surfaces to be formed in creating the
crack to the elastic strain energy released in the vicinity of the crack
surfaces, for the particular stress distribution which he calculates. From
these considerations he derives the following condition for crack formation:

$$n > \frac{12\,\gamma}{b\,\tau_a}\,,\tag{1}$$

where n is she number of piled-up dislocations in the slip plane in which
slip has been arrested, τ_a is the applied shear stress causing the slip,
γ is the specific surface energy of the material and b is the slip vector of
the dislocations.

PETCH [2] has made use of KOEHLER's result to rationalize the observed effect of grain size on she brittle fracture stresses of polycrystalline
zinc and iron by equating the maximum normal stress computed from
KOEHLER's equations to the estimated theoretical cleavage strength of
the material. PETCH does not make use of the actual numerical value of
the theoretical cohesive strength but assumes it to be a constant independent of the grain size. In the course of his derivation PETCH derives
the following relation between the grain size, d, the shear stress acting

on the slip plane, τ, the number of piled up dislocations, n, required to cause fracture and the elastic and crystallographic constants of the material:

$$n = \frac{l\tau}{4A}, \tag{2}$$

where $A = \frac{Gb}{2\pi(1-\nu)}$, with G the shear modulus, b the slip vector and ν POISSON's ratio; τ is now $(\tau_{applied} - \tau_0)$, τ_0 being a constant representing the resistance so dislocation motion arising from structural defects.

From the inequality (1) or equation (2) one may estimate the number of dislocations in a stopped array which would be required to initiate fracture according to either modification of this particular model if the shear stress for brittle fracture is known. The results of such estimates, using observed fracture stresses at $-196°$C for polycrystalline iron and zinc, are shown in Table 1.

Table 1

Material	Mean Grain Diameter	Stress to Cause Brittle Fracture[1] (dynes/cm²)	Number of blocked dislocations estimated from:	
			(1)	(2)
Polycrystalline Iron	0.1 mm.	$3.7 \cdot 10^9$	> 225	100
Polycrystalline Zinc	0.1 mm.	$5.5 \cdot 10^8$	> 1000	~ 1000

[1]. Expressed as a shear stress = One half normal stress for fracture at $-196°$C. Data for iron from PETCH [2] or LOW [23] data for zinc from PETCH [29] or GREEN-WOOD and QUARRELL [13].

The very large values shown in Table 1 for the number of dislocations in a stopped array required to explain the observed fracture stresses immediately raises a question as to the validity of the model. If the normal stresses in the region of the obstacle which has stopped the dislocation array are very high, is must be remembered that the shear stresses are also. For example, ESHELBY, FRANK and NABARRO [8] and COTTRELL [12] find that the shear stress on the leading dislocation will be of she order of $n\tau$, where τ is the applied shear stress and n the number of dislocations in the array. Thus, for the numbers of dislocation shown in Table 1, the shear stress on the leading dislocation becomes of the order of G, the shear modulus. Under these circumstances; it becomes difficult to imagine a mechanism for blocking dislocation motion which would be effective in holding up the dislocation array until the normal stresses could rise to a sufficiently high value to cause fracture. It seems more likely that slip will occur ahead of the blocked array and that this model is more appropriate for the propagation of slip in a polycrystalline aggregate, as FRANK [7] proposes, than for the nucleation of fracture.

The possible obstacles to slip which have been suggested are hard particles of a second phase [5], grain boundaries [2, 13], or dislocations of opposite sign on neighboring slip planes [14]. Of these, only the first might reasonably be expected to have a sufficiently high shear strength to permit the pile-up of the required number of dislocations to nucleate fracture.

FRANK [7] estimates that for an array of dislocations held up by a grain boundary, the applied shear stress necessary to nucleate slip in an adjacent grain would be of the order of $G/50\,n$ so $G/20\,n$ where n is the number of dislocations in the array and G the shear modulus. The applied shear stress required to activate existing FRANK-READ sources in the blocking grain would be considerably less. For the observed shear stresses at fracture given in Table 1 this means that only 10 to 100 dislocations would be required to start slip in a new grain and thus relieve the stress concentration at the end of the array. This estimate is for the favorable situation where the slip plane and direction in the blocking grain are nearly parallel to slip plane in the first grain, but even if this were not true the shear stresses on other planes might be expected to be great enough to start slip rather than fracture, especially in the case of the cubic metals which have a multiplicity of slip systems available. A possible exception could conceivably occur in the case of zinc with only a single slip plane where a 90° mis-orientation of slip planes across a boundary is possible.

Another possible obstacle to slip suggested by MOTT [14] might be the interaction force between two dislocations of opposite sign moving in opposite directions on closely spaced slip planes. The maximum interaction force per unit length opposing dislocation motion is:

$$0.25\,\frac{G\,b^2}{2\,\pi\,(1-v)\,y}\,,$$

where y is the distance separating the two slip planes [15]. The force per unit length acting on the leading dislocation in an array of n dislocations for an applied shear stress τ, is $n\tau b$. Thus, for an applied shear stress of the order of 10^9 dynes/cm² and the minimum possible spacing of the slip planes $(y = b)$, n could not be expected to exceed about 50, with wider spacings of the slip planes this number would be correspondingly reduced.

Finally, such a model fails to account for the observed brittle fracture stresses of single crystals of the metals such as iron or zinc. Iron single crystals fracture under applied shear stresses of the order of 10^9 dynes/cm² at $-196°$C while for zinc single crystals the value may be as low as 10^7 dynes/cm². In the case of single crystals it is particularly difficult to conceive of obstacles of sufficient strength to resist the shear stresses at the leading edge of an array of dislocations and to permit the pile-up of the number of dislocations required to nucleate fracture.

The relationship between grain size and fracture stress, i. e., that which has been cited as evidence in support of this particular model for fracture nucleation [2, 13] appears more likely to result from a relationship between the stress for slip propagation in a polycrystalline aggregate and the grain size and the fact that slip is required to initiate fracture.

Another model for fracture nucleation by the action of moving dislocations may be based on the observation, first made by SEITZ [16] and by READ [17], that if a screw dislocation moving in a slip plane cuts another screw dislocation, oblique to the slip plane, then a line of vacancies will be generated in the slip plane from the crossing point onward. MOTT [18] has pointed out that if several successive screw dislocations pass through the same crossing-point, then a sheet of vacancies, i. e., a crack nucleus, may be generated by this mechanism. FISHER [19] has suggested this mechanism as a possible means of nucleating fractures but points out that a sheet of vacancies should be unstable, and collapse to a ring dislocation unless a sufficient normal stress acts across the plane of the crack nucleus. He estimates that for a crack nucleus several hundred inter-atomic distances in width, i. e., several hundred moving screw dislocations passing through a single crossing-point, the stress to prevent collapse should be about 10^{10} dynes/cm^2, several times larger than the observed brittle fracture stresses for iron, and one or two orders of magnitude greater than the observed fracture stresses for zinc.

One may estimate the lower limit of the size of the crack nucleus it would be necessary to produce by the vacancy generation mechanism, from the GRIFFITH criterion for crack propagation. For example, if the observed brittle fracture stress for an iron single crystal at $-196°$C is taken as $4 \cdot 10^9$ dynes/cm^2, then from the known surface energy $(1.5 \cdot 10^3$ ergs/cm$^2)$ and the elastic modulus $(10^{12}$ dynes/cm$^2)$ one estimates from the GRIFFITH relation: $\sigma_c = \sqrt{\dfrac{2\,E\gamma}{\pi\,c}}$, that C, the crack size, must be at least $2 \cdot 10^{-4}$ cm, about 10^3 inter-atomic distances.

The above estimate for the number, n, of screw dislocations crossing a single intersecting screw dislocation again leads to an unreasonably large value for n. However, it is to be expected that in any real crystal, the moving dislocations will encounter large number of screw dislocations intersecting the glide plane so that a crack of sufficient size to be self propagating might be rather readily produced by this mechanism.

A low angle twist boundary might well provide a sufficient number of crossing-points close enough to each other so that a relatively small number of moving dislocations on the glide plane could generate a large enough crack nucleus. If, as shown in fig. 3, d is the distance between screw dislocations intersecting a glide plane in an array such as a twist boundary, then if $n = d/b$ moving dislocations cut the boundary, the vacancy sheets generated at each crossing-point would merge to form a much larger crack, limited only by the length of the boundary having a twist character. The spacing of screw dislocations in a pure twist boundary is given by $d = b/\Theta$ [15, 17] and if for example, $\Theta = 1°$, $d = 60\,b$ and n need be no greater than 60 to generate a crack having the dimensions of the total slip path in one direction and the boundary length in the other direction.

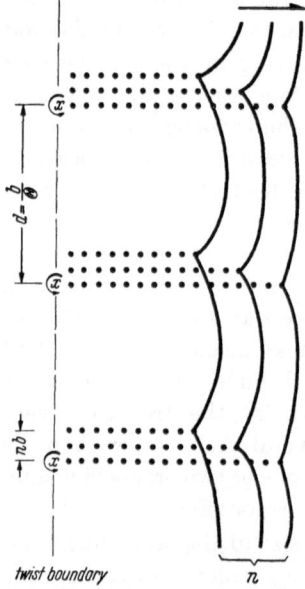

Fig. 3. Schematic representation of crack generation by a series of n screw dislocations cutting a low angle twist boundary. Θ = angle of twist

The validity of such a model can hardly be decided finally on the basis of experimental observations available at present. However, certain requirements of the model may be examined qualitatively. The structural requirement that dislocation arrays such as a low angle twist boundary shall exist in real crystals seems not unlikely since low angle boundaries may be readily observed by the etch-pit technique on single crystals grown either from the melt, or by the strain anneal technique [20, 21]. An array of screw dislocations intersecting an active glide plane may be expected to offer considerable resistance to the motion of other screw dislocations in the glide plane so that the first dislocation from a source arriving at the low angle boundary might be expected to be held up for a time. However, the pile-up of other dislocations of the same sign behind the arrested dislocation should provide the necessary local shear stress to force it through the array and begin the vacancy generation process. Perhaps the most critical consideration will be whether or not the vacancy sheets can be prevented from collapsing to edge dislocations before they can grow to a sufficient size to merge with each other and produce a macroscopic crack. At present no reliable way of estimating the stress required to hold two planes of atoms at one interatomic distance from each other suggests itself.

3. Crack propagation in metals. While the Gʀɪғғɪᴛʜ theory of crack propagation has had some success [11] in explaining the observed fracture strengths of non-deformable solids, such as glasses at room temperature, it has been repeatedly pointed out that the theory fails in the case of metals. As Oʀᴏᴡᴀɴ [22] has observed, brittle crack propagation in metals differs from that in glasses in that in the case of metals there is evidence for appreciable local plastic deformation accompanying the propagation of brittle cracks. After finding X-ray diffraction evidence for deformation at the fracture surface, OROWAN estimated the amount of such deformation after brittle fracture of a steel at low temperatures. From this estimate he concluded that the energy absorbed in separating the material, i.e., in crack propagation, must have been several orders of magnitude larger than the surface energy of the material. He suggested that the GRIFFITH equation relating the critical stress for crack propagation and the energy should be modified and that the surface energy term of the order of 10^3 ergs/cm^2 be replaced by a term representing the energy of local plastic deformation at the crack surface, of the order of 10^6 ergs/cm^2. Experimental measurements which permit the calculation of this energy for polycrystalline iron or steel fractured at low temperatures, one by Low [23] and the other by WELLS [24], both give values of the order of magnitude estimated by OROWAN.

There are two principal ways in which localized plastic deformation may occur during crack propagation. The first of these is the localized plastic deformation which may occur due to the shear stresses at the tip of a crack if the velocity of crack propagation is not so great that slip cannot be initiated in the stressed region. This aspect of the mechanism of crack propagation was suggested by MOTT [25] and has been investigated to some extent by HALL [26], and more generally by GILMAN [27]. A second method by which plastic deformation may occur and by which energy may be absorbed in crack propagation is in the highly localized deformation which occurs at the "cleavage-steps" or "tearlines" which are observed on the fracture surface of metals and other solids which have failed by brittle fracture.

Fig. 4. Typical cleavage step pattern on cleavage surface of a single crystal of 3% silicon iron. Note that low angle tilt boundary does not affect pattern. X 600

An example of these cleavage steps is shown in fig. 4 which is a photo-micrograph of the fracture surface of a single crystal of silicon—iron cleaved at $-196°$C. These patterns of lines on the cleavage surface represent steps in the plane of the crack as it propagated through the crystal and the manner in which localized plastic deformation may arise at one of these steps is shown schematically in fig. 5.

Fig. 5. Schematic representation of successive stages in cleavage step formation and localized plastic deformation at cleavage step. View is looking toward advancing edge of cleavage crack. Compare fig. 7

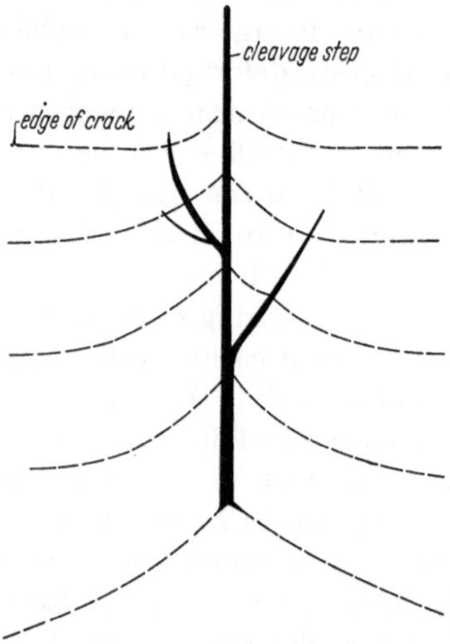

Fig. 6. Schematic representation of origin of "river" pattern of cleavage steps. Compare fig. 4 and 8

Such a cleavage step can be expected to retard the crack front locally so that a cusp would be produced in the leading edge of the crack. Subsequent cleavage steps in the region immediately adjacent to the first step will be produced in a direction normal to the crack front so that steps tend to run together producing the characteristic "river-like" pattern shown in fig. 4. Fig. 6 is a schematic representation of the manner in which this characteristic cleavage step pattern is produced. If the steps which join together in this way are of the same sign, then as they come together a larger and larger step will be produced, leading to very large level differences and local plastic deformation in a larger volume of metal as shown in fig. 7.

Gilman [28] has studied the formation of cleavage steps in zinc single crystals and describes a number of ways in which these steps may originate. He found that cleavage steps may originate at screw dislocation cut by the advancing crack and at regions of internal stress produced by inhomogeneous plastic deformation. He also observed that large amounts of slip prior to cleavage increased the number of cleavage steps as did raising the temperature of cleavage.

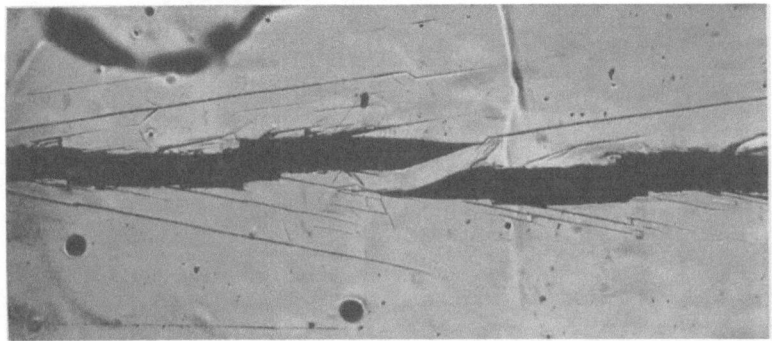

Fig. 7. Vertical section through a large cleavage step in a 3 % silicon iron single crystal. Compare fig. 5. X 100

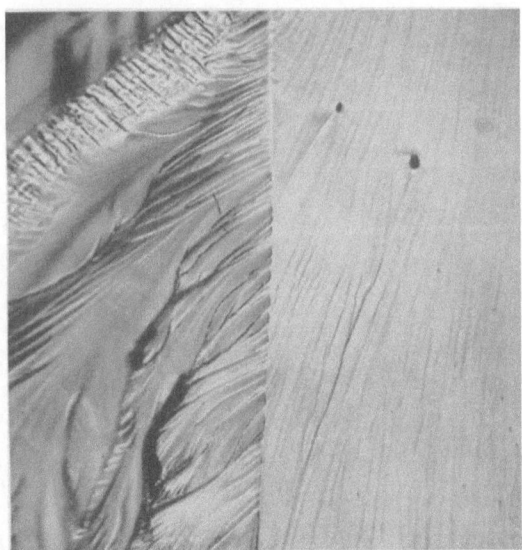

Fig. 8. Cleavage steps originating at a low angle twist boundary and at second phase particles in 3 % silicon iron crystal. Low angle boundary lies in a (110) plane with a 3° twist about an axis normal to the boundary. Direction of crack propagation upper right to lower left. X 500

There are a number of other ways in which cleavage steps may arise. For example, as might be expected, a low angle twist boundary such as that shown in fig. 8 leads to a great many new cleavage steps, all of the same sign, and consequently a great deal more energy is required for crack propagation once the boundary is crossed. Fig. 8 also shows another source of cleavage steps, i. e., particles of a second phase may cause local

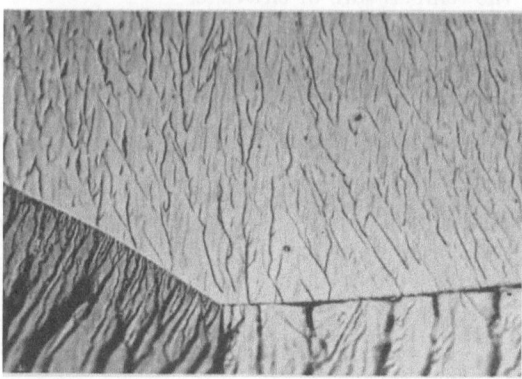

Fig. 9. Cleavage steps originating at two high angle grain boundaries. Left boundary 10° tilt, right boundary 10° tilt, plus 3° twist. Direction of crack propagation top to bottom (3% silicon iron). X 250

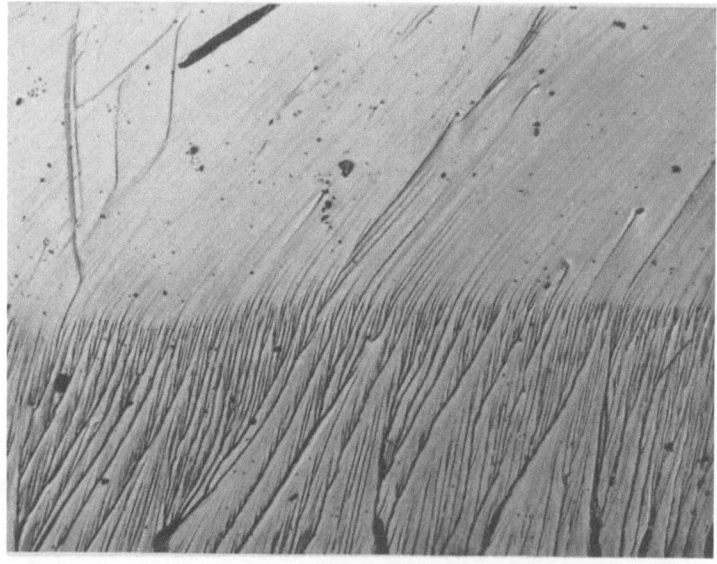

Fig. 10. Cleavage steps originating at point where cleavage crack was stopped and then started again immediately at −195°C. Direction of crack propagation upper right to lower left. 3% silicon iron crystal. X 150

level differences in the plane of the crack. Thus one might expect that a high density dispersion of such particles should increase the resistance to brittle crack propagation.

High angle grain boundaries, requiring large changes in the orientation of the crack plane, lead to large cleavage steps as may be seen in fig. 9. The increased resistance to crack propagation caused by the presence of such boundaries undoubtedly accounts for much of the increase in ductility and fracture stress which is observed as the grain size of a metal is decreased.

Finally, cleavage steps may be generated by plastic deformation ahead of an advancing crack if the crack velocity can be reduced to the point where deformation by slip has time to occur. This effect is illustrated in fig. 10 which shows the point at which a cleavage crack in a silicon iron single was stopped and then immediately started again, all at $-195°$ C. The only interpretation of this very striking cleavage step pattern which seems possible, is that, as the crack slowed down and as it was started up again, plastic flow occurred ahead of the leading crack to a defect structure or an internal stress pattern which caused the crack to change levels locally, thus producing the observed cleavage steps. This result immediately suggests that the velocity of crack propagation relative to the rate at which slip may be initiated is the determining factor in whether or not brittle behavior is observed in a metal. A slowly moving crack will produce regions of deformation ahead of it which in turn will produce a stepped cleavage surface, tending to increase the energy required for crack propagation and to further reduce the crack velocity. This concept was first suggested by MOTT [25] and has been examined recently in some detail by GILMAN [27].

It is a pleasure to acknowledge many helpful discussions with my colleagues at the General Electric Company Research Laboratory and in particular those with J. C. FISHER, J. J. GILMAN and E. W. HART. Much of the experimental work described herein could not have been performed without the very able assistance of J. S. ERICKSON.

References

[1] HALL, E. O.: Proc. phys. Soc., Lond., Ser. B **64**, 747 (1951).
[2] PETCH, N. J.: Iron Steel Inst. **174**, 25 (1953).
[3] CRACKNELL, A. and N. J. PETCH: Acta Metall. **3**, 186 (1955).
[4] ZENER, C.: Phys. Rev. **69**, 128 (1946).
[5] ZENER, C.: Trans. Amer. Soc. Met. **40A**, 3 (1948).
[6] KOCHENDÖRFER, A.: Arch. Eisenhüttenw. **25**, 351 (1954).
[7] FRANK, F. C.: Symposium on the Plastic Deformation of Crystalline Solids, p. 89. Pittsburgh: May 1950.
[8] ESHELBY, J. D., F. C. FRANK and F. R. N. NABARRO: Phil. Mag. **42**, 351 (1951).
[9] KOEHLER, J. S.: Phys. Rev. **85**, 480 (1952).
[10] STROH, A. N.: Proc. Roy. Soc., Lond., Ser. A **223**, 404 (1954).
[11] GRIFFITH, A. A.: First International Congress of Applied Mechanics, Delft 1924, p. 55.

[12] Cottrell, A. H.: Progress in Metal Physics I, chap. ii. London 1949.
[13] Greenwood, G. W. and A. G. Quarrel: J. Inst. Met. 82, 551 (1954).
[14] Mott, N. F.: Proc. Roy. Soc., Lond., Ser. A 220, 1 (1953).
[15] Cottrell, A. H.: Dislocations and Plastic Flow in Crystals. Oxford 1953.
[16] Seitz, F.: Advances in Physics 1, 43 (1952).
[17] Read, T. W.: Dislocations in Crystals. New York 1953.
[18] Mott, N. F.: Proc. Phys. Soc., Lond., Ser. B 64, 733 (1951).
[19] Fisher, J. C.: Acta Metall. 3, 109 (1955).
[20] Vogel, F. L. et al.: Phys. Rev. 90, 489 (1953).
[21] Dunn, C. G.: Trans. Amer. Soc. Met. 45A, 170 (1953).
[22] Orowan, E.: Rep. Progr. Physics, Physical Society London 12, 185 (1949).
[23] Low, jr., J. R.: Trans. Amer. Soc. Met. 46A, 163 (1954).
[24] Wells, A. A.: Welding Res. 7, 34r (1953).
[25] Mott, N. F.: Cambridge University Conference on Brittle Fracture in Mild Steel Plates, October 26, 1945.
[26] Hall, E. O.: J. Mech. Phys. Solids 1, 227 (1953).
[27] Gilman, J. J.: Propagation of Cleavage Cracks in Crystals (to be published J. Appl. Phys.).
[28] Gilman, J. J.: J. Metals 7, 1252 (1955).
[29] Petch, N. J.: Progr. Metal Phys. 5, 1 (1954).

The Origin of Dislocations in Crystals Grown from the Melt

By **F. C. Frank,** Bristol

With 3 figures

There is no doubt that the phenomena of plasticity are explicable in terms of dislocations, and that dislocations are present in almost all crystals. The exceptions ("whiskers") prove the rule. The behaviour of dislocations is now understood in considerable detail, but their origin is still understood very imperfectly. We know of the ways in which deformation can multiply the dislocation content of a crystal. We can also see that the recovery processes by which the dislocation content diminishes again will become extremely slow when the content of dislocations becomes small. For the driving forces are inversely proportional to radii of curvature of dislocation lines or of dislocation grids, which in turn may be reckoned proportional to the general scale of the dislocation structure: the frictional forces impeding the motion could be the intrinsic "PEIERLS-NABARRO force", but in metals of presently available purity are almost certainly "COTTRELL forces" of interaction between dislocations and impurity atoms. What determines the dislocation content of the new grains when annealing involves a solid recrystallization is something of which we know very little indeed. The experimental facts are meagre, and to make progress with this problem we must probably look for extensions of experimental technique like the HEDGES-MITCHELL method of revealing dislocations in the interior of transparent crystals [1], or to further applications of X-ray microscopy to the study of the "wavy" lattice curvatures characteristic of crystals grown by the strain anneal method [2]. It is possible that dislocations are generated in a somewhat similar way to that discussed here, the extra volume in grain boundaries providing an effective source of vacancies. In the case of crystals grown from the melt, however, we do have a probable understanding of the way in which their initial dislocation content is established.

Firstly let us remark that fast grown crystals are dendritic, and slight bending of the dendrite arms ultimately produces dislocations at the meeting surface between material which crystallizes on different arms. The seed on which a crystal grows is either a pre-existing crystal, dislocated as all ordinary crystals are, or a nucleus formed in supercooled liquid, which will first grow dendritically and so acquire dislocations. The crystal which grows on a dislocated seed will itself grow with dislocations in it, though it might progressively lose many of them by mutual annihilation and combination. Observations show however that this is not what happens, but rather that we must look for processes making new dislocations which must operate during crystal growth.

The basic idea [3] is due to TEGHTSOONIAN, a co-worker of CHALMERS. Crystallization from the melt is necessarily performed in a travelling temperature gradient. The vacancy content of the newly formed solid may be assumed to have its equilibrium value at the melting point. Somehow it must finally diminish to the equilibrium value belonging to a lower, and usually much lower, final temperature. It is TEGHTSOONIAN's idea that these vacancies condense into discs, the faces of the discs collapsing together to make rings of edge dislocation (a situation first envisaged by NABARRO). The evidence for this process is the peculiar nature of the disorientations in the "lineage" structure of crystals grown from the melt, in which rotations about the growth axis greatly predominate. No alternative explanation for this phenomenon has been advanced. We shall here develop TEGHTSOONIAN's idea in greater detail.

Let us first briefly summarize the facts. Crystals grown slowly from the melt almost invariably have a substructure in the form of cells, of the order of a mm. across and very much elongated along the growth axis, disoriented from each other by relative rotations varying from a few minutes to a few degrees, depending on the substance and conditions of growth. This is not a consequence of dendritic growth, which is easily recognized and avoided. It is also distinct from another recognizable substructure on a scale 10 times smaller or less, not directly connected with disorientations, but attributable to traces of impurity (see, for example, SMIALOWSKI [4], RUTTER and CHALMERS [5]), which we shall not further consider here, though when present it no doubt modifies the development of the disorientation sub-structure. The latter occurs with qualitative similarity in non-metals, such as sodium chloride, as well as metals. For quantitative information about it in the metals we are largely indebted to the work of CHALMERS and his associates, and a detailed study was made by TEGHTSOONIAN and CHALMERS [3] for the case of tin. The cells were roughly square in cross-section, with average widths decreasing from 1.2 mm. to 0.25 mm. as the growth rate increased from less than 1 mm. per minute to 10 mm. per minute. The average disorientation between neighbours decreased at the same time from 2.8° to 1.2°. Disorientations as large as 5° were observed. Successive disorientations were not random, but tended to alternate in sign so that the orientation anywhere remained within about 4° of the mean. The direction of the subboundaries was near to the growth axis so long as the growth rate was small, but deviated towards crystallographycally indicated directions at higher growth rates. At a place where the crystal growth front widens, new sub-boundaries are not produced immediately, but develop in a distance of a few centimetres, initially with a very small disorientation which progressively increases during the next centimetre or two. The rate of increase for various boundaries in the same specimen is about the same, 1° per cm. for example. Thus there is a continuous twist in each cell at this stage, of the order of $^1/_2$° per cm.: but there is some variation in the final value at which this disorientation ceases to increase Similar phenomena have been observed in lead, zinc, copper, and silver by CHALMERS et al., and the present writer has observed them in sodium chloride. There does not seem to be any great difference, even quantitatively, between various materials. We shall now try to use TEGHTSOONIAN's basic suggestion to account for the above facts.

At any temperature T the fraction of lattice sites vacant should be, in thermal equlibrium, approximately

$$n = e^{-U/kT}$$

where U is the formation energy for a vacancy. U is in no case very reliably known, but it seems very probable that self-diffusion in most metals and many other solids proceeds by a vacancy mechanism, in which case U should be a moderate fraction of the activation energy for self-diffusion. The measurements of KAUFFMANN and KOEHLER [6] for gold, which appear to be rather satisfactory, give $U = 1.28$ eV, and Q (the activation energy for diffusion of vacancies) $= 0.68$ eV, their sum being in excellent agrement with the observed activation energy for self-diffusion of gold, 1.965 eV. With these figures, at the melting point

$$n = e^{-10.67} = 2.5 \cdot 10^{-5}.$$

Activation energies for diffusion in metals and probably for solids generally, are fairly strongly correlated with their melting points, T_M. For the purposes of numerical estimate we shall assume in general

$$\frac{U}{kT_M} \sim 10,$$

according to which the fraction of vacant sites at the melting point is about $5 \cdot 10^{-5}$. (Some estimates of this quantity in other materials are as high as 10^{-3}.) The corresponding fraction at room temperature will in general be negligibly small compared with this. We may assume that the vacancy concentration maintains its equilibrium value at the solid-liquid interface, and have to enquire how this vacancy excess is removed as the temperature falls behind the growth front. Consistently with the above estimate for U/kT, the diffusion coefficient for vacancies, D_{vac}, is about 10^{-6} cm^3/sec. at the melting point (and less, if we have underestimated n).

Then we may easily show that diffusion to the free surface is not an adequate process for their removal in specimens of say 1 cm. diameter. As LE CLAIRE [7] has pointed out, there is a thermal diffusion drift of vacancies in a temperature gradient, with a drift velocity $D_{vac} Q' (dT/dx)/kT^2$, in which Q', the heat of transport, is approximately equal to Q, the activation energy for diffusion of vacancies. If we assume $T_M/(dT/dx) \sim 50$ cm., this drift velocity is only about 10^{-7} cm./sec. and therefore negligible compared with usual rates of crystal growth in the laboratory.

Hence the vacancies must in some way condense out in the interior of the specimen. They may do this on dislocations or other condensation nuclei already present. If there are not enough of these, the supersaturation of vacancies will rise to such a value, say 10 fold supersaturation, that spontaneous nucleation of large vacancy aggregates can occur. These may be of two kinds: pores, and flat aggregates which are loops of edge dislocation in a plane normal, or approximately normal, to their BURGERS vectors. These will grow by edgewise addition of vacancies. They will lie in various planes, but only those in a plane near to parallelism with the growth axis will be able to grow on, following up the growth front. The pores, and the dislocation loops in other planes, will be left behind. This process will develop and select a system of dislocation loops, the long arms of which are edge dislocations approximately parallel to the growth axis. This should be the case whether the condensation occurs on new-formed loops or dislocations pre-existing in the crystal seed. The elastic interactions between edge dislocations should cause them to assemble into parallel arrays of the same kind, as observed in "polygonisation", the resulting configuration being that indicated in fig. 1. This growing

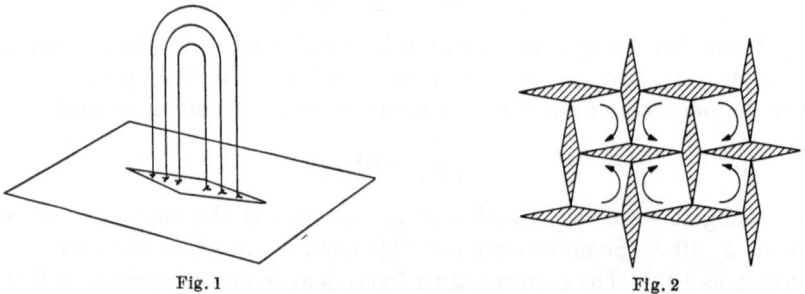

Fig. 1   Fig. 2

system of loops removes from every plane of the crystal a narrow rhombus of lattice sites, also indicated in fig. 1. If we enquire how a number of these systems of loops can grow without developing a severe stress, we see that they must be arranged essentially as in fig. 2 and that the removal of the rhombic prisms of material will rotate alternate prismatic regions between them in opposite directions about the growth axis. The angle of rotation is equal to n, the fraction of lattice sites removed. The nature of the rotation corresponds to what is observed, but the observed rotation is not constant, progressively increasing, and reaching values such as $2°$, certainly much larger than the fraction of lattice sites vacant at the melting point. This is also explicable, because the ends of the loops are in an unstable position: they are attracted to the interface with the liquid, so that when there are just enough of them to remove the necessary number of vacancies, the configuration will not persist, but the leading loops will grow ahead and reach the interface (fig. 3 a, b). Then, however,

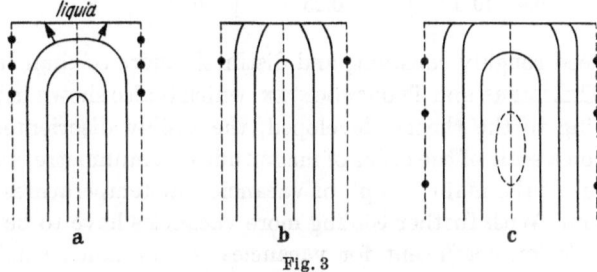

Fig. 3

they are no longer effective for removing vacancies merely by lengthening, and the vacancy excess will drive them apart, developing a tensile region in the gap. A new dislocation loop will then preferentially form in this region (fig. 3 c) so that the dislocation number and the amount of rotation at the lineage boundaries will continually increase, until the rate of production of new loops is balanced by the rate of annihilation of dislocations by the meeting of opposite pairs.

We may also verify that the transverse scale of these lineage cells is of the correct order of magnitude. The equilibrium concentration of vacancies must fall by a factor e in a distance of approximately $kT^2/U (dT/dx)$, and accordingly in a time $kT^2/v U (dT/dx)$ where v is the growth velocity. This time suffices for diffusion over a distance

$$ l \sim \left[\frac{D_{\text{vac}} k T^2}{v U (dT/dx)} \right]^{1/2} $$

which is about 1 mm. when v is 10^{-3} cm./sec., the other numerical estimates being those made earlier. With sub-boundaries significantly further apart than this, vacancies in the interior of the cells would not be sufficiently rapidly removed, and further dislocation loops would nucleate within the cells. Thus the order of magnitude of the observed scale of lineage is satisfactorily accounted for. We may also check the proportionality of lineage scale to $v^{-1/2}$, from CHALMERS and TEGHTSOONIAN's results shown in table 1.

Table 1

Growth rate, v mm./min.	Average striation spacing, l mm.	$l\,v^{1/2}$	$l\,v$
0— 1.0	1.20	0.85	0.6
0.5— 1.5	1.05	1.05	1.05
1.0— 2.0	0.85	1.04	1.27
1.5— 2.5	0.75	1.06	1.50
2.0— 3.0	0.60	0.95	1.50
2.5— 3.5	0.50	0.87	1.50
3.0— 4.0	0.45	0.84	1.67
3.5— 4.5	0.40	0.80	1.60
5.5— 6.5	0.35	0.86	2.10
9.0—10.0	0.25	0.77	2.37

$l\,v^{1/2}$ is indeed roughly constant, and distinctly more so than $l\,v$, as suggested by CHALMERS and TEGHTSOONIAN, which is also shown in the table.

According to the theory developed, the visibly disoriented lineage structure, on a scale of the order of magnitude of a millimetre, arises from the removal of the main "crop" of vacancies, at temperatures near the melting point. With further cooling more vacancies have to be removed and the diffusion coefficient for vacancies is now much smaller. This should cause a further sub-division of the lineage cells with a much sparser distribution of dislocations on a finer scale. The mechanism which caused a progressive increase of the lineage disorientation is here absent. As formed, this latter crop of dislocations should have a similar arrangement to those of the visible lineage, but being sparser it will be more easily shifted by external stresses, such as those due to thermal contraction, so that the final arrangement of these dislocations may actually be very different.

References

[1] HEDGES and MITCHELL: Phil. Mag. (VII) 44, 223 (1953).
[2] SCHULZ: J. Metals 6, 1082 (1954).
[3] TEGHTSOONIAN and CHALMERS: Canad. J. Phys. 29, 370 (1951); 30, 388 (1952).
[4] SMIALOWSKI: Z. Metallkde. 29 133 (1937).
[5] RUTTER and CHALMERS: Canad. J. Phys. 31, 15 (1953).
[6] KAUFFMANN and KOEHLER: Phys. Rev. 92, 555 (1955).
[7] LE CLAIRE: Phys. Rev. 93, 344 (1954).

Magnetoresistivity of Plastically Deformed Metals

By **H. G. van Bueren**, Eindhoven

With 3 figures

1. Introduction. By plastic deformation vacancies, interstitials and dislocations are introduced into a metal. To study the dependence on strain of their number, mechanical, optical, electrical and X-ray methods have been used. However, no satisfactory agreement exists between the results. This is mainly due to three reasons:

1. It is not possible to separate the influences of the different defects, than by applying methods (e.g. temperature treatment) that may well affect the concentrations to be determined in an unknown way.

2. The absolute magnitude of the expected influence of a known concentration of defects is not known.

3. The measured concentrations may represent only a fraction of the true concentrations.

The difficulties are particularly outspoken in the results of the electrical (resistivity) measurements [*1*]. Deformation of a metal at a very low temperature results in an appreciable resistivity increase (20% strain at 20° K doubles the resistivity of 99.998% copper), part of which disappears again on heating up the wire. The additional resistivity anneals out in various recovery processes [*2*] that are associated with diffusion processes of the various defects. Analysis of the recovery seems to point to the conclusion that, in copper at least, about half of the additional resistivity is caused by dislocations, as it is this part of the resistivity that remains after heating the deformed wire to such temperatures where the point defects should have disappeared, by diffusion, completely from the metal.

There are several arguments, however, that point against this conclusion. No conclusive evidence has yet been found [*3*] for *anisotropy* of the additional resistance, which might be present if dislocations caused a large part of it. Moreover, experiments especially designed to obtain information on the *relative influence of point defects*, yield results that can most easily be explained by assuming that dislocations contribute only to about 10% of the resistivity [*4*]. Most important is the disagreement which turns up between experiment and theory, when the recovery results are interpreted in the way indicated. The theoretical scattering cross-section of dislocations has been derived by HUNTER and NABARRO [5]. From this and the measured resistivity, dislocation densities can be derived ($\approx 10^{11}$ cm^{-2} after 20% strain) that are too high by a factor of 10 if compared with the dislocation densities as indicated by various X-ray and optical experiments on deformed metals: $\leq 10^{10}$ cm^{-2} after 20% strain. The latter value follows also from quite general theoretical arguments [6]. One would therefore expect the dislocations not to produce more than a few percents of the measured additional resistivity at low temperatures.

In order to obtain more information on this point, a method was sought by which one could determine directly the (relative) concentration of dislocations alone, without the disturbing influence of point defects. Such a method was discovered in measuring the change of resistivity of a low temperature deformed wire in a transverse magnetic field.

2. Experimental. All experiments were carried out in apparatus designed by Mr. JONGENBURGER of this laboratory and in close collaboration with him. Wires of pure polycrystalline copper (99.998%, obtained from JOHNSON and MATTHEY), 0.5 mm. thick, were extended in a low temperature tensile machine described before [4], at 20°K (liquid H$_2$). The wire was placed between the poles of an electromagnet producing a transverse field essentially homogeneous over the length of the wire. All wires were first annealed for two hours at 550°C in vacuum. After extension by 5, 10 and 20% respectively, the resistivity of the wire was measured in the magnetic field, the fieldstrength being varied in steps from zero to 19.000 Oe. The results were mutually compared in a so-called *reduced* or KOHLER-diagram, in which the relative change of the resistivity in a field H, $[\varrho(H) - \varrho(0)]/\varrho(0)$, is plotted (either logarithmically or linearly) against the "reduced" field strength $H/\varrho(0)$. In this way the effect of variations of the resistivity in zero field that arise from the presence of impurities are eliminated [$H/\varrho(0)$ being effectively a measure for the ratio between mean free path of the conduction electrons and the radius of the classical orbit they describe in the magnetic field]. That the impurity content had indeed no effect is shown in figs. 1 and 2, in which the same curve applies to the undeformed pure metal and to a very much less pure comparison wire.

Straining the wire plastically resulted in appreciable deviations from this curve as is also shown in figs. 1 and 2. The relative magnetoresistivity *increases* by the strain, and for not too high fields the relative increase is in first approximation *proportional to the strain* and *independent of the fieldstrength*. The ratio between magnetoresistivity and ordinary resistivity is found to be *independent* of the strain. At high field strengths the effect of deformation diminishes and even seems to disappear.

Subsequent warming up the wire to room temperature for sufficient time to affect a complete room temperature recovery of the ordinary resistivity (only 60% of the additional resistivity was left), had *no effect* upon the magnetoresistivity as measured again at 20°K. That is, the

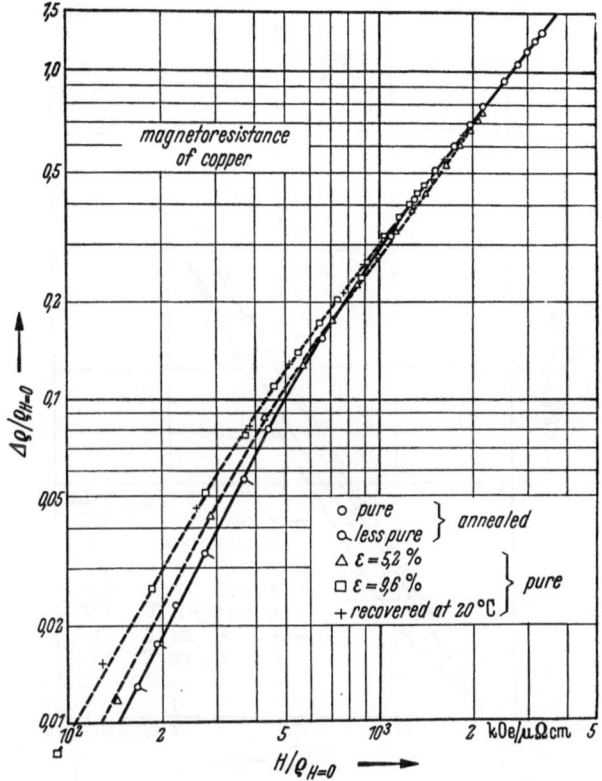

Fig. 1. Magnetoresistance in copper deformed at 20°K, plotted logarithmically

effect of the previous deformation in the magnetoresistance curve was still fully present. Even annealing the wire at 200° C for one hour, resulting in a further decrease by 10% of the additional ordinary resistivity, did not alter the form and place of the magnetoresistance curve. Only by heating above the recrystallization temperature, and thereby eliminating effectively all effects of the previous deformation, both ordinary and magnetoresistivity returned to their original values, pertaining to the undeformed wire (fig. 2).

This behaviour of the magnetoresistance can only be understood by concluding that vacancies and interstitials have no influence on it, as is shown by the lack of effect of annealing. That is, only *dislocations* affect the magnetoresistance. A further conclusion is that the additional resistance left after annealing at sufficiently high temperature, is intimately connected with dislocations and not with point defects.

From the first point it can be deduced that measuring the magneto-resistivity is indeed an effective way of observing dislocations. The second

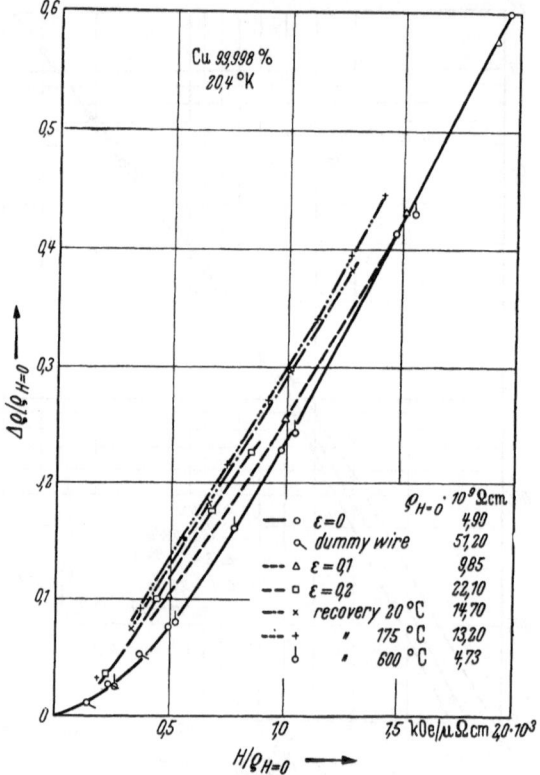

Fig. 2. Linear plot of data on magnetoresistance in deformed copper at 20°K

statement points very strongly to the conclusion that dislocations are indeed responsible for about half of the additional resistance caused by cold-work, although the possibility can not be ruled out that other defects, such as vacancy conglomerations, may play a part. This seems rather improbable however. It follows then that the theoretical estimates of the dislocation scattering are probably to low by a factor of 10.

This seems a serious criticism. In order to base it not only on experimental but also on theoretical evidence, we have computed the effect of dislocations on the magnetoresistance in an approximate way.

3. The magnetoresistance of a dislocated metal. Lattice defects can affect the magnetoresistance as apparent in the reduced diagram in three ways, viz. by scattering the conduction electrons anisotropically, by deforming the shape of the FERMI-surface (due to the elastic strains around them) and by producing so-called size effects [7] when they are present in more or less ordered arrays. It is not to be expected that point defects could make their presence known in either way, as they scatter isotropically, produce only short range strains and cannot make up efficiently reflecting layers when present in densities as encountered in practice. The same holds true for chemical impurities in unordered alloys, and as has been already signalled, experiment confirms this.

Dislocations however produce appreciable anisotropic scattering. They moreover have long range strains associated with them, and often occur in ordered arrays (subboundaries). It is possible that at low temperatures size effects might be associated with dislocations, as dislocation walls, if present, may easily locally be separated by distances of the order of the mean free path or less, that is 10^{-3} cm or less at 20°K. However, the dependence of magnetoresistance on field strength should then be expected to be quite different from that observed.

To estimate the effect of the lattice strain on the FERMI-surface, in principle the method of HUNTER and NABARRO [5] could be followed. In view of the huge theoretical difficulties that then arise, and the inherent uncertainty present in their treatment, as shown by the disagreement with observations, no estimate has been obtained. The only phenomenon that yields to semi-quantitative treatment is the influence of the *anisotropic scattering* on the magnetoresistance. We shall only deal with that here.

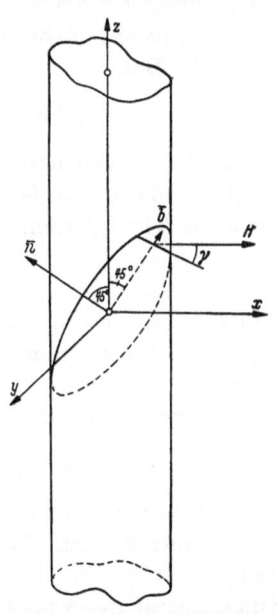

Consider a metal wire in a transverse magnetic field H, containing N parallel edge dislocations per cm². When the current j runs in an arbitrary direction, the contribution to the resistivity caused by these dislocations is given by

$$\Delta \varrho_{\text{disl}} = N \varrho_1 \cos^2(b\,j) + N \varrho_2 \cos^2(n\,j), \quad (1)$$

where b is the BURGERS vector and n the unit vector normal to the glide plane of each dislocation. In order to simplify the computation somewhat, we shall restrict the possible orientations of the dislocations as follows

Fig. 3. Geometry of the dislocated state assumed in the computation

(fig. 3). We take the angle between glide planes and wire axis to be 45°, and assign the same value to the angle between the BURGERS vectors (slip direction) and wire axis. In polycrystalline metals of small grainsize these seem plausible assumptions. Then we can specify the orientation of the dislocation set by one angle γ, viz. the angle between the axes of the dislocations and a given direction in a plane perpendicular to the wire axis, for which we take the direction of the magnetic field strength H. Introducing rectangular coordinates x along H, y perpendicular to H and to the wire axis and z along this axis and indicating the current direction with respect to this system by the polar angles ϑ and φ (the z-axis being taken as polar axis), equation (1) becomes:

$$\Delta\varrho_{\mathrm{disl}} = N \frac{\varrho_1 + \varrho_2}{2} [\cos^2\vartheta + \sin^2\vartheta \cos^2(\varphi - \gamma)] +$$
$$+ N(\varrho_1 - \varrho_2) \sin\vartheta \cos\vartheta \cos(\varphi - \gamma). \qquad (2)$$

Assuming Matthiesen's law to be valid, the resistivity of the dislocated metal is simply given by

$$\varrho = \varrho_0 + \Delta\varrho_{\mathrm{disl}}, \qquad (3)$$

where ϱ_0 is the resistivity of the undeformed wire.

The resistivity of a metal can be found by solving the so-called Boltzmann equation that governs the equilibrium shape of the distribution funktion $f(k)$ in momentum space of the conduction electrons. An often used approximation to facilitate the solution is to describe the influence of the collisions of the electrons with the lattice by a relaxation time τ. Then the Boltzmann equation has the simple form

$$-\frac{e}{\hbar}\left(F + \frac{1}{\hbar c}[\nabla_k E \times H]\right) \cdot \nabla_k f + \frac{f - f_0}{\tau} = 0. \qquad (4)$$

Here $E(k)$ is the energy of electrons of wavevector k, F and H are the electric and magnetic field strengths and f_0 is the undisturbed distribution function. The equation holds only when

$$\frac{eH\tau}{mc} < 1, \qquad (5)$$

otherwise the deviations, due to the magnetic field, of the electrons from their path become comparable to the mean free path and quantum effects come in.

Solving (4) for f, the current j is given by

$$j = \frac{e}{4\pi^3\hbar} \int\!\!\int\!\!\int \nabla_k E f\, dk_x\, dk_y\, dk_z, \qquad (6)$$

where the integration is over the whole of momentum space.

The introduction of τ can only be defended rigorously in very simplified models, where τ is essentially independent of k. We now make the fundamental assumption that equation (4) will describe, at least approximately, also the state of affairs in a dislocated metal. Due to the anisotropy of the resistance, τ must now necessarily depend on k, and the use of such a τ in Boltzmann's equation is open to criticism. However, we may expect that the results of this assumption will give us at least some insight in the problem under consideration. An exact treatment with the aid of anisotropic transition probabilities would lead to forbidding mathematical difficulties.

The relaxation time must be chosen in such a way that on computing the current in the direction of the applied electric field from (4) and (6), the correct anisotropic resistance (2) is obtained. We shall write $\tau = \tau_0 - \tau_1(\boldsymbol{k})$ where τ_0 applies to the undislocated metal (and is thus independent of $\boldsymbol{k}$). Denoting the polar coordinates of $\boldsymbol{k}$ in the coordinate system defined earlier by ϑ' and φ', we try the assumption that τ_1 can be expanded in terms of spherical harmonies in ϑ' and φ' of orders zero and two. The 6 coefficients of the expansion can then be determined by the procedure mentioned above, viz. comparing the ordinary resistivity according to (2) with the expression found from (4) and (6), containing these six coefficients. The comparison is straightforward when the energy surface is spherical and yields the unambiguous result:

$$\tau = \tau_0 \left\{ 1 + N \frac{\varrho_1 + \varrho_2}{2\varrho_0} - \frac{5}{4} N \frac{\varrho_1 + \varrho_2}{\varrho_0} [\cos^2 \vartheta' + \sin^2 \vartheta' \cos^2 (\varphi' - \gamma)] - \right.$$
$$\left. - \frac{5}{2} N \frac{\varrho_1 - \varrho_2}{\varrho_0} \sin \vartheta' \cos \vartheta' \cos (\varphi' - \gamma) \right\} . \qquad (7)$$

We shall use this expression, derived formally, in the following discussion.

We are inserested in the change of resistance by a transverse magnetic field when the current flows in the direction of the wire axis ($\vartheta = 0$) and the only existing field components are $H_x = H$, F_y and F_z. A similar problem has been studied by DAVIS [8], and his results can directly be applied here. For the coefficient of transverse magnetoresistivity the following expression holds:

$$B_t = \frac{\varrho(H) - \varrho(0)}{\varrho(0) H^2} = - \left(\frac{e}{\hbar^2 c} \right)^2 \frac{I_2 I_4 + I_3^2}{I_1 I_2} , \qquad (8)$$

where $\varrho(H)$ and $\varrho(0)$ stand for the resistivities of the dislocated metal in fields H and zero respectively, and I_1, I_2, I_3 and I_4 are integrals in $\boldsymbol{k}$-space, defined as

$$\left. \begin{aligned} I_1 &= \iiint \frac{\partial f_0}{\partial E} \tau \left(\frac{\partial E}{\partial k_z} \right)^2 dk_x \, dk_y \, dk_z = \frac{\pi \hbar^2}{e^2} \varrho^{-1}(0) , \\ I_2 &= \iiint \frac{\partial f_0}{\partial E} \tau \left(\frac{\partial E}{\partial k_y} \right)^2 dk_x \, dk_y \, dk_z , \\ I_3 &= \iiint \frac{\partial f_0}{\partial E} \tau \frac{\partial E}{\partial k_z} \Omega \left(\tau \frac{\partial E}{\partial k_y} \right) dk_x \, dk_y \, dk_z , \\ I_4 &= \iiint \frac{\partial f_0}{\partial E} \tau \frac{\partial E}{\partial k_z} \Omega \left[\tau \Omega \left(\tau \frac{\partial E}{\partial k_z} \right) \right] dk_x \, dk_y \, dk_z . \end{aligned} \right\} \qquad (9)$$

The operator Ω is defined as

$$\Omega = \frac{\partial E}{\partial k_y} \frac{\partial}{\partial k_z} - \frac{\partial E}{\partial k_z} \frac{\partial}{\partial k_y} . \qquad (10)$$

Making use of the fact that $\partial f_0 / \partial E$ only differs appreciably from zero near the FERMI-surface $E(\boldsymbol{k}) = E_0$, the integrals can be evaluated exactly when this surface is assumed to be *spherical*. We then find the, at first sight somewhat peculiar, result that the coefficient of magnetoresistivity depends only *quadratically* on the dislocation density:

$$B_t = \frac{N^2}{(n\,e\,c)^2\,\varrho_0^2}\left[a\left(\frac{\varrho_1 + \varsigma_2}{\varrho_0}\right)^2 + b\left(\frac{\varrho_1 - \varrho_2}{\varrho_0}\right)^2\right], \qquad (11)$$

where n is the number of conduction electrons per atom, and a and b are numerical factors of the form $\alpha + \beta \cos^2\gamma$, α and β being of order unity. Also terms of higher order exist, but they have been neglected.

This result can be understood by a deeper study of the effect of the underlying assumption, viz. the description of the dislocated metal by adding an anisotropic term to the relaxation time. We shall not go into this here.

The supposition of a spherical energy surface implies that the magneto-resistance of the undeformed metal should be zero. To take account of the presence of very appreciable magnetoresistance in undeformed copper, a more intricate model for the energy surfaces must be used. This model must still have the property that only the effect of an *anisotropic* increase of the ordinary resistivity of the metal shall be detectable in the reduced magnetoresistivity coefficient, as obviously in the Kohler-diagram isotropic contributions to τ_0 have no effect. The simplest model that obeys this condition is the so-called two-band model [8]. It is supposed that the current is carried by electrons and holes, characterized by the concentrations n_1 and n_2 and effective masses m_1 and m_2. To simplify the computations and to avoid introduction of another undefined parameter, the plausible assumption is made that the same anisotropic relaxation time is associated with both bands. Assuming that the influence on it of dislocations is also the same, we have:

$$\varrho(0) = \frac{\pi\,\hbar^2}{e^2}\left(I_1^{(1)} + I_1^{(2)}\right)^{-1}, \qquad (12)$$

$$B_t = -\left(\frac{e}{\hbar^2 c}\right)^2 \frac{(I_2^{(1)} + I_2^{(2)})(I_4^{(1)} + I_4^{(2)}) - (I_3^{(1)} - I_3^{(2)})^2}{(I_1^{(1)} + I_1^{(2)})(I_2^{(1)} + I_2^{(2)})}, \qquad (13)$$

where the superscripts denote the two bands in which the integrals must be evaluated.

The results of a numerical evaluation, assuming both bands to be spherical and the dislocation resistance to be small compared to the bulk resistance, are:

$$\sigma(0) = e^2\,\tau_0\left(\frac{n_1}{m_1} + \frac{n_2}{m_2}\right)\left(1 - \frac{1}{2}\,N\,\frac{\varrho_1 + \varrho_2}{\varrho_0} + O\right), \qquad (14)$$

$$B_t = \frac{e^2\,\tau_0^2}{c^2}\,\frac{\dfrac{n_1}{m_1}\,\dfrac{n_2}{m_2}\left(\dfrac{1}{m_1} + \dfrac{1}{m_2}\right)^2}{\left(\dfrac{n_1}{m_1} + \dfrac{n_2}{m_2}\right)^2}\left(1 - \frac{1}{2}\,(1 + \cos^2\gamma)\,N\,\frac{\varrho_1 + \varrho_2}{\varrho_0} + O'\right), \qquad (15)$$

where O and O' stand for terms containing higher powers of N than the first, which moreover depend in an intricate way on the band constants. The order of magnitude of O' can be inferred from equation (11).

What we are interested in, is the relative increase of the ordinates of the KOHLER-curve, $[\varrho(H) - \varrho(0)]/\varrho(0)$, at a constant value of $H/\varrho(0)$ caused by the dislocations; that is, we are interested in the quantity

$$\beta = \frac{[B_t \varrho^2 (0)]_N - [B_t \varrho^2 (0)]_0}{[B_t \varrho^2 (0)]_0} , \qquad (16)$$

where the subscript N denotes the dislocated state, the subscript 0 the undeformed state. From (14) and (15) it follows that

$$\beta = \frac{1}{2} \sin^2 \gamma \, N \frac{\varrho_1 + \varrho_2}{\varrho_0} + O''. \qquad (17)$$

The effect on the magnetoresistivity is thus zero when the dislocation lines are parallel to $\boldsymbol{H}$, maximum when they are perpendicular to the field. Up to linear terms in N the effect does not depend on the anisotropy in the resistivity perpendicular to the dislocation axis, and a similar result should therefore also hold for screw dislocations.

When the orientations are arbitrarily distributed except for the conditions imposed on them in the beginning of this section, as should be the case in polycrystalline material, one obtains by averaging

$$\beta = \frac{1}{4} \, N \frac{\varrho_1 + \varrho_2}{\varrho_0} + O''. \qquad (18)$$

The relative effect of dislocations on the ordinary resistivity, $\alpha = (\varrho_N - \varrho_0)/\varrho_0$, follows from (14) to become

$$\alpha = \frac{1}{2} \, N \frac{\varrho_1 + \varrho_2}{\varrho_0} + O, \qquad (19)$$

thus

$$\frac{\beta}{\alpha} = \frac{1}{2} \qquad (20)$$

in the model used.

4. Discussion. Equations (16) to (20) have been derived under various simplifying assumptions, of which two impose restrictions on the experimental conditions. The first is that expressed by (5), that can also be written as

$$\frac{H}{\varrho(0)} < n e c; \qquad (5\,\text{a})$$

the second is the assumption that the dislocation resistivity is relatively small. Both assumptions are verified in our case for copper strained less than 10% and observed in field strengths less than 10.000 Oe at liquid H_2. The results obtained experimentally are given in table 1.

Table 1

Observed values of α and β measured at $H/\varrho_0 = 300\ \text{kOe}/\mu\Omega$ cm in polycrystalline copper deformed at $20°\text{K}$, and critical reduced field strength $[H/\varrho\,(0)]_{cr}$

Wire no.	Extension %	α %	β %	β/α	$[H/\varrho\,(0)]_{cr}$
1	20	164	58	0.35	—
2	20	165	65	0.39	$450\ \text{kOe}/\mu\Omega$ cm
2	10	(49)	31	(0.63)	350
3	10	79	50	0.63	500
4	10	71	48	0.68	350
3	5	(26)	(32)	(1.10)	400
4	5	(27)	19	(0.70)	350

The quantity α was measured as the relative increase at $20°\text{K}$ in resistivity, after deformation at $20°\text{K}$ and annealing as $200°\text{C}$. β was derived from the individual Kohler curves at the constant value $H/\varrho(0) = 300\ \text{kOe}/\mu\,\Omega$ cm, that is in that region of the diagram where the observations are most reliable. It is seen that the observed values of β/α compare reasonably well with the theoretical value 0.50. Better agreement could hardly be expected when it is realized that the adopted dislocation arrangement is rather a favourable one as regards the magneto-resistance; in practice, instead of pure edge dislocations perpendicular to the wire axis, certainly dislocations of mixed type and oriented more at random will occur. This will result in a decrease of β/α. The use of an anisotropic relaxation time also prevents of course an exact numerical comparison between theory and experiment.

A certain critical reduced field strength $[H/\varrho(0)]_{cr}$ could be estimated from our experiments. At this field strength the relative effect of dislocations, essentially constant below it, begins to abate. As is also demonstrated in table 1, this critical reduced field strength is found to be fairly independent of the resistivity in field zero, and equal to about $400\ \text{kOe}/\mu\,\Omega$ cm. This constancy gives rise to the obvious conclusion that the existence of a critical field strength is closely associated with the condition expressed in equations (5) and (5a). Expressed numerically in the units used, (5a) becomes

$$\frac{H}{\varrho\,(0)} < 512\ \text{kOe}/\mu\Omega\,\text{cm}, \tag{5b}$$

and is therefore indeed in close numerical agreement with the observed critical field strength. A completely different theoretical treatment is necessary, however, to explain the influence of dislocations in high fields.

The conclusion seems pretty well warranted that the observational evidence on the influence of dislocations on the magnetoresistance can be satisfactorily explained by the anisotropic scattering of these defects.

Theory and experiment agree that the magnetoresistance of deformed metals reflects clearly the behaviour of dislocations in them.

The conclusion reached in section 2, viz. that the dislocation scattering cross section is in fact larger by about a factor of 10 than the theoretical estimate, seems thus verified.

It is a pleasure to express my gratitude to my colleague JONGEN-BURGER who is largely responsible for the experimental part and to Mr. AALBERTS for his assistance in the measurements. I am much obliged to Drs POLDER and SMIT of this Laboratory for valuable help and criticism.

5. Summary. The magnetoresistivity at $20°K$ has been measured of polycrystalline pure copper wires deformed at this temperature. In the reduced (KOHLER) diagram the deformation results in an increase of the relative coefficient of magnetoresistivity. Subsequent annealing at temperatures up to $200°C$ have no effect on this, only by heating above $400°C$ the magnetoresistivity can be brought back to its original value. This seems to point to dislocations as the cause of the magnetoresistivity increase. When it is assumed that the effect is due to the anisotropic scattering caused by the dislocations, substantial agreement between theory and experiment is reached. The magnitude of the magnetic effect is a constant fraction of the relative ordinary resistivity caused by dislocations. From this, conclusions can be drawn about the scattering cross-section of dislocations. In strong magnetic fields the dislocation effect diminishes, the reason for which is briefly discussed.

References

[1] BROOM,T.: Advances in Physics **3**, 26 (1954). — H. G.van BUEREN: Z. Metallkde **46**, 272 (1955).

[2] BUEREN, H. G. VAN: Z. Metallkde **46**, 272 (1955). — A. SEEGER: Z. Naturforsch. **10a**, 251 (1955). — W. M. LOMER and A. H. COTTRELL: Phil. Mag. **46**, 711 (1955).

[3] KOEHLER, J. S.: Contribution to Birmingham Conference on the Mechanical Properties of Solids, 1954.

[4] BUEREN, H. G. VAN and P. JONGENBURGER: Nature, Lond. **175**, 544 (1955).

[5] HUNTER, S. C. and F. R. N. NABARRO: Proc. roy. Soc., Lond., Ser. A **220**, 542 (1953). Other workers have obtained more or less accurate estimates of the same order of magnitude.

[6] BUEREN, H. G. VAN: Acta Metall. **3**, 519 (1955).

[7] SONDHEIMER, E. H.: Advances in Physics **1**, 1 (1952).

[8] DAVIS, L.: Phys. Rev. **56**, 93 (1939). — See also A. H. WILSON: The theory of Metals, Chapter 8, 2nd edition. Cambridge 1953.

Neuere mathematische Methoden
und physikalische Ergebnisse zur Kristallplastizität

Von **A. Seeger,** Stuttgart

Mit 13 Abbildungen

1. Elastische und plastische Verformung; innere Spannungen. Die Theorie der Plastizität umfaßt gegenwärtig zwei große Teilgebiete, nämlich die phänomenologische Theorie, auch mathematische Plastizitätstheorie genannt, und die atomistische Theorie, deren Ziel es ist, die plastischen Eigenschaften von Ein- und Vielkristallen mit Hilfe der Eigenschaften von Versetzungen und anderen Störungen des regelmäßigen Kristallbaus quantitativ zu deuten. Obwohl beide Teilgebiete sich in jüngster Zeit rasch entwickelt haben, ist die Verbindung zwischen ihnen noch außerordentlich lose.

Dies mag vor allem zwei Gründe haben:

1. Die phänomenologische Theorie bezieht sich ausschließlich auf Vielkristalle. Im Vergleich zu den an Einkristallen vorliegenden Experimenten sind die an Vielkristallen gewonnenen experimentellen Resultate, die sich für eine versetzungstheoretische Deutung eignen, verhältnismäßig spärlich.

2. Die bisherige Entwicklung der Versetzungstheorie der plastischen Verformung benützte vor allem atomistische Modelle und darauf aufbauende Rechnungen. Es fehlte bis jetzt eine Formulierung der Versetzungstheorie, die einen einfachen und ungezwungenen Übergang zu den makroskopischen Begriffen der phänomenologischen Plastizitätstheorie gestattet.

In jüngster Zeit hat E. KRÖNER [1] eine Formulierung der Versetzungstheorie[1] gegeben, die einen geeigneten Ausgangspunkt für die zukünftige Verbindung der beiden Zweige der Plastizitätstheorie abgeben dürfte. Obwohl diese Verbindung selbst bis jetzt noch nicht hergestellt ist, erschien im Rahmen dieses Kolloquiums ein Bericht über den derzeitigen Stand dieser Überlegungen und über deren Zusammenhang mit anderen neueren Arbeiten angebracht.

---

[1] Wegen eines Überblicks über die hier benutzten Ergebnisse der Theorie der Kristallversetzungen vgl. [2].

Vom atomistischen Standpunkt aus kann man die elastische und die plastische Verformung von Kristallen dadurch gegeneinander abgrenzen, daß letztere durch die *Bewegung von Versetzungen* erfolgt, erstere dagegen nicht. Für den Zusammenhang mit der makroskopischen Beschreibung ist jedoch darüber hinaus die *Charakterisierung des verformten Zustandes* wesentlich. In voller Allgemeinheit lassen sich die Verhältnisse nur dann überblicken, wenn man nicht, wie in der Elastizitätstheorie üblich, den symmetrischen Verzerrungstensor ε^S und die Drehungen der Volumelemente[1] unabhängig voneinander betrachtet, sondern sie in bestimmter Weise zu einem unsymmetrischen Tensor ε zusammenfaßt. Dieser, neun wesentliche Komponenten enthaltende „*Distorsionstensor*" setzt sich gemäß

$$\varepsilon = \varepsilon^S + \varepsilon^A \tag{1}$$

aus dem symmetrischen Anteil ε^S und dem antisymmetrischen Anteil ε^A zusammen. Letzterer ist bekanntlich gemäß[2]

$$\varepsilon^A = I \times \vec{\Phi} \equiv I \times \left(\vec{\Phi}' + \vec{\Phi}'' \right) \tag{2}$$

durch einen axialen Vektor $\vec{\Phi}$ eindeutig bestimmt (I = Einheitstensor zweiter Stufe).

Wir werden in diesem Abschnitt an Hand von Beispielen die verschiedenen Verformungsarten besprechen und ihre charakteristischen Züge hervorheben. Auf den Zusammenhang mit der Theorie des unsymmetrischen Distorsionstensors ε und auf einige Anwendungen in der Plastizitätstheorie werden wir in Abschnitt 2 eingehen.

Abb. 1.
Ursprünglich gerader Stab

a) *Elastische Biegung eines geraden Stabes* (Abb. 1 und 2). Wie bei jeder elastischen Verformung kann man bei der elastischen Biegung angeben, wohin die materiellen Punkte des Körpers elastisch verschoben wurden. Der gesamte Verformungszustand ist durch die Kenntnis des eindeutig bestimmten Verschiebungsfeldes $\dot{s}$ bestimmt. Die Forderung, daß ein eindeutiges Verschiebungsfeld existieren soll, legt dem Feld des Verzerrungstensors ε^S eine gewisse Einschränkung auf. ε^S wird ein sog. Deformator

Abb. 2.
Elastisch gebogener Stab unter der Wirkung der Momente $\pm M$

$$\varepsilon^S = \text{Def } \dot{s} \equiv \frac{1}{2} \left(\nabla \dot{s} + \dot{s} \nabla \right). \tag{3}$$

---

[1] Wir beschränken uns hier auf infinitesimale Drehungen (und Verzerrungen), so daß die Drehungen durch einen Vektor beschrieben werden können.

[2] Auf die Zerlegung

$$\vec{\Phi} = \vec{\Phi}' + \vec{\Phi}'' \tag{2 a}$$

werden wir in Abschn. 2, a) zurückkommen.

Der Verzerrungstensor hat in diesem Falle die sog. *Kompatibilitätsbedingungen*

$$\text{Ink } \boldsymbol{\varepsilon}^S \equiv \nabla \times \boldsymbol{\varepsilon}^S \times \nabla = 0 \tag{4}$$

zu erfüllen. Die in (4) verwendete Schreibweise für die Kompatibilitätsbedingungen soll ausdrücken, daß ein bestimmter, dem Verzerrungsfeld zugeordneter symmetrischer Tensor zweiter Stufe, die sog. Inkompatibilität

$$\eta = \text{Ink } \boldsymbol{\varepsilon}^S \tag{5}$$

für Deformatoren, insbesondere also für elastische Verzerrungen, Null ist.

Ein elastisches Verzerrungsfeld besitzt insgesamt drei Freiheitsgrade[1], z. B. die drei Komponenten des Verschiebungsvektors oder die sechs Komponenten des symmetrischen Tensors zweiter Stufe $\boldsymbol{\varepsilon}^S$, die jedoch noch drei aus (4) folgenden Nebenbedingungen genügen müssen[2]. Mit den elastischen Verzerrungen ist eine durch den Drehvektor

$$\vec{\omega} = \frac{1}{2} \text{rot } \mathfrak{s} = -\vec{\varPhi}' \tag{6}$$

beschriebene Drehung der Volumelemente verbunden. Da jedoch $\vec{\omega}$ (bzw. $\boldsymbol{\varepsilon}^A = -I \times \vec{\omega}$) durch $\mathfrak{s}$ eindeutig bestimmt ist, bringt diese Drehung der Volumelemente keine weiteren Freiheitsgrade mit sich. Die leicht zu verifizierende Darstellung

$$\boldsymbol{\varepsilon} = \boldsymbol{\varepsilon}^S + \boldsymbol{\varepsilon}^A = \text{Def } \mathfrak{s} - \frac{1}{2}(\text{rot } \mathfrak{s}) \times I \equiv \nabla \mathfrak{s} \equiv \text{Grad } \mathfrak{s} \tag{7}$$

für den Distorsionstensor $\boldsymbol{\varepsilon}$ bei elastischer Verformung macht dies deutlich.

Zur Aufrechterhaltung elastischer Biegung sind äußere Momente notwendig[3]. Sieht man von der sog. neutralen Faser ab, so herrschen überall im Stab Zug- oder Druckspannungen. Macht man in der unter Zugspannungen stehenden Seite einen Schnitt, so klafft der Stab unter der Wirkung dieser Spannungen, wie in Abb. 3 angedeutet, auf. In den so entstandenen keilförmigen Einschnitt kann man (z. B. auf der Druckseite entnommenes) Material einfüllen und alsdann den mechanischen Zusammenhalt wieder herstellen. Führt man diese Operationen in der gesamten Probe in atomaren Dimensionen aus, so

Abb. 3. Aufklaffen des elastisch gebogenen Stabes beim Anbringen einer Kerbe

---

[1] Unter der Zahl der „Freiheitsgrade" einer physikalischen Größe verstehen wir hier die Zahl ihrer unabhängigen Komponenten.

[2] Es ist

$$\text{Div } \eta \equiv \nabla \cdot \eta = 0, \tag{5a}$$

so daß (4) nur drei und nicht sechs einschränkende Bedingungen für $\boldsymbol{\varepsilon}^S$ liefert.

[3] Allgemein gesprochen können elastische Verzerrungen in einem einfach zusammenhängenden, dem Hookeschen Gesetz gehorchenden Körper nur dann bestehen, wenn äußere Kräfte oder Momente wirken. Siehe z. B. [3].

kann man die beim Biegen auftretenden Zug- und Druckspannungen nahezu vollständig beseitigen und einen Zustand erreichen, bei dem zur Aufrechterhaltung der Biegung keine äußeren Momente notwendig sind. Die Verformung ist in diesem Falle *bleibend*. Derselbe Verformungszustand kann im Prinzip durch *plastische Verformung* erreicht werden; das wesentliche Kennzeichen hierbei ist, daß die Verformung nicht auf Null zurückgeht, wenn die die Verformung erzwingenden äußeren Kräfte und Momente aufhören zu wirken. Man sieht ohne weiteres ein, daß man für einen derartigen Verformungszustand nicht mehr in allgemeingültiger Weise ein Verschiebungsfeld definieren kann.

b) *Plastische Verformung.* Die plastische Verformung von Kristallen erfolgt, zumindest bei nicht allzu hohen Temperaturen, durch das Gleiten von Versetzungen. Es ist ein empirisches Ergebnis, daß fast die gesamte nicht elastisch erfolgte Verformung bleibend ist[1]. Wie sogleich näher ausgeführt werden wird, ist der durch plastische Biegung zustande kommende Verzerrungszustand jedoch wesentlich komplizierter als der durch die oben geschilderte Operation erreichte Zustand. Letzterer läßt sich am einfachsten durch eine Verteilung von Stufenversetzungen eines einzigen Typs beschreiben. Jede eingefügte oder herausgenommene Netzebene wird durch eine Versetzung begrenzt, wobei sich die in Abb. 4 angedeutete Verteilung von Stufenversetzungen ergibt. Bei der plastischen Biegung eines Stabes wird sich jedoch im allgemeinen die „*Idealverteilung*" von Abb. 4 *nicht* einstellen; es werden vielmehr Versetzungen von entgegengesetzten Vorzeichen vorhanden sein, deren Wirkungen auf die Krümmung des Stabes sich teilweise aufheben. Dies hat zur Folge, daß die Gesamtzahl der Versetzungen in einem plastisch verformten Kristall

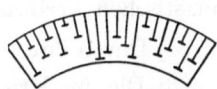

Abb. 4. „Idealverteilung" von Stufenversetzungen in einem plastisch gebogenen Stab. Der lange Strich im Zeichen $\perp$ deutet die eingeschobenen Netzebenen an, der kurze Strich gibt die Gleitebene der Stufenversetzungen an. Die Stufenversetzung befindet sich am Berührungspunkt der beiden Striche

Abb. 5. Schematische Darstellung der plastischen Dehnung eines Stabes. a) Form des Stabes vor, b) nach der Dehnung. Es hat ein Materialtransport und eine Vergrößerung der Kristalloberfläche stattgefunden

wesentlich größer ist, als es für die Aufrechterhaltung der während der Verformung erzwungenen Netzebenenkrümmung notwendig wäre.

Die Richtigkeit der eben genannten Behauptung kann man sehr einfach am Beispiel eines plastisch gezogenen Kristalls oder Polykristalls einsehen (Abb. 5). Zur plastischen *Dehnung* ist (wiederum bei Beschrän-

---

[1] Bei mittleren und höheren Temperaturen tritt nach plastischer Verformung und Aufhören der Beanspruchung ein gewisses „Rückwärtskriechen" auf. Es ist jedoch sehr oft vernachlässigbar gering.

kung auf hinreichend tiefe Temperaturen) die *Bewegung* von Versetzungen notwendig, da ja Material aus dem Kristallinnern an die Kristalloberfläche transportiert werden muß. Im Gegensatz zu der in Abb. 4 dargestellten Biegung sind jedoch zur *Aufrechterhaltung* der Verformung ohne äußere Kräfte keine Versetzungen notwendig. Es ist z. B. denkbar, daß alle Versetzungen an der Kristalloberfläche aus dem Kristall ausgetreten sind.

Der empirische Befund ist, daß dies im allgemeinen nicht der Fall ist. Man findet vielmehr, daß der verformte Kristall eine größere Fließspannung aufweist als der unverformte, also während der plastischen Verformung „verfestigt" wurde. Diese Verfestigung wird im wesentlichen durch die inneren Spannungen bewirkt, die die Versetzungen im Kristall hervorrufen[1].

Aus dem Vorstehenden geht hervor, daß die Versetzungen bei der plastischen Verformung drei verschiedene Rollen spielen.

α) Durch ihre Bewegung rufen sie eine *Scherung* der Kristalle hervor.

β) Die im Kristall nach der Verformung zurückbleibenden Versetzungen stabilisieren die durch die plastische Verformung entstandenen *Gitterkrümmungen*, nachdem die äußeren Kräfte und Momente zu wirken aufgehört haben.

γ) Die Versetzungen rufen innere Spannungen (sog. Eigenspannungen) hervor, die sich z. B. in der *Verfestigung* plastisch verformter Kristalle äußern.

Wir diskutieren nunmehr diese drei Rollen und die zwischen ihnen bestehenden Zusammenhänge:

α) Als Idealfall ist α) ohne das Auftreten von β) und γ) denkbar. Dies zeigt das Beispiel der plastischen Dehnung eines Stabes (Abb. 5). Wenn die die Verformung bewirkenden Versetzungen alle an der Kristalloberfläche austreten, so bleiben weder Krümmungen noch innere Spannungen zurück[2]. Letzteres gilt nur, wenn die Versetzungen reine Gleitbewegungen, auch konservative Bewegungen genannt, ausgeführt haben. Waren an der Verformung auch nichtkonservative, d. h. mit Volumenänderungen verbunden Bewegungen von Versetzungen beteiligt (was in Wirklich-

---

[1] Wenn sich zwei nichtparallele Versetzungslinien im Laufe ihrer Bewegung kreuzen, so trägt neben dem Spannungsfeld der Versetzungen auch noch die Störung der Kristallstruktur im Versetzungszentrum zur makroskopisch zu beobachtenden Fließspannung bei, z. B. bei der Bildung von sog. Sprüngen. In manchen Fällen (vgl. Abschn. 8) kann man diesen Beitrag jedoch vernachlässigen.

[2] Wie Rieder [5] neuerdings zeigen konnte, kann prinzipiell jede plastische Formänderung in der Weise erfolgen, daß am Ende des Verformungsvorganges keine Versetzungen im Medium mehr zurückbleiben. Wegen der zahlreichen der Versetzungsbewegung entgegenstehenden Hindernisse kann diese Verformungsart in Wirklichkeit nie allein auftreten.

keit wohl immer der Fall ist), so bleiben im verformten Kristall Gitterlücken und Zwischengitteratome zurück. Diese bilden vom Standpunkt
der Kontinuumstheorie aus Quellen innerer Spannungen, auf die wir
unter γ) näher eingehen werden.

β) Das in a) behandelte Beispiel zeigt, daß man in gewisser Näherung
Versetzungsverteilungen haben kann, die zwar Gitterkrümmungen, jedoch keine inneren Spannungen hervorrufen. Die Näherung besteht darin,
daß man die Versetzungsstärke b (= Betrag des BURGERS-Vektors) nicht
als eine durch die Kristallstruktur bestimmte endliche Größe auffaßt,
sondern den Grenzübergang $b \to 0$ durchführt. Damit die vorgegebenen
Krümmungen erhalten bleiben, muß man auch den Abstand der Versetzungen gegen Null gehen lassen, also kontinuierliche Versetzungsverteilungen betrachten. In diesem Grenzfall ist es in der Tat möglich, Ver-
setzungsanordnungen ohne innere Spannungen zu haben. BILBY, BUL-
LOUGH und SMITH [4] haben im einzelnen untersucht, wie man die Grund-
begriffe der Versetzungstheorie, z.B. BURGERS-Umlauf und BURGERS-
Vektor, im Falle einer kontinuierlichen Verteilung von Versetzungen zu
definieren hat. Den Zusammenhang zwischen der Krümmung der Gitterebenen und einer auf die oben geschilderte Weise erhaltenen kontinuierlichen Versetzungsverteilung ohne innere Spannungen hat NYE [6] angegeben. Die wesentliche Größe ist dabei der Tensor α der Versetzungsdichte[1], der ein im allgemeinen unsymmetrischer Tensor zweiter Stufe
ist und durch die Gleichung

$$d\mathfrak{b} = d\mathfrak{f} \cdot \alpha \qquad (8)$$

oder auch, gleichwertig mit (8), durch

$$\alpha = \frac{d\mathfrak{b}}{d\mathfrak{f}} \qquad (8\,\text{a})$$

definiert wird. Dabei bedeutet $d\mathfrak{b}$ den resultierenden BURGERS-Vektor
der durch das gerichtete Flächenelement $d\mathfrak{f}$ hindurchtretenden Versetzungen.

Neben den von NYE betrachteten *räumlichen* kontinuierlichen Verset-
zungsverteilungen sind für manche Anwendungen auch noch *flächenhafte*
Versetzungsverteilungen ohne Spannungsfelder von großem Interesse.
Sie bilden das kontinuumsmäßige Analogon des BURGERS-BRAGGschen
Korngrenzenmodells. Diese „flächenhaften" Versetzungen, die von
BILBY [7] näher diskutiert worden sind, bilden die Grenzflächen zwischen
Bereichen, die um einen endlichen Winkel gegeneinander verdreht worden
sind. Wegen weiterer Einzelheiten sei auf die zitierte Arbeit von BILBY
verwiesen.

---

[1] NYE bezeichnet $-\tilde{\alpha}$, den zu $-\alpha$ transponierten Tensor, als Versetzungsdichte.
Die hier benützte Vorzeichenwahl ist die in der Kristallphysik übliche (siehe [2],
Ziff. 3 u. 34).

Aus (8) folgt durch Integration über eine Fläche F

$$\mathfrak{b} = \iint_F d\mathfrak{b} = \iint_F d\mathfrak{f} \cdot \alpha, \tag{9}$$

woraus sich für eine geschlossene Fläche F mit Hilfe des Gaussschen Satzes wegen der Erhaltung des Burgers-Vektors [6]

$$\text{Div } \alpha = 0 \tag{9a}$$

ergibt. Man kann also das Integral (9) durch ein Randintegral über die Berandung der Fläche F ersetzen und gemäß der Gleichung

$$\alpha = -\text{Rot } \beta \equiv -\nabla \times \beta \tag{10}$$

einen (im allgemeinen asymmetrischen) Tensor zweiter Stufe β einführen, auf dessen physikalische Bedeutung wir zurückkommen werden.

Die Krümmung der Gitterebenen eines Kristalls wird zweckmäßigerweise durch den Krümmungstensor $\varkappa$ beschrieben. Dieser im allgemeinen ebenfalls unsymmetrische Tensor zweiter Stufe gibt gemäß der Gleichung

$$\overrightarrow{d\vartheta} = \varkappa \cdot d\mathfrak{r} \tag{11}$$

an, welche (infinitesimale) relative Drehung $\overrightarrow{d\vartheta}$ zwei Volumelemente erfahren, deren Ortsvektoren $\mathfrak{r}$ sich um den infinitesimalen Vektor $d\mathfrak{r}$ voneinander unterscheiden. Wie Nye gezeigt hat, sind in dem hier betrachteten Falle einer Versetzungsverteilung ohne innere Spannungen die beiden Tensoren $\varkappa$ und α durch die Gleichungen

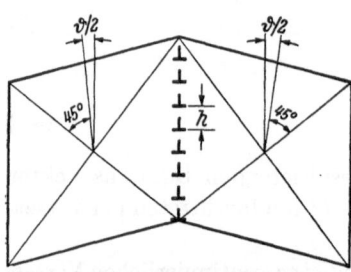

$$\varkappa = -\alpha + \frac{1}{2}\alpha_I I, \tag{12a}$$

$$\alpha = -\varkappa - \varkappa_I I \tag{12b}$$

Abb. 6. Burgers-Bragg-Korngrenze zwischen zwei um den Winkel ϑ gegeneinander geneigten Körnern. Die Drehachse ist parallel zu den in der Korngrenze liegenden Stufenversetzungen

verknüpft, die es gestatten, bei vorgegebener Krümmung die zugehörige Versetzungsverteilung auszurechnen [α_I und $\varkappa_I$ bedeuten die ersten Skalare (Spuren) der betreffenden Tensoren].

Die Verhältnisse lassen sich besonders einfach am Burgers-Braggschen Korngrenzenmodell [8, 9] für die Korngrenze zwischen zwei um den Winkel ϑ gegeneinander geneigten Kristallen illustrieren. Im Falle kleiner Winkel ϑ besteht zwischen der Versetzungsstärke b und dem Abstand h der Versetzungen der Zusammenhang (vgl. Abb. 6)

$$\vartheta = \frac{b}{h}. \tag{13}$$

Man sieht, daß der Grenzübergang $b \to 0$, $h \to 0$ zu einer kontinuierlichen Verteilung von Versetzungen infinitesimaler Stärke unter Erhaltung der Verdrehung ϑ möglich ist. Da sich, wie die nähere Diskussion zeigt [8, 10], das Spannungsfeld einer derartigen Korngrenze nur bis zu einem Abstand von der Größenordnung h auf beiden Seiten der Korngrenze erstreckt, wird es durch diesen Grenzübergang vollständig eliminiert.

An vorliegendem Beispiel sieht man ferner, daß zur Erzeugung der hier besprochenen Krümmungen im allgemeinen die in α) behandelten Gleitbewegungen notwendig sind. Zerlegt man nämlich den ursprünglich geraden Kristall durch einen Schnitt in zwei Hälften (Abb. 7 a) und dreht diese um die Winkel $-\vartheta/2$ und $+\vartheta/2$, so lassen sie sich nur dann ohne elastische Spannungen zusammenfügen, wenn die in Abb. 7 b angedeuteten plastischen Scherungen stattgefunden haben. Andererseits gibt es auch Fälle, in denen „innere Drehungen" ohne innere Spannungen und ohne resultierende Scherungen möglich sind. Ein Beispiel hierfür ist ein aus dem elastischen Körper her-
ausgeschnittener Kreiszylinder,
der um einen beliebigen Winkel
um seine Achse verdreht wird
und dann wieder mit seiner Umgebung verschweißt wird.

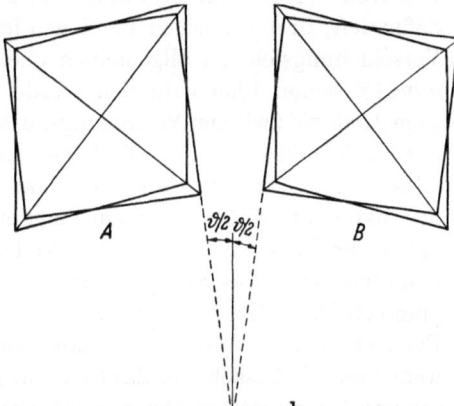

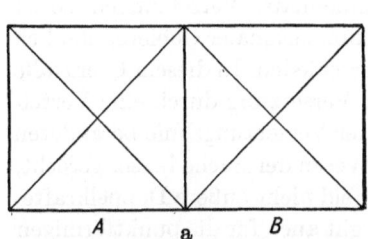

Abb. 7. Erzeugung der in Abb. 6 gezeichneten Korngrenze durch Drehung und Scherung. a) Ein Einkristall wird in zwei Hälften A und B zerschnitten. b) Die beiden Kristalle A und B werden um die Winkel $\pm \vartheta/2$ gedreht und anschließend ohne weitere Drehung so geschert, daß sie sich ohne elastische Verspannung, wie in Abb. 6 gezeichnet, zusammenfügen lassen

γ) Wie schon oben betont wurde, hat man nach der plastischen Verformung eines Kristalls keineswegs die nach den Gleichungen (12) durch die makroskopische Krümmung bestimmten Versetzungsverteilungen. Es bleibt vielmehr ein Überschuß von Versetzungen beider Vorzeichen, die weitreichende innere Spannungsfelder hervorrufen, welche auch über größere Volumenteile gemittelt nicht verschwinden. Um wenigstens näherungsweise die „idealen" Versetzungsverteilungen zu erhalten, muß man den verformten Kristall bei höheren Temperaturen anlassen. Viele der Versetzungen entgegengesetzten Vorzeichens annihilieren sich dabei. Für die zurückbleibenden Versetzungen gelten näherungsweise die Glei-

chungen (12). Nach plastischer Biegung ist diese: Vorgang (sog. Polygonisierung) mit verschiedenen Methoden sichtbar gemacht worden.

Wir diskutieren nun die mit den nichtidealen Versetzungsverteilungen
verknüpften Spannungen. Die zugehörigen Deformationen sind inkompa-
tibel, da sie nicht durch elastische Verformung eines einfach zusammenhängenden Körpers erzeugt werden können [11]. Weitere Beispiele für
derartige Spannungen, die auch Eigenspannungen genannt werden, bilden
Wärmespannungen, wie sie z.B. beim ungleichförmigen Abkühlen von
Werkstoffen entstehen können, magnetostriktive Spannungen sowie die
von Zwischengitteratomen und Fremdatomen hervorgerufenen Spannungen. In allen diesen Fällen sind im elastischen Medium Gebiete vorhanden, in denen die Kompatibilitätsbedingungen (4) verletzt sind. In
jenen Bereichen, in denen der Inkompatibilitätstensor

$$\eta = \text{Ink } \boldsymbol{\varepsilon}^S \equiv \nabla \times \boldsymbol{\varepsilon}^S \times \nabla \tag{14}$$

von Null verschieden ist, kann man überhaupt kein Verschiebungsfeld
definieren; in den übrigen Bereichen ist im Falle von Versetzungen das
Verschiebungsfeld im allgemeinen unendlich vieldeutig. Wie von Burgers [8] schon 1939 gefunden wurde, besteht in dieser Hinsicht eine
enge Analogie zwischen Versetzungslinien und Wirbellinien in der Theorie
der idealen Flüssigkeiten. Die Versetzungslinien bilden also den *Sitz von*
„Inkompatibilitäten". Betrachtet man linienhafte Versetzungen, so ist
der Inkompatibilitätstensor nur längs eindimensionaler Gebiete, nämlich
gerade der Versetzungslinien, von Null verschieden. In diesem Grenzfalle
kann man zwar das Spannungsfeld einer Versetzung durch eine Verteilung von Doppelkräften längs einer von der Versetzungslinie beranderten
Fläche beschreiben, doch wird man dem Wesen der Sache besser gerecht,
wenn man als Ursache für das Spannungsfeld nicht äußere Doppelkräfte,
sondern Inkompatibilitäten ansieht. Dies gilt auch für die punktförmigen
elastischen Singularitäten, die häufig als Modelle für Gitterlücken, Zwischengitteratome und Fremdatome in Kristallen verwendet werden.

Das Spannungsfeld einer Versetzung ist stets von den unter α) und β)
diskutierten Distorsionen begleitet. Man kann nämlich eine Versetzung
nicht ohne Gleitbewegungen (oder ohne sog. nichtkonservative Bewegungen) von außen in einen Kristall einführen. Wie das Burgers-
Braggsche Korngrenzenmodell zeigt, rufen Versetzungen auch immer
innere Drehungen hervor. Ohne diese Nebenerscheinungen kann man
inkompatible Deformationen durch den *Einbau von Zwischengitteratomen*
(oder auch Fremdatomen) in einen Kristall erzeugen.

Man denke sich hierzu, ohne Material zu entfernen, durch Anwendung
äußerer Kräfte einen Hohlraum im Kristall erzeugt, der das Zwischengitteratom gerade aufnehmen kann. Nachdem man das Zwischengitteratom, das man sich z.B. als dreiachsiges Ellipsoid vorstellen kann, ein-

gebaut hat, werden die äußeren Kräfte entfernt. Das Zwischengitteratom bildet dann eine Quelle innerer Spannungen. Man sieht, daß außerhalb des Atoms die Kompatibilitätsbedingungen erfüllt sind, innerhalb dagegen nicht[1]. Der so erzeugte Spannungszustand hat insgesamt drei Freiheitsgrade, drei aufeinander senkrecht stehenden *Doppelkräften*[2] entsprechend (Abb. 8).

2. Der allgemeine Distorsionszustand[3]. a) *Mathematische Analyse*[4].

Im Abschnitt 1 haben wir die elastische Verformung, eindeutig gekennzeichnet durch die Angabe des Verschiebungsvektors $\mathfrak{s}$ als Funktion des Ortsvektors $\mathfrak{r}$, den drei verschiedenen Arten nicht-elastischer Verformungen, die in $b\ \alpha$), $b\ \beta$) und $b\ \gamma$) behandelt wurden, gegenübergestellt. Das Unterscheidungsmerkmal zwischen diesen beiden Gruppen ist, daß bei der elastischen Verformung benachbarte Punkte immer benachbart bleiben, während bei den drei anderen Verformungsarten die Nachbarschaftsverhältnisse gestört werden, z. B. durch Gleitbewegungen, durch innere Verdrehungen (Beispiel des

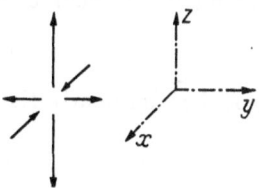

Abb. 8. Inkompatible Spannungen ohne innere Drehungen können durch „Doppelkräfte ohne Moment" hervorgerufen werden

gedrehten Kreiszylinders) oder durch Herausnehmen oder Einfügen von Materie (nichtkonservative Bewegungen von Versetzungen, Einbau von Fremdatomen usw.).

Als Ausgangspunkt für die mathematische Beschreibung [*14*] dieser verschiedenen Verformungsarten benützen wir, wie schon eingangs erwähnt, den allgemeinen nichtsymmetrischen Distorsionstensor ε. Er läßt sich immer in der Form

$$\varepsilon = \mathrm{Grad}\,\mathfrak{s} + \mathrm{Rot}\,\zeta \equiv \nabla\mathfrak{s} + \nabla \times \zeta \qquad (15)$$

schreiben. Der erste Anteil gibt, wie (7) zeigt, die elastische Distorsion wieder; der zweite Anteil mit einem unsymmetrischen Tensor zweiter Stufe ζ muß den nichtelastischen Distorsionsarten entsprechen. Man erhält diese, indem man Rot ζ in seine invarianten Bestandteile zerlegt. Hierbei ist es zweckmäßig, ζ mit Hilfe eines im allgemeinen ebenfalls unsymmetrischen Tensors zweiter Stufe ι und eines Vektors $\mathfrak{a}$ in der Form

$$\zeta = \iota \times \nabla + \mathfrak{a}\,\nabla \qquad (16)$$

---

[1] Im Außengebiet können im Gegensatz zum Innengebiet elastische Verschiebungen in eindeutiger Weise definiert werden.

[2] Es handelt sich hier um sog. „Doppelkräfte ohne Moment". Siehe hierzu [*12*].

[3] In dieser Arbeit verwenden wir „Distorsionen" als Oberbegriff zu „Verzerrungen und Drehungen", womit der enge Zusammenhang mit VOLTERRAS Distorsionen [*13*] zum Ausdruck gebracht wird. Letztere werden durch singuläre Versetzungslinien erzeugt, während wir hier kontinuierliche Versetzungsverteilungen im Auge haben.

[4] Die folgenden Ausführungen sind sehr knapp gehalten, da demnächst eine ausführliche Darstellung von E. KRÖNER [*16*] erscheinen wird.

darzustellen. Mit dem Ansatz (16) erhält man

$$\text{Rot } \zeta = \text{Ink } \iota + \text{Def} (\text{rot } \mathfrak{a}) + \frac{1}{2} [(\text{rot } \mathfrak{a}) \nabla - \nabla (\text{rot } \mathfrak{a})] . \tag{17}$$

Es werde nun der Vektor

$$\mathfrak{u} = \text{rot } \mathfrak{a} , \tag{18}$$

der der Nebenbedingung

$$\text{div } \mathfrak{u} = 0 \tag{18a}$$

genügt, eingeführt. Außerdem soll ι in seinen symmetrischen Anteil ι^S und in seinen antisymmetrischen Anteil ι^A aufgeteilt werden. Für letzteren gilt, wie man leicht nachrechnet [15],

$$\text{Ink } \iota^A = - \left(\text{grad div } \vec{\iota^A} \right) \times I , \tag{19}$$

wobei $\vec{\iota^A}$ der dem antisymmetrischen Tensor ι^A gemäß der Gleichung

$$\iota^A \equiv I \times \vec{\iota^A} \tag{19a}$$

zugeordnete axiale Vektor ist. Führt man noch die skalare Größe

$$\lambda \equiv - \text{div } \vec{\iota^A} \tag{20}$$

ein, so bekommt man die folgende Zerlegung in invariante Bestandteile:

$$\text{Rot } \zeta = \text{Ink } \iota^S + \text{Def } \mathfrak{u} + \left(\text{grad } \lambda + \frac{1}{2} \text{rot } \mathfrak{u} \right) \times I . \tag{21}$$

Die beiden ersten Summanden in (21) geben einen Beitrag zum Verzerrungstensor ε^S, die beiden letzten Glieder geben einen Beitrag

$$\vec{\Phi}'' = \frac{1}{2} \text{rot } \mathfrak{u} + \text{grad } \lambda \tag{22}$$

zum antisymmetrischen Anteil

$$\varepsilon^A = I \times \vec{\Phi} = I \times \left(\vec{\Phi}' + \vec{\Phi}'' \right) \tag{23}$$

des Distorsionstensors [vgl. Gl. (2)].

Die physikalische Bedeutung der symmetrischen Anteile in (21) ist leicht einzusehen. Da die Inkompatibilität eines Deformators Null ist, gibt nur Ink ι^S einen Beitrag zur Inkompatibilität des Verzerrungstensors. Ink ι^S stellt die mit den unter 1 b γ) besprochenen Eigenspannungen verbundenen Verzerrungen dar und besitzt wegen den mit jeder Inkompatibilität verknüpften Nebenbedingungen [vgl. (5a)] drei unabhängige Komponenten. Def $\mathfrak{u}$ stellt den spannungsfreien, durch das Gleiten von Versetzungen bewirkten Anteil der Verzerrung dar; Gleichung (18a) drückt die Tatsache aus, daß mit dem Gleiten von Versetzungen keine Volumänderung verbunden ist.

Die antisymmetrischen Glieder geben gemäß (22) Drehungen von Volumelementen wieder. Nur der Anteil grad λ ist von den oben diskutierten Anteilen des Verformungstensors unabhängig. Er beschreibt die verzerrungslose Verdrehung von Volumelementen, die in Abschnitt 1 am Ende von 1 b β) erwähnt worden war und entspricht einem Freiheitsgrad. 1/2 rot u ist eine Drehung, die mit Def u immer gekoppelt ist (von trivialen Ausnahmefällen abgesehen) und die dafür sorgt, daß der Zusammenhang der Volumelemente erhalten bleibt. Beide Glieder ergeben zusammen zwei Freiheitsgrade. Sie beschreiben plastische Scherungen. Wir haben sie bisher immer der Bewegung von Versetzungen zugeschrieben. Selbstverständlich können diese Scherungen auch durch triviales, versetzungsloses Gleiten benachbarter Netzebenen, wie es vielleicht in Kristallen mit ausgeprägter Schichtstruktur (Graphit) auftritt, und durch Zwillingsbildung entstehen. Die Vektoren ŝ und u unterscheiden sich rein geometrisch dadurch voneinander, daß die mit elastischen Verschiebungen ŝ verknüpften Drehungen im Gegensatz zu „inneren Drehungen" die Nachbarschaftsverhältnisse ungeändert lassen. Dies wird an anderer Stelle genauer besprochen werden [16].

Betrachtet man den allgemeinen nichtsymmetrischen Distorsionstensor ε, so muß man sich nach einer geeigneten Verallgemeinerung der Gleichung (5) umsehen, die die Grundgleichung für die nicht durch äußere Kräfte hervorgerufenen Verzerrungen ist. Wie die Diskussion in Abschnitt 1 gezeigt hat, gibt es Versetzungsverteilungen, die nichts zur Inkompatibilität η beitragen, so daß die neue Grundgleichung einen Zusammenhang zwischen der Versetzungsdichte α und dem Distorsionstensor ε geben muß. Der in (10) eingeführte Tensor β hat die Dimension von ε; eine Analyse der verschiedenen Möglichkeiten zeigt, daß zwischen β und ε der einfachste mögliche Zusammenhang besteht und daß

$$\beta = \varepsilon \qquad (24)$$

ist. Da der Zusammenhang zwischen dem Spannungstensor σ, der selbstverständlich symmetrisch ist, und der Dichte f der äußeren Volumkräfte durch die eingeführten Verallgemeinerungen nicht geändert wird, hat man das folgende Paar von Grundgleichungen:

$$\text{Div } \sigma + f = 0, \qquad (25\,a)$$

$$\text{Rot } \varepsilon + \alpha = 0. \qquad (25\,b)$$

Wegen der Identität Rot Grad $\equiv 0$ tragen die elastischen Distorsionen (7) nichts zur Versetzungsdichte bei. Der Zusammenhang zwischen der Inkompatibilität η und der Versetzungsdichte α ist durch

$$\eta = Sym \{\text{Rot } \tilde{\alpha}\} \equiv -\frac{1}{2}(\alpha \times \nabla - \nabla \times \tilde{\alpha}) \qquad (26)$$

gegeben, wobei $Sym \{\ \}$ den symmetrischen Anteil des in geschweiften Klammern stehenden Tensors bedeutet. Wie man leicht nachprüft, ergeben die von NYE betrachteten Versetzungsverteilungen in der Tat keinen Beitrag zu der durch (26) definierten Inkompatibilität η. Für $\eta = 0$ wird andererseits (25b) mit der NYEschen Gleichung (12b) identisch, da geometrische Überlegungen zeigen, daß der NYEsche Drehvektor $\vec{\vartheta}$ von (11) mit grad λ + rot $\mathfrak{u}$ zu identifizieren ist. Man muß dabei beachten, daß Rot Def $\mathfrak{u}$ = Rot [1/2 (rot $\mathfrak{u}$)×I] gilt. Im allgemeinen Fall kann man demnach (25b) auch Rot [Ink ι^S + I × $\vec{\vartheta}$] = − α schreiben, in welcher Form sie ursprünglich von KRÖNER [1] abgeleitet wurde.

In das HOOKEsche Gesetz und in die Gleichung für die Energiedichte hat man den symmetrischen Teil des Distorsionstensors einzusetzen. Die Energiedichte lautet z. B.

$$U = \frac{1}{2}\,\boldsymbol{\sigma} \cdot \cdot \,\boldsymbol{\varepsilon}^S \,, \tag{27}$$

wo

$$\boldsymbol{\varepsilon}^S = \mathrm{Def}\,(\mathfrak{s} + \mathfrak{u}) + \mathrm{Ink}\,\iota^S \,. \tag{28}$$

ist. Die Begründung für das Auftreten von $\mathfrak{u}$ in diesen Gleichungen wird an anderer Stelle [16] gegeben werden.

Im Falle verschwindender äußerer Kräfte, also

$$\mathrm{Div}\,\boldsymbol{\sigma} = 0\,, \tag{29}$$

wird die Berechnung der von den Versetzungen hervorgerufenen Spannungsenergie sehr erleichtert durch die Einführung eines Spannungsfunktionstensors φ gemäß

$$\boldsymbol{\sigma} = -\,\mathrm{Rot}\,\boldsymbol{\varphi}\,. \tag{30}$$

Mit Hilfe des zu φ transponierten Tensors $\widetilde{\varphi}$ läßt sich die Energiedichte bei unendlich ausgedehntem Medium als

$$U = \frac{1}{2}\,\widetilde{\boldsymbol{\varphi}} \cdot \cdot\,\boldsymbol{\alpha} = \frac{1}{2}\,\boldsymbol{\varphi} \cdot \cdot\,\widetilde{\boldsymbol{\alpha}} \tag{31}$$

schreiben, so daß $\widetilde{\varphi}$ die Bedeutung eines „*Versetzungspotentials*" hat. Es geht aus den MAXWELL-MORERAschen Spannungsfunktionen[1] χ durch Differentiation hervor und ist im Falle eines homogenen unendlich ausgedehnten Mediums (Schubmodul G, POISSONsche Konstante ν) durch die Differentialgleichung

$$\Delta\,\boldsymbol{\varphi} = \frac{2\,G}{1-\nu}\,(\widetilde{\boldsymbol{\alpha}}^l - \nu\,\boldsymbol{\alpha}^l) \tag{32}$$

mit

$$\boldsymbol{\alpha}^l = -\,\mathrm{Rot}\,\boldsymbol{\varepsilon}^l \equiv -\,\mathrm{Rot}\,\mathrm{Ink}\,\iota^S \,, \tag{32a}$$

also auf verhältnismäßig einfache Weise, bestimmt.

---

[1] Siehe hierzu und wegen der Anwendung der Spannungsfunktionen zur praktischen Berechnung der Spannungsfelder von Versetzungslinien [17]. Wegen der Randbedingungen an der Trennfläche zwischen zwei Medien bzw. an freien Oberflächen vgl. man [14].

b) *Physikalische Anwendungen.* Wie schon eingangs erwähnt wurde, ist zu erwarten, daß sich die im vorstehenden besprochene Analyse des allgemeinen Distorsionszustands und der inneren Spannungen als wertvolles Hilfsmittel bei der Deutung der plastischen Erscheinungen in Vielkristallen erweisen wird. Wir wollen einige der sich bietenden Möglichkeiten und ihre Zusammenhänge mit den Untersuchungen in Abschnitt 3 kurz skizzieren.

Die nicht mit inneren Spannungen verknüpften Verformungsarten gestatten es einem duktilen Ein- oder Vielkristall, einem äußeren Zwang mit viel geringerem Energieaufwand nachzugeben, als dies bei rein elastischer Verformung möglich wäre. Bei der Zugverformung von *Einkristallen* handelt es sich dabei überwiegend um plastische Scherungen. Innere Drehungen treten z. B. bei der Bildung der sog. Deformationsbänder und bei deren Polygonisierung beim Anlassen auf; im Gesamtbild der Erscheinungen spielen diese Drehungen jedoch nur eine Nebenrolle.

Ganz anders liegen in dieser Hinsicht die Verhältnisse bei der *plastischen Verformung von Vielkristallen.* Hier bildet sich während der Verformung bei nicht zu tiefer Temperatur innerhalb der Körner eine *Zellstruktur* [18,19] aus. Die Zellwände sind im wesentlichen BURGERS-BRAGG-Korngrenzen oder — genauer ausgedrückt — Gebiete, in denen geeignete Anordnungen von Versetzungen inkompatible Drehungen hervorrufen. Bei Einkristallen tritt diese Zellstruktur unter vergleichbaren Bedingungen nicht oder nur mit sehr kleinen Orientierungsunterschieden zwischen benachbarten Zellen auf [20]. Die Zellbildung ist somit wohl eine Folge der *inhomogenen* Verformung (z. B. Biegung), welche die einzelnen Körner bei der plastischen Verformung eines Vielkristalls erleiden.

Die Zellgrenzen sind nach einer Verformung bei tiefen Temperaturen (z. B. von Al bei $-183°$ C) [21] viel weniger scharf ausgeprägt als nach Verformung bei höheren Temperaturen (z. B. Al bei Raumtemperatur). Ein thermisch aktivierter Prozeß spielt somit bei der Bildung der Zellgrenzen eine wesentliche Rolle. Bei den kubischen Metallen gibt das empirische Bild hinsichtlich der Zellbildung einen ähnlichen Eindruck wie bei der *Gleitbandbildung* (s. Abschnitt 3). Hier wie dort scheint die Quergleitung von Schraubenversetzungen der maßgebende Prozeß zu sein. Aluminium, in dem die Quergleitung wegen der geringen Aufspaltung vollständiger Versetzungen in Halbversetzungen ziemlich leicht erfolgen kann [22], zeigt nach Verformung bei Raumtemperatur wohlausgeprägte Zellgrenzen. α-Fe, das sich wegen der fehlenden Aufspaltung in Halbversetzungen hinsichtlich der Quergleitung ähnlich wie Al verhalten sollte, weist nach Verformung bei Raumtemperatur fast ebenso scharfe Zellgrenzen wie Al auf [19]. Dieser Befund läßt sich *nicht* auf Grund der an und für sich naheliegenden Hypothese klären, daß das Klettern von Stufenversetzungen die Hauptrolle spielt, da bei Raumtemperatur dieser Pro-

zeß in α-Eisen wegen mangelnder Beweglichkeit der Leerstellen ziemlich
sicher nicht stattfinden kann[1]. Mit der Quergleitungshypothese ist im
Einklang, daß die Metalle Cu, Ag, Au und Ni, die eine starke Aufspaltung
in Halbversetzungen aufweisen [23], nach der Verformung bei Raumtemperatur wesentlich weniger wohldefinierte Zellwände besitzen[2] [24].

Faßt man die vorstehenden Erörterungen und die in Abschnitt 3 zu
besprechenden Erscheinungen bei der Gleitbandbildung zusammen, so
erhält man folgendes Bild für den Mechanismus der Zellbildung: Mit
wachsender Spannung ist eine zunehmende Zahl von Schraubenversetzungen in einzelnen Gleitlinien in der Lage, diese Gleitlinien mit Hilfe
thermischer Energie zu verlassen und seitwärts „auszuweichen". Die
dafür erforderlichen Mindestspannungen liegen bei Al und α-Fe wesentlich niedriger als bei Cu, Ag, Au und Ni. Diese Versetzungen können als
neue FRANK-READ-Quelle wirken und damit zur Ausbildung neuer Gleitlinien (und bei mehrfacher Wiederholung des Vorgangs zur Gleitbandbildung) führen. Dies ist der hauptsächliche Prozeß in Einkristallen, in
denen der Kristall den äußeren Kräften durch eine Scherung nachzugeben
versucht. Bei Polykristallen tritt jedoch wegen der Anisotropie der elastischen und plastischen Eigenschaften der Körner eine inhomogene Verformung auf, die im Anfangsstadium der Verformung wohl überwiegend
elastischer Natur ist. Diese *elastischen* Verzerrungen und Drehungen
können durch Bildung von Zellwänden mit geeigneten Versetzungsanord-
nungen abgebaut werden. Es erscheint natürlich, daß ein Teil der aus den
einzelnen Gleitlinien „befreiten" Versetzungen dazu verwendet wird,
derartige Zellwände zu bilden.

Die hier entwickelte Auffassung, daß die in die Gleitbänder und in
die Zellgrenzen gehenden Versetzungen an einer begrenzten Anzahl von
Stellen gemeinsam aus einzelnen, besonders starken Gleitlinien befreit
werden, erklärt zwanglos den empirischen Befund [25], daß die Zellgröße
bei großer Verformung („limiting particle size") und der Minimalabstand
der Gleitbänder bei allen untersuchten kubischen Metallen etwa gleich
sind.

**3. Gleitmechanismus und Oberflächenerscheinungen in dichtest ge-
packten Kristallen.** a) *Allgemeines.* Auf dem Gebiet der Versetzungstheorie
der plastischen Verformung von Einkristallen hat während der letzten
beiden Jahre besonders bei den kubisch-flächenzentrierten und bei den
hexagonal dichtest gepackten Metallen eine rasche Entwicklung eingesetzt. Als ein geeigneter Ausgangspunkt erwies sich dabei das gemein-

---

[1] Es erscheint jedoch wahrscheinlich, daß bei Zn und Cd, die beide während
einer Verformung bei Raumtemperatur eine sehr ausgeprägte Zellstruktur ent-
wickeln, das Klettern von Versetzungen wesentlich ist.

[2] Auf die empirische Korrelation zwischen Versetzungsaufspaltung und Zellbildung hat auch P. B. HIRSCH (private Mitteilung) hingewiesen.

same Bauprinzip der beiden Strukturen, die beide durch Aufeinanderstapeln dichtest gepackter Ebenen erzeugt werden können. Dies hat zur
Folge, daß die in den dichtest gepackten Ebenen der beiden Strukturen
(hexagonale Basisebene bzw. {111}-Ebenen) liegenden Versetzungen gemeinsame Eigenschaften besitzen, die vergleichende Studien zulassen. Eine
derartige gemeinsame Eigenschaft stellt die Fähigkeit vollständiger Versetzungen (mit BURGERS-Vektoren $\mathfrak{b} = 1/3 < 2\bar{1}\bar{1}0 >$ bzw. $\mathfrak{b} = 1/2 < 110 >$)
dar, in diesen Ebenen in sog. Halbversetzungen aufzuspalten [26]. Während die unaufgespaltene Versetzung als eine linienhafte Störung des
regelmäßigen Kristallbaus aufgefaßt werden kann, muß man sich eine
aufgespaltene Versetzung als ein flaches, in der (0001)- bzw. {111}-Gleitebene liegendes Band vorstellen, das auf beiden Seiten von je einer Halbversetzung begrenzt wird. Zwischen diesen beiden Halbversetzungen erstreckt sich ein sog. Stapelfehler, der im Falle der hexagonalen dichtesten
Kugelpackung ein Gebiet kubisch-flächenzentrierter Struktur und im
Falle des kubisch-flächenzentrierten Gitters ein Gebiet hexagonal dichtest gepackter Struktur ist. Je nachdem, ob in einem Metall die spezifische Energie je Flächeneinheit eines Stapelfehlers, die sog. Stapelfehlerenergie γ, verhältnismäßig groß oder verhältnismäßig klein ist, ist das
Stapelfehlerband innerhalb einer Versetzung schmal oder breit. Da man
weder die Aufspaltungsweite noch die Stapelfehlerenergie direkt messen
kann, ist man bei der quantitativen Diskussion auf indirekte Daten angewiesen. Diese weisen darauf hin, daß bei den Metallen Al, Cd, Zn, Mg
die beiden Halbversetzungen um 1 bis 2 Atomabstände voneinander
getrennt sind, während bei Cu (und auch bei Au, Ag und Ni) die Aufspaltung, die hier ziemlich stark vom Charakter der Versetzungen abhängt, zwischen etwa 5 Atomabständen (Schraubenversetzung) und etwa
12 Atomabständen (Stufenversetzung) variiert [23].

Diese Verschiedenheit der Aufspaltung macht sich in allen jenen Prozessen bemerkbar, in denen die Aufspaltung ganz oder teilweise rückgängig gemacht werden muß, z.B. beim Überkreuzen zweier Versetzungslinien oder bei der sog. Quergleitung von Schraubenversetzungen.
Die Aktivierungsenergien für diese Prozesse können berechnet werden
und ergeben sich als wesentlich größer für Cu als für Al [27].

Auf Grund der eben erwähnten Vorstellungen und Rechnungen
konnte die Temperaturabhängigkeit der Fließspannung der oben aufgeführten Metalle in quantitativen Einzelheiten gedeutet werden [28, 29].
Während die Theorie im Falle von Cu verhältnismäßig kompliziert ist
und wir im Rahmen dieses Berichts nicht näher darauf eingehen können[1],
liegen die Verhältnisse bei den Metallen mit großen Stapelfehlerenergien
einfach. Für die Temperaturabhängigkeit der kritischen Schubspannung
(bzw. der Fließspannung nach Vorverformung) ergibt die Theorie den in

---

[1] Wegen den experimentellen Daten siehe [30].

Abb. 9 wiedergegebenen Verlauf. (Dabei ist von einigen kleineren Korrekturen abgesehen.) Er stimmt mit den experimentellen Resultaten an Zn, Cd, Mg und Al gut überein. Oberhalb der Temperatur $T_0(\dot{a})$, die von der Abgleitungsgeschwindigkeit abhängt und mit wachsendem $\dot{a}$ zunimmt, wird die Fließspannung dieser Metalle durch das Spannungsfeld τ_G bestimmt, das von den im Kristall verteilten Versetzungen herrührt und im wesentlichen temperaturunabhängig ist. Unterhalb von T_0 trägt auch der sog. „Wald" der Versetzungen, die mit Hilfe der thermischen Energie von den gleitenden Versetzungen durchschnitten werden, zur Fließspannung bei, so daß diese mit fallender Temperatur linear ansteigt.

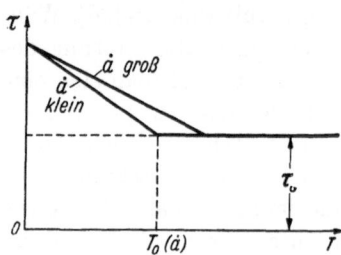

Abb. 9. Temperatur- und Geschwindigkeitsabhängigkeit der Fließspannung τ von hexagonal dichtest gepackten und kubischflächenzentrierten Metallen hoher Stapelfehlerenergie (T = absolute Temperatur; $\dot{a}$ = Abgleitungsgeschwindigkeit)

b) *Oberflächenerscheinungen und ihre Interpretation.* Die Kenntnis der für die Fließspannung, also eine makroskopisch meßbare, das ganze Kristallvolumen betreffende Größe, maßgebenden atomistischen Vorgänge ermöglicht es, folgendes Problem zu behandeln: Sind die auf der Oberfläche von verformten Kristallen sichtbaren Gleitspuren für die Vorgänge im Kristallinnern typisch und kann man aus Oberflächenbeobachtungen Rückschlüsse auf den Gleitmechanismus ziehen?

Obwohl die Ergebnisse von Oberflächenbeobachtungen zur Entwicklung von Theorien über die Verfestigung usw. benützt wurden [*31*], und obwohl wiederholt Versuche gemacht wurden, die oben aufgeworfene Frage experimentell zu beantworten [*32*], wurde in der Literatur immer wieder die Möglichkeit erörtert, daß der Kristalloberfläche hinsichtlich der plastischen Eigenschaften eine Sonderstellung zukomme. Es liegt im Wesen des vorliegenden Problems, daß man nicht ein für allemal die Äquivalenz der Oberflächenerscheinungen und der Gleitvorgänge im Innern beweisen kann. Es ist jedoch möglich, die Oberflächenbeobachtungen mit den Daten zu vergleichen, die man aus Messungen der Verfestigungskurve ableiten kann. Wir werden hier kurz über zwei Fälle berichten, die vorzügliche Übereinstimmung ergeben haben. Diese Übereinstimmung läßt es als fast sicher erscheinen, daß man bei einfachen Metallen aus dem Oberflächenbild auf die Verhältnisse im Kristallinnern schließen darf. Aus jüngster Zeit liegen zwei experimentelle Arbeiten über Oberflächenerscheinungen vor, die im Hinblick auf derartige vergleichende Untersuchungen unternommen worden waren. Müller und Leibfried [*33*] haben Al-Einkristalle um 12% gedehnt, und zwar bei Raumtemperatur mit Dehngeschwindigkeiten von $\dot{\varepsilon} = 5 \cdot 10^{-7}$ sec^{-1} bis $\dot{\varepsilon} = 5 \cdot 10^2$ sec^{-1}, also über neun Zehnerpotenzen hinweg, und in flüssiger Luft mit Dehngeschwindigkeiten von $\dot{\varepsilon} = 3 \cdot 10^{-3}\,\text{sec}^{-1}$ bis $\dot{\varepsilon} = 5 \cdot 10^{2}\,\text{sec}^{-1}$. Beobachtet wurden u. a. Dichte, Breite, Länge und Häufigkeitsverteilung der Gleitbänder sowie die verschiedenen Störungen des regelmäßigen Gleitlinienbildes als Funktion der Gleitgeschwindigkeit $\dot{\varepsilon}$. Verfestigungskurven wurden nicht gemessen; da jedoch das Verhalten der kritischen Schubspannung und der Verfestigungskurve von Aluminium bei normalen Gleitgeschwindigkeiten als Funktion der Temperatur verhältnismäßig gut bekannt ist, kann man den hier angestrebten Vergleich dennoch durchführen. Wir beschränken uns hier auf ein besonders geeignetes Beispiel.

Abb. 10 zeigt die Abhängigkeit der Gleitbandlänge bei Raumtemperatur von der Verformungsgeschwindigkeit $\dot{\varepsilon}$. Die Tatsache, daß die Gleitbandlänge über den größten Teil der untersuchten Dehnungsgeschwindigkeit hinweg praktisch nicht von der Dehnungsgeschwindigkeit abhängt, zeigt, daß die Hindernisse, die das Längenwachstum von Gleitbändern verhindern, nicht mit Hilfe thermischer Energie durchbrochen werden können.

Dies ist ein bemerkenswertes Ergebnis, da man sich bei diesen Versuchen im Bereich III der Verfestigungskurve (s. unten, Abb. 12) befindet, in dem der Verfestigungsanstieg ziemlich stark von der Temperatur und damit auch von der Gleitgeschwindigkeit abhängt. Bei normalen Verformungsgeschwindigkeiten[1] (Ab-

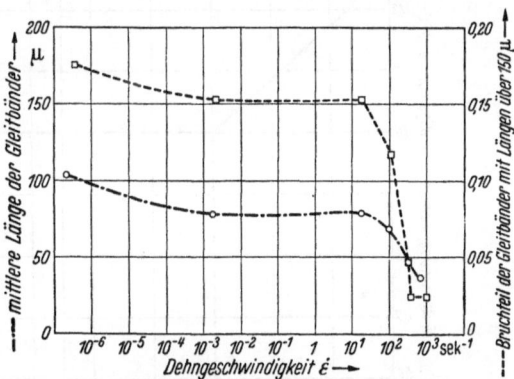

Abb. 10. Länge der Gleitbänder von Aluminium-Einkristallen, die bei Zimmertemperatur mit verschiedenen Dehngeschwindigkeiten $\dot{\varepsilon}$ um 12% gedehnt worden sind

gleitungsgeschwindigkeit $\dot{a} \approx 10^{-4}\,\text{sec}^{-1}$) liegt die Temperatur $T_0(\dot{a})$ für Aluminium unterhalb der Zimmertemperatur, und zwar etwa bei $-100^{\circ}\,\text{C}$. Erhöht man die Abgleitungsgeschwindigkeit, so steigt $T_0(\dot{a})$ an und erreicht schließlich Zimmertemperatur. Wird die Abgleitungsgeschwindigkeit noch weiter erhöht, so stellt der Versetzungswald bei Raumtemperatur ein wirkungsvolles Hindernis für die Versetzungsbewegung dar, so daß die Versetzungen z. T. im Versetzungswald stecken bleiben und die durchschnittliche Länge der Gleitlinien geringer wird, wie dies in Abb. 10 zum Ausdruck kommt. Die kritische Abgleitgeschwindigkeit,

---

[1] Wir verwenden hier sowohl die Dehngeschwindigkeit $\dot{\varepsilon}$ als auch die Abgleitungsgeschwindigkeit $\dot{a}$. $\dot{a}$ ist etwa um einen Faktor 2,2 größer als $\dot{\varepsilon}$.

bei der dieser Abfall in der Gleitbandlänge einsetzt, ist im vorliegenden Falle

$$\dot{a}_{\mathrm{krit}} = 50\,\mathrm{sec}^{-1}. \tag{33}$$

Man kann hieraus die für das Durchschneiden einer einzelnen Versetzungslinie im Versetzungswald benötigte Aktivierungsenergie nach der
Formel [*34, 28*]

$$U_0 = kT_0\,(\dot{a})\ln\left(\frac{NFb\,\nu_0}{\dot{a}}\right) \tag{34}$$

berechnen. Hierbei bedeutet ν_0 eine Frequenz von der Größenordnung
der DEBYE-Frequenz, b den Betrag des BURGERS-Vektors der schnei-
denden Versetzung, N die Zahl der Stellen je Volumeinheit, an denen die
Versetzungen im „Wald" aufgehalten werden, und F die Fläche, die im
Durchschnitt nach dem thermischen Überwinden des Hindernisses von
dieser Versetzung überstrichen wird. Setzt man plausible Werte [*28*] ein,

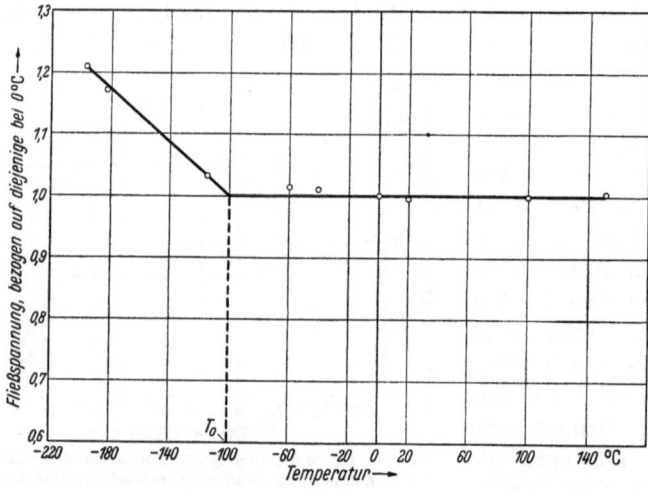

Abb. 11. Temperaturabhängigkeit der Fließspannung von Aluminium-Einkristallen als Funktion der
Temperatur nach COTTRELL und STOKES. An den Meßwerten wurde eine Korrektur angebracht, die
die Temperaturabhängigkeit der elastischen Konstanten von Aluminium-Einkristallen nach P. M.
SUTTON: Physic. Rev. **91**, 816 (1953) berücksichtigt

z. B. $NFb\,\nu_0 = 10^{10}\,\mathrm{sec}^{-1}$, so erhält man mit $T_0 = 293°\,\mathrm{K}$ aus Gl. (33) und
(34) als Aktivierungsenergie in den Messungen von MÜLLER und LEIB-
FRIED

$$U_0 = 0{,}49\,\mathrm{eV}. \tag{35}$$

Aus der von COTTRELL und STOKES [*35*] gemessenen Temperaturabhängigkeit der Fließspannung von Aluminiumkristallen ergibt sich, nachdem
eine Korrektur für die Temperaturabhängigkeit des Schubmoduls an-
gebracht wurde (Abb. 11), $T_0 = -100°\,\mathrm{C}$ (für $\dot{a} = 5 \cdot 10^{-5}\,\mathrm{sec}^{-1}$), woraus
man

$$U_0 = 0{,}48\,\mathrm{eV} \tag{36}$$

ableitet. Die Übereinstimmung der aus Spannungsmessungen und aus Oberflächenbeobachtungen ermittelten Werte ist verblüffend gut und eine starke Stütze für die hier vertretene Auffassung, daß beide eng miteinander zusammenhängen. Da beide Messungen bei mittleren Verformungsgraden ausgeführt wurden, dürfte $\ln(N F b \nu_0)$ in beiden Fällen gleich sein. Eliminiert man dieses Glied aus (34) mit Hilfe der Messungen bei zwei Gleitgeschwindigkeiten $\dot{a}_1$ und $\dot{a}_2$, so erhält man die Aktivierungsenergie

$$U_0 = \frac{k \, T_0(\dot{a}_1) \, T_0(\dot{a}_2)}{T_0(\dot{a}_2) - T_0(\dot{a}_1)} \ln \frac{\dot{a}_2}{\dot{a}_1} \qquad (37)$$

aus lauter bekannten bzw. gemessenen Größen. Man findet, indem man die angegebenen Meßwerte in (37) einsetzt,

$$U_0 = 0{,}49 \text{ eV}. \qquad (38)$$

Schon bevor experimentelle Daten vorlagen, aus denen man mit einiger Genauigkeit U_0 bei Al ermitteln konnte, war es möglich gewesen, die verschiedenen in U_0 enthaltenen Beiträge theoretisch zu berechnen [36]. Wir fügen hier einige kurze Angaben hierüber ein, wegen deren ausführlicher Begründung auf die Originalarbeit verwiesen sei.

U_0 ist die Aktivierungsenergie für das Schneiden von Versetzungen in einer $\{111\}$-Ebene, die eine überwiegende Stufenkomponente haben, mit Versetzungen in einer anderen $\{111\}$-Ebene, welche überwiegend Schraubencharakter haben. Da beide bei einem derartigen Schneidprozeß beteiligten Versetzungen etwas in Halbversetzungen aufgespalten sind, enthält U_0 zwei Beiträge, die von der Bildung je einer Einschnürung in diesen Versetzungen herrührt. Sie betragen 0,21 eV (Stufenversetzung) und 0,11 eV (Schraubenversetzung). Außerdem wird in der Stufenversetzung noch ein sog. Sprung („jog") gebildet, wobei es zwei etwa gleichhäufige Möglichkeiten gibt. Im einen Fall ($\{110\}$-Sprung) kommt zu obigem noch ein Beitrag von 0,13 eV, im anderen ($\{100\}$-Sprung) ein Beitrag von 0,19 eV hinzu. Man erhält also $U_0 = 0{,}45$ eV und $U_0 = 0{,}51$, so daß der experimentelle Wert zwischen den beiden theoretischen Werten liegt. Die Existenz zweier diskreter Aktivierungsenergien erklärt sehr wahrscheinlich die Beobachtung von COTTRELL und STOKES, daß unmittelbar oberhalb der Temperatur T_0 (Abb. 11) noch eine geringe Temperaturabhängigkeit vorhanden ist.

Die zweite der oben erwähnten experimentellen Untersuchungen [37] befaßt sich mit dem Zusammenhang zwischen der Gleitbandbildung und dem Verlauf der Verfestigungskurve. Es ist schon seit langer Zeit bekannt, daß die Gleitbänder bei Aluminium bei höheren Temperaturen stärker als bei niedrigen Temperaturen ausgebildet sind [38]. Andererseits hängt die Verfestigungskurve kubisch-flächenzentrierter Metalle wie in Abb. 12 angegeben von der Temperatur ab [39]. Es ist ein naheliegender Gedanke, die Temperaturabhängigkeit der Gleitbandbildung und der

Verfestigungskurve miteinander in Verbindung zu bringen und damit einen Zusammenhang zwischen den Oberflächenerscheinungen und den Vorgängen im Kristallinnern herzustellen. Frühere derartige Versuche haben aus folgendem Grunde nicht das gewünschte Ergebnis geliefert: Die Überlegungen stützten sich sämtlich auf die Verhältnisse bei Aluminium, insbesondere bei Raumtemperatur. Nun nimmt aber Aluminium bei Raumtemperatur eine Sonderstellung ein, die durch seine im Vergleich zu anderen kubisch-flächenzentrierten Metallen hohe Stapelfehlerenergie bedingt ist und die darin besteht, daß die Übergangspunkte α und β in Abb. 12 zusammenfallen.

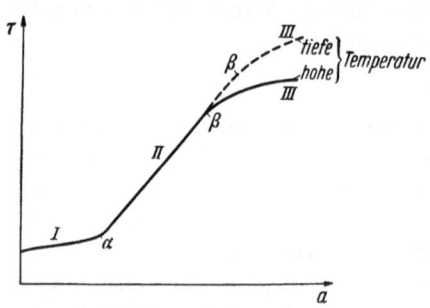

Die typische kubisch-flächenzentrierte Verfestigungskurve erhält man bei der Verformung von Aluminiumkristallen bei der Temperatur der flüssigen Luft [40; 41].

Aus diesem Grunde haben DIEHL, MADER und SEEGER Untersuchungen bei Zimmertemperatur *und* bei der Temperatur der flüssigen Luft sowie außer an Aluminiumkristallen auch an Kupfer- und Nickeleinkristallen durchgeführt. In allen untersuchten Fällen ergab sich, daß die Gleitbandbildung am Punkt β, also am Beginn des temperaturabhängigen

Abb. 12. Allgemeiner Verlauf und Temperaturabhängigkeit der Verfestigungskurve kubisch-flächenzentrierter Metalleinkristalle (schematisch). In den Bereichen I und II ist der Verfestigungsanstieg praktisch temperaturunabhängig. Mit steigender Temperatur rückt der Übergangspunkt β zwischen den Bereichen II und III zu kleineren Spannungen und Abgleitungen

Bereichs III der Verfestigungskurve, einsetzt. Da dieses Zusammentreffen unter ganz verschiedenen Bedingungen kein Zufall sein kann, muß geschlossen werden, daß die Temperaturabhängigkeit des Verfestigungsanstiegs und die Gleitbandbildung bei kubisch-flächenzentrierten Kristallen eine gemeinsame Ursache haben. Von den zahlreichen hierfür in Frage kommenden Prozessen konnten alle bis auf einen auf Grund theoretischer Überlegungen, z. T. unter Zuhilfenahme weiterer experimenteller Ergebnisse, ausgeschieden werden. Es wurde geschlossen, daß der am Punkt β einsetzende thermisch aktivierte Prozeß die Quergleitung von Schraubenversetzungen ist, die schon von KOEHLER [42] und später von LEIBFRIED und HAASEN [43] als Ursache für die Gleitbandbildung vorgeschlagen worden war. Die Quergleitung besteht im kubisch-flächenzentrierten Gitter darin, daß eine Schraubenversetzung von einer {111}-Gleitebene auf eine andere nicht parallele {111}-Gleitebene überwechselt. Wegen der damit verbundenen teilweisen Rückbildung der Aufspaltung in Halbversetzungen hängt die Aktivierungsenergie dieses Vorgangs stark von der Stapelfehlerenergie des betreffenden Metalls ab und erklärt damit zwanglos die beobachteten Unterschiede zwischen Aluminium und den anderen kubisch-flächenzentrierten Metallen [22].

Den theoretischen Überlegungen liegt folgendes Modell für den Gleitmechanismus bei kubisch-flächenzentrierten Metallen zugrunde: Im Bereich I sind die Laufwege L der Versetzungen vergleichbar mit dem Durchmesser der üblicherweise verwendeten Kristalle, also von der Größenordnung mm. Dies geht aus der Existenz eines Dickeneffektes [44] sowie aus lichtmikroskopischen Beobachtungen an Al-Einkristallen [45] hervor. Das Ende des Bereichs I wird nach DIEHL und SEEGER [46] dadurch bestimmt, daß in einem der latenten Gleitsysteme die wirksame Schubspannung (äußere Schubspannung minus latente Verfestigung) die kritische Schubspannung erreicht und Gleiten in diesem System in merklichem Maße erfolgt. Die Wechselwirkung zwischen den Versetzungen der beiden Gleitsysteme führt zu einer Verminderung des Versetzungslaufweges und damit zu dem in Punkt α zu beobachtenden Verfestigungsanstieg. Infolge der damit verbundenen Vergrößerung der äußeren Spannung erreicht die wirksame Schubspannung rasch die kritische Schubspannung weiterer Gleitsysteme und führt zur Erzeugung von Versetzungen in mehreren Gleitsystemen, so daß die Bildung von LOMER-COTTRELLschen unbeweglichen Versetzungen [47] nicht nur zwischen dem Hauptgleitsystem und einem Nebengleitsystem, sondern auch zwischen zwei Nebengleitsystemen stattfindet. Diese LOMER-COTTRELL-Versetzungen haben zwei hauptsächliche Wirkungen: Sie tragen einerseits zur Stabilität des verformten Zustandes gegenüber einem Zurückgleiten bei Entlastung bei und vermindern andererseits den Laufweg L mit zunehmender Abgleitung a, und zwar nach einem Gesetz

$$L \sim \frac{1}{a}. \tag{39}$$

Rührt die Verfestigung nicht von einer „Erschöpfung von FRANK-READ-Quellen", sondern von den inneren Spannungsfeldern der im Kristall aufgestauten Versetzungen her [48], so kann man für Abschätzungszwecke die Verfestigung umgekehrt proportional zum mittleren Abstand der Versetzungen und damit umgekehrt proportional zur Wurzel aus ihrer Dichte setzen. Mit dem Gesetz (39) erhält man dann eine lineare Verfestigungskurve, wie sie im Bereich II mit großer Genauigkeit beobachtet wird [49]. Die Verfestigung in diesem Bereich wird hauptsächlich von einer immer dichter werdenden Erfüllung des Kristalls mit der besonders von WILSDORF und KUHLMANN-WILSDORF [50] studierten Elementarstruktur bewirkt.

Eine Abnahme der Gleitlinienlänge (als Maß für L) etwa nach dem Gesetz (7) ist zuerst an Kupfereinkristallen von BLEWITT, COLTMAN und REDMAN [51] und an Aluminiumeinkristallen von KELLY [45] beobachtet

worden. Die Gültigkeit von Gl. (39) im Bereich II konnte von Diehl und
Rebstock [52] mit ziemlich großer Genauigkeit an Kupfereinkristallen
verifiziert werden. Diehl und Rebstock konnten ferner die oben beschriebene Rolle der sekundären Gleitsysteme experimentell bestätigen.
Eine Zwischentorsion vermindert L um einen festen Betrag, läßt aber die
Variation von L mit der Abgleitung ungeändert (Abb. 13). Die Betätigung
zusätzlicher Gleitsysteme bei der Torsion gibt zur Bildung einer großen
Zahl unbeweglicher Versetzungen
und zur Verminderung des Ver-
setzungslaufwegs im Hauptgleitsystem Anlaß.

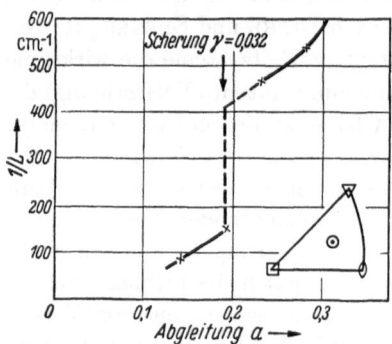

Abb. 13. Abhängigkeit der während einer kleinen
Zusatzverformung neu entstandenen Gleitband-
länge L von der Abgleitung a sowie Einfluß einer
Zwischentorsion (Scherung $\gamma = 0{,}032$) nach
Diehl und Rebstock

Neben der Elementarstruktur
beobachtet man in Abschnitt 2
einzelne starke Gleitlinien [37] (ge-
legentlich auch Paare von solchen).
An der Spitze dieser Gleitlinien
(bzw. am Rande der betreffenden
Gleitzone) herrscht eine große
Spannungskonzentration. Sind an-
gelegte Spannung und Temperatur
hoch genug, so beginnen an ihnen
Schraubenversetzungen durch
Quergleitung zu entweichen. Be-
findet sich eine andere wohlausgebildete Gleitzone mit Schraubenversetzungen entgegengesetzten Vorzeichens in der Nähe, so annihiliert sich
ein Teil der entweichenden Schraubenversetzungen mit diesen und gibt
zu den bekannten Bildern mit Quergleitung zwischen lichtmikroskopisch
sichtbaren Gleitlinien Anlaß [53]. Ein anderer Teil wird unter der Wirkung des lokalen Spannungsfeldes durch „doppelte Quergleitung" auf
eine zur ursprünglichen Gleitebene parallele Gleitebene umgelenkt und
bildet dort Versetzungsquellen. Letztere können zur Ausbildung neuer,
eng benachbarter Gleitlinien und damit zur Bandbildung führen.

Unter den Mechanismen, die als Ursache für die Gleitbandbildung
(bzw. die Temperaturabhängigkeit der Verfestigungskurve) vorgeschlagen worden sind und die man auf Grund der empirischen Ergebnisse
ausschließen kann, befindet sich auch die von Friedel [54] und Cottrell[1] diskutierte Möglichkeit eines thermisch aktivierten Zusammen-
brechens von Lomer-Cottrell-Versetzungshindernissen. Da in diesem
Fall das Längenwachstum der Gleitbänder thermisch aktiviert wäre,
müßte die Gleitbandlänge in starkem Maße von der Gleitgeschwindigkeit
abhängen. Abb. 10 zeigt, daß das Gegenteil der Fall ist; die bei sehr hohen
Verformungsgeschwindigkeiten auftretende starke Geschwindigkeits-

---

[1] Siehe den Beitrag von A. H. Cottrell in diesem Band, S. 33, sowie [35].

abhängigkeit konnte oben in befriedigender Weise erklärt werden. Dies bedeutet, daß die Hindernisse für die Gleitbandausbreitung nicht durchbrochen, sondern — wie es der hier vertretene Mechanismus verlangt — umgangen werden. Neben den bisher überwiegend benutzten negativen Argumenten gibt es überzeugende direkte experimentelle Beweise für die Richtigkeit des Quergleitungsmechanismus. Nach dem oben Gesagten muß am Punkt β der Verfestigungskurve die lichtmikroskopisch sichtbare Quergleitung einsetzen. KELLY [45] konnte an Aluminiumkristallen zeigen, daß dies in der Tat der Fall ist. Verformt man einen Kristall bei flüssiger Luft soweit im Bereich II, daß man beim Anwärmen auf Zimmertemperatur in den Bereich III kommt, so beobachtet man beim Weiterverformen bei der höheren Temperatur das Phänomen des ,,work-softening'' [55]. KELLY [45] fand an Aluminiumeinkristallen, daß in diesem Falle die neugebildeten Gleitbänder lang und wellig sind. Bei Vergrößerung der Auflösung erkennt man, daß sie aus kurzen geraden Stücken zusammengesetzt sind, welche durch Quergleitung miteinander verbunden sind. Die Länge der geraden Stücke ist etwa gleich der Länge der isolierten Gleitlinien, die vor dem Aufwärmen vorhanden waren. Dieser Befund zeigt direkt, daß bei dem für das ,,work-softening'' maßgebenden Prozeß, der natürlich auch für die Temperaturabhängigkeit des Bereichs III der Verfestigungskurve verantwortlich ist, die Versetzungshindernisse nicht durchbrochen, sondern durch Quergleitung umgangen werden[1].

Das Ergebnis der vorstehenden Diskussion ist, daß ein großer Teil der Oberflächenerscheinungen in verformten kubisch-flächenzentrierten Metallen aufs innigste mit den Vorgängen im Kristallinnern verknüpft werden konnte. Es wird sich sicher auch in der Zukunft als wertvoll erweisen, Oberflächenerscheinungen in großem Umfange zur Deutung der atomistischen Vorgänge bei der Kristallplastizität heranzuziehen.

Die Abfassung des vorstehenden Berichts wurde durch Rat und Tat von den Angehörigen des Instituts für theoretische und angewandte Physik der Technischen Hochschule Stuttgart unterstützt. Besonders zu

---

[1] *Zusatz bei der Korrektur:* Ähnliche lichtmikroskopische Beobachtungen an ,,verformungs-entfestigten'' Aluminiumeinkristallen wie in [45] finden sich auch bei COTTRELL und STOKES [35] (insbes. Abb.16). In neuester Zeit konnte S. MADER (unveröffentlichte Ergebnisse) zeigen, daß sowohl die bei Verformungsentfestigung im Lichtmikroskop sichtbare Quergleitung als auch die bei Kupfereinkristallen im Bereich III der Verfestigungskurve auftretende Fragmentierung der Gleitlinien [51] im Elektronenmikroskop als Quergleitung zu sehen sind. Wenn die lichtmikroskopischen Gleitspuren unkristallographisch sind oder nicht genau der Spur der Quergleitebene folgen, so erscheinen sie im Elektronenmikroskop als treppenförmig aus Quergleitung und Gleitung im Hauptgleitsystem aufgebaut. Durch diese Beobachtungen sind die in dieser Arbeit dargelegten Anschauungen über die Bedeutung der Quergleitung bei kubisch-flächenzentrierten Metallen vollständig bestätigt worden.

Dank verpflichtet fühle ich mich hierbei den Herren Prof. U. DEHLINGER, Dr. J. DIEHL, Dipl.-Phys. E. KRÖNER und Dipl.-Phys. H. REBSTOCK. Die Ausführungen über den Mechanismus der Zellbildung in Vielkristallen verdanken sehr viel ausführlichen Diskussionen mit Herrn Prof. Dr. N.F. MOTT und Herrn Dr. P. B. HIRSCH. Weiteren Dank schulde ich Herrn Dr. A. H. COTTRELL, Herrn Dr. G. LEIBFRIED, Herrn Dr. A. KELLY und Herrn Dipl.-Phys. H. MÜLLER für die Mitteilung ihrer experimentellen Ergebnisse vor der Veröffentlichung. Ein Teil der hier referierten experimentellen und theoretischen Arbeiten wurde von der Deutschen Forschungsgemeinschaft unterstützt, wofür ebenfalls herzlich gedankt sei.

Literatur

[1] KRÖNER, E.: Z. Phys. 142, 463 (1955); Z. Naturforsch. 11 a, 95 (1956).

[2] SEEGER, A.: Handbuch der Physik, Bd. VII/1. Berlin/Göttingen/Heidelberg 1955.

[3] BIEZENO, C. B. und R. GRAMMEL: Technische Dynamik, Bd. 1, S. 56 ff. Berlin/ Göttingen/Heidelberg 1953.

[4] BILBY, B. A., R. BULLOUGH, and E. SMITH: Proc. roy. Soc., Lond., Ser. A 231, 263 (1955).

[5] KRÖNER, E. u. G. RIEDER: Z. Phys. 145, 424 (1956).

[6] NYE, J. F.: Acta Metall. 1, 153 (1953).

[7] BILBY, B. A.: Report of a Conference on Defects in Crystalline Solids, p. 124. London: Physical Society 1955.

[8] BURGERS, J. M.: Proc., nederl. Akad. Wetensch. 42, 293 (1939).

[9] BRAGG, W. L.: Proc. phys. Soc., Lond. 52, 54 (1940).

[10] SEEGER, A.: Naturwiss. 38, 526 (1951).

[11] Siehe zu den hier behandelten Fragen die ausführlichen Darstellungen von E. KRÖNER: Z. angew. Phys. 7, 249 (1955). — J. D. ESHELBY: Phil. Trans. Roy. Soc., Lond., Ser. A 244, 87 (1951). — F. R. N. NABARRO: Advances in Physics 1, 269 (1952).

[12] LOVE, A. E. H.: A Treatise on the Mathematical Theory of Elasticity, 4. Aufl., p. 186. Cambridge 1927.

[13] VOLTERRA, E.: Ann. Ecole norm. sup. [3], 24, 401 (1907).

[14] KRÖNER, E.: Z. Naturforsch. 11a, 95 (1956).

[15] KRÖNER, E.: Z. Phys. 139, 175 (1954), Anhang.

[16] KRÖNER, E.: Dissertation Stuttgart 1956, Veröffentlichung demnächst.

[17] KRÖNER, E.: Z. Phys. 139, 175 (1954); 141, 386 (1955); Proc. phys. Soc., Lond., Ser. A 68, 53 (1955).

[18] WOOD, W. A. and W. A. RACHINGER: J. Inst. Met. 76, 237 (1949). — P. B. HIRSCH and J. N. KELLAR: Acta crystallogr., Lond. 5, 163 (1952). — P. B. HIRSCH: Acta crystallogr., Lond. 5, 168, 172 (1952). — P. GAY and A. KELLY: Acta crystallogr., Lond. 6, 165 (1953). — P. GAY, P. B. HIRSCH, and A. KELLY: Acta crystallogr., Lond. 7, 41 (1955).

[19] P. B. HIRSCH: Progr. Metal Phys. 6 (Zusammenfassung).

[20] BALL, C. J. (Cavendish Laboratory, Cambridge): unveröffentlichte Untersuchungen.

[21] KELLY, A. and W. T. ROBERTS: Acta Metall. 3, 96 (1955).

[22] Schöck, G. and A. Seeger: Report of Conference on Defects in Crystalline Solids, p. 340. London: Physical Society 1955.

[23] Seeger, A.: Report of a Conference on Defects in Crystalline Solids, p. 328. London: Physical Society 1955. — A. Seeger: Phil. Mag. 46, 1194 (1955). — A. Seeger u. G. Schöck: Acta Metall. 1, 519 (1953).

[24] Gay, P. and A. Kelly: Acta crystallogr., Lond. 6, 165 (1953). — P. B. Hirsch, A. Kelly and J. W. Menter: Proc. Phys. Soc. B 68, 1132 (1955).

[25] Gay, P., P. B. Hirsch, and A. Kelly: Acta crystallogr., Lond. 7, 41 (1954).

[26] Heidenreich, R. D. and W. Shockley: Report of a Conference on the Strength of Solids, p. 57. London: Physical Society 1948. — Siehe hierzu auch den Beitrag von A. H. Cottrell in diesem Band sowie die schon erwähnte Zusammenfassung [2].

[27] Seeger, A.: Z. Naturforsch 9a, 856 (1954). — G. Schöck and A. Seeger: Report of a Conference on Defects in Crystalline Solids, p. 340. London: Physical Society 1955.

[28] Seeger, A.: Z. Naturforsch. 9a, 870 (1954).

[29] Seeger, A.: Phil. Mag. 46, 1194 (1955).

[30] Adams, M. A. and A. H. Cottrell: Phil. Mag. 46, 1187 (1955).

[31] Mott, N. F.: Phil. Mag. 43, 1151 (1952); 44, 742 (1953). — G. Leibfried and P. Haasen: Z. Phys. 137, 67 (1954).

[32] Brown, A. F.: Advances in Physics 1, 427 (1953).

[33] Müller, H. u. G. Leibfried: Z. Phys. 142, 87 (1955). — H. Müller: Diplomarbeit Göttingen 1954.

[34] Seeger, A.: Phil. Mag. 45, 771 (1954).

[35] Cottrell, A. H. and R. J. Stokes: Proc. roy. Soc., Lond. A 233, 17 (1955).

[36] Seeger, A.: Report of a Conference of Defects in Solids, p. 391. London: Physical Society 1955.

[37] Diehl, J., S. Mader u. A. Seeger: Z. Metallkde. 46, 650 (1955).

[38] Brown, A. F.: J. Inst. Met. 80, 115 (1951/52).

[39] Andrade, E. N. da C. and C. Henderson: Phil. Trans. Roy. Soc. A 244, 177 (1951). — T. H. Blewitt, R. R. Coltman, and J. K. Redman: Report of a Conference on Defects in Solids, p. 369. London: Physical Society 1955.

[40] M. Heinzelmann: Diplomarbeit Stuttgart 1948/49.

[41] Staubwasser, W.: Dissertation Göttingen 1954. In der Al-Verfestigungskurve bei −185°C in Abb. 2 der Arbeit von Diehl, Mader u. Seeger ist leider ein Zeichenfehler unterlaufen. Der Beginn des Bereichs III liegt etwa bei $\tau = 1{,}4$ kp/mm^2.

[42] Koehler, J. S.: Imperfections in Nearly Perfect Crystals, p. 146. Wiley: New York 1952; Phys. Rev. 86, 52 (1952).

[43] Leibfried, G. u. P. Haasen: Z. Phys. 137, 67 (1954).

[44] Andrade, E. N. da C. and C. Henderson: Phil. Trans. Roy. Soc. A 244, 177 (1951). — T. Suzuki [zitiert bei A. Seeger: Z. Naturforsch. 9a, 758 (1954), und bei P. Haasen u. G. Leibfried: Fortschr. Phys. 2, 73 (1954), insb. S.163]. — M. S. Paterson: Trans. Amer. Inst. Mining metallurg. Engr. 203, 696 (1955); Acta Metall. 3, 491 (1955). — J. Garstone, R. W. K. Honeycombe, and G. Greetham: Acta Metall. 4 (im Druck).

[45] Kelly, A.: Veröffentlichung demnächst.

[46] Siehe F. Röhm u. J. Diehl: Z. Metallkde. 43, 126 (1952), sowie A. Seeger: Z. Naturforsch. 9a, 758, 870 (1954).

[47] Cottrell, A. H.: Dislocations and Plastic Flow in Crystals, p. 171. Oxford 1952.

[48] Seeger, A.: Z. Naturforsch. 9a, 758 (1954).

[*49*] Wegen der Genauigkeit, mit der bei Cu die Verfestigungskurve in Bereich II linear ist, vgl. man J. Diehl: Diss. Stuttgart 1955; Z. Metallkde **47**, 331, (1956).

[*50*] Wilsdorf, H. u. D. Kuhlmann-Wilsdorf: Naturwiss. **38**, 502 (1951); Z. angew. Phys. **4**, 361, 409, 418 (1952). — D. Kuhlmann-Wilsdorf u. H. Wilsdorf: Acta Metall. **1**, 394 (1953).

[*51*] Blewitt, T. A., R. R. Coltman and J. K. Redman: Report of a Conference on Defects in Solids, p. 369. London: Physical Society 1955.

[*52*] Diehl, J. u. H. Rebstock: Z. Naturforsch. **11a**, 169 (1956). Die Torsionsversuche wurden an dünnwandigen Hohlkristallen durchgeführt, so daß es trotz der inhomogenen Verformung gerechtfertigt ist, aus Oberflächenbeobachtungen Schlüsse auf die Vorgänge im Kristallinnern zu ziehen.

[*53*] Cahn, R. W.: J. Inst. Met. **79**, 129 (1951). Diese Art der Quergleitung, die für Aluminium typisch ist, ist in ihrem Wesen verschieden von der auf α-Messing zu beobachtenden „Quergleitung" [R. Maddin, C. H. Mathewson, W. R. Hibbard: Trans. Amer. Inst. Mining metallurg. Engr. **175**, 86 (1948); **185**, 529 (1949)], wie schon R. Maddin: Bull. Inst. Met. **1**, 105 (1952), bemerkt hat. Letztere entsteht durch eine nicht thermisch aktivierte Betätigung von Versetzungsquellen im Quergleitungssystem und sollte eigentlich durch eine andere Bezeichnung von derjenigen in Aluminium unterschieden werden.

[*54*] Friedel, J.: Phil. Mag. **46**, 1169 (1955).

[*55*] Stokes, R. J. u. A. H. Cottrell: Acta Metall. **2**, 341 (1954).

Les mécanismes de rupture des métaux

Par **C. Crussard**, **J. Plateau** et **Y. Morillon**, St.-Germain-en-Laye

Avec 8 figures

1. Introduction. Dans des corps fragiles, comme les verres, le mécanisme des ruptures est relativement simple : une rupture s'amorce sur une fissure préexistante (appelée fissure de GRIFFITH) et se propage par décohésion, sans déformation des parties adjacentes à la fissure.

Dans les métaux, on n'a jamais observé de rupture qui ne fût accompagnée de déformation, au moins dans les zones avoisinant les surfaces de rupture. D'autre part, on n'a aucune preuve de l'existence de fissures analogues aux fissures de GRIFFITH, sauf dans des cas spéciaux comme des corrosions intergranulaires ou des pores créés par une déformation de fluage, par exemple.

Ces simples observations soulèvent aussitôt une foule de questions. Les déformations accompagnant la rupture en sont-elles la cause ou la conséquence? La déformation est-elle nécessaire à la propagation des fissures? ou à l'amorçage des ruptures? ou aux deux? crée-t-elle des «fissures de GRIFFITH» dans le métal? etc ...

En ce qui concerne la propagation, il existe un cas où la déformation crée manifestement la condition favorable à la rupture: ce sont les *ruptures ductiles* (aussi dites «à nerfs») et les *ruptures fibreuses*, qui surviennent après des déformations importantes, de l'ordre de 100%, et produisent des surfaces mates. D'autres modes de rupture surviennent après des déformations beaucoup moindres: ce sont les ruptures *fragiles* ou «*à grains*», à l'aspect caractéristique, formé de facettes brillantes; mais même dans ce cas il y a déformation: une déformation d'ensemble, faible ou très faible, précédant la rupture, et une déformation assez forte localisée dans les couches avoisinant les surfaces de rupture.

On ne sait pas encore à l'heure actuelle si, dans les métaux usuels, à grains fins, cette dernière déformation est une condition essentielle de la propagation de la rupture, ou une conséquence secondaire[1]. Ce point est

---

[1] Dans le clivage de cristaux uniques ou de masses à gros grains, il semble bien que la déformation soit une conséquence secondaire de la propagation de la fissure.

encore à l'étude. Aussi ne parlerons-nous ici que d'essais faits à l'Irsid pour éclaircir le mécanisme de *l'amorçage des ruptures*.

2. Méthodes expérimentales. Les méthodes employées ont été des plus classiques. Comme essais mécaniques, pour faire apparaître des ruptures fragiles dans des conditions variées, nous nous sommes adressés à des essais de résilience et de traction à basse température:

1. Essais de résilience sur mouton-pendule, à températures variées allant de $-160°$C à $+40°$C, sur des éprouvettes à entailles d'acuités différentes (entaille en V, entaille UF, à rayon de 1 mm ou entaille à grand rayon de 3 mm). On mesure l'énergie de rupture, ou l'angle de pliage avant rupture, facile à déterminer en rapprochant les surfaces de rupture des deux moitiés rompues, après avoir meulé les liserés « à nerfs ».

2. Essais de traction dans l'azote liquide, à $-196°$C sur éprouvettes lisses (microéprouvettes Chévenard, de 1,5 mm de diamètre). Du fait de la petitesse de ces éprouvettes, les effets d'entailles sont faibles. On mesurait l'allongement et la tension de rupture.

En plus des grandeurs mécaniques déterminées dans ces essais, nous avons fait des examens des surfaces de rupture, à l'oeil, à la loupe binoculaire et au microscope électronique. Nous avons fait aussi quelques coupes avec attaques micrographiques appropriées pour observer les zones déformées plastiquement.

3. Essais de résilience: bimodalité d'amorçage. On a reconnu depuis longtemps, sur les courbes résilience-température des aciers doux, une zone de températures où les valeurs de la résilience présentent une forte dispersion, ou plus exactement tendent à se grouper autour de deux courbes situées à deux niveaux d'énergie différents [1].

Cette bimodalité, ou coexistence à une même température de deux groupes de valeurs de l'énergie de rupture, a été étudiée statistiquement et interprétée au cours d'études [2] dont nous résumons ici l'essentiel, en insistant sur un aspect nouveau des conclusions auxquelles nous sommes arrivés.

1. *Etude statistique.* Dans une série d'aciers doux et extra-doux, des essais de résilience avec entaille UF ont montré que dans le domaine de température correspondant à la partie basse de la transition, en prenant à chaque température un nombre suffisant d'éprouvettes (nous avons été jusqu'à 150), on peut montrer avec certitude que la distribution est bimodale, et groupée autour d'une valeur haute et d'une valeur basse de l'énergie. Dans la suite, pour simplifier, nous appellerons *ruptures ductiles* celles du groupe supérieur, *ruptures fragiles* celles du mode inférieur.

En recollant, comme il a été dit, les deux moitiés d'éprouvettes de telle sorte qu'on applique l'une sur l'autre les surfaces de rupture à grain, l'angle que l'on observe mesure la déformation atteinte au moment où s'amorce la rupture fragile. On constate qu'à chaque température, la distribution des valeurs de cet angle est également bimodale.

Lorsque la température varie, les valeurs correspondant aux ruptures ductiles et celles correspondant aux ruptures fragiles varient également, croissant en général avec la température, mais surtout la *proportion des deux types de rupture change:* le pourcentage de ruptures ductiles croît depuis 0 à la limite inférieure du domaine de température en question, jusqu'à 100 à la limite supérieure. Ceci signifie que la probabilité pour une éprouvette de se rompre selon le «mode ductile» croît en même temps de 0 à 1, ou que celle d'une rupture sur le «mode fragile» décroît de 1 à 0.

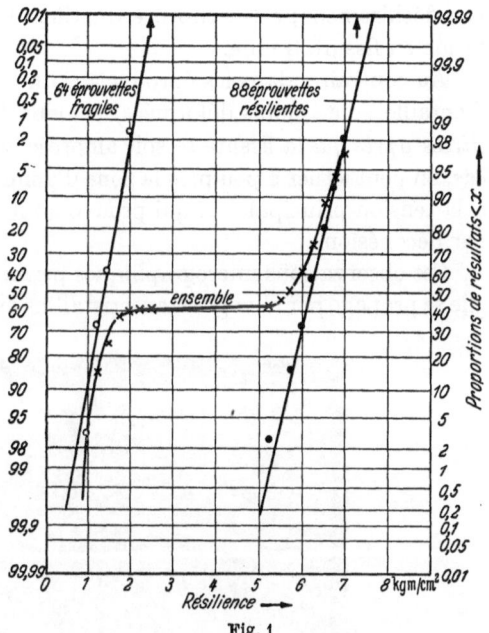

Fig. 1

On peut donner de ces résultats statistiques diverses représentations, parmi lesquelles nous donnerons celle des *fréquences cumulées* (fig. 1), car elle nous servira dans la suite: pour toutes les éprouvettes rompues à une certaine température, on représente, pour chaque valeur de l'énergie, le pourcentage des éprouvettes dont l'énergie de rupture est inférieure ou égale à la valeur considérée. Ce pourcentage étant porté en coordonnées galtoniennes, si la distribution est normale, les points doivent s'aligner sur une droite; s'ils ne le font pas, la distribution d'ensemble n'est pas normale: il en est ainsi dans le cas présent, mais les distributions correspondant à chacun des deux modes de rupture sont normales.

Nous avons retrouvé cette bimodalité, non seulement sur des aciers doux, mais sur des aciers de toutes nuances, et ce avec tous les types d'entailles essayés. Mais le saut de résilience (différence moyenne d'énergie entre les ruptures du groupe ductile et du groupe fragile) est d'autant plus faible que l'entaille est plus aigüe. Il est faible également pour certains aciers extra-doux, ou au contraire pour des aciers durs alliés (par exemple aciers auto-trempants au Cr—Ni, aussi bien à l'état tenace qu'à l'état fragilisé de revenu). Dans ce cas, il faut un grand nombre d'éprouvettes pour mettre en évidence la bimodalité.

2. *Mode d'amorçage de la rupture.* En étudiant des micrographies de coupes d'éprouvettes rompues, on s'aperçoit que, dans le cas des ruptures ductiles, les zones situées au fond de l'entaille sont très fortement déformées, et leur forme même indique que le métal s'est déchiré progressivement à partir de la surface du fond de l'entaille. Les surfaces de rupture à cet endroit ont un aspect mat, à nerf, qui passe rapidement à l'aspect brillant et granuleux des ruptures fragiles. Il y a donc *amorçage ductile*, par déchirement; cette rupture ductile, en s'approfondissant, se transforme en fissure fragile.

Au contraire, dans les éprouvettes du type fragile, les zones du fond d'entaille sont moins déformées, et leur forme prouve *qu'elles ont été étirées après que* la fissure se soit amorcée en profondeur. Dans quelques cas, on peut situer à peu près la zone d'amorçage: la fissure y est normale à la tension principale, ce qui prouve qu'il s'agit là d'une *amorce fragile*, par décohésion.

Ces observations micrographiques permettent de voir si la fissure fragile se propage par clivages transcristallins ou par décohésions intergranu-

Fig. 2

laires. Nous avons rencontré les deux cas extrêmes, avec souvent des cas mixtes; celà dépend de la nature du métal, du traitement qu'il a subi, et de la température de l'essai. Dans un acier particulier, bien que les éprouvettes examinées soient peu nombreuses, on observe un phénomène curieux: à très basse température, on observe presque uniquement des clivages; c'est aussi le cas à limite supérieure de la zone de transition (température la plus élevée à laquelle on observe une rupture fragile, qui dans l'acier en question correspondait à une résilience déjà très élevée);

mais entre les deux, pour une déformation préalable faible comme on en rencontre à la limite inférieure de la zone de transition, on observe une *proportion notable de décohésions intergranulaires.*

La bimodalité des ruptures dans la zone de transition provient donc de ce que deux modes d'amorçage sont possibles.

Mais pourquoi ce saut dans l'énergie de rupture, quand on passe d'un mode à l'autre? On conçoit que pour l'amorçage ductile, une forte déformation locale, donc une forte énergie soit nécessaire. Mais pourquoi la déformation de l'éprouvette au moment où débute la portion fragile de la cassure (cas d'un amorçage fragile) reste-t-elle en général faible, et ne peut-elle croître jusqu'à rejoindre celle de l'amorçage ductile? Pour comprendre ce point, il faut voir comment se localisent les déformations dans une éprouvette de résilience.

Dans ce but, nous avons sectionné des éprouvettes de rupture fragile par un plan médian, et après traitement et attaque convenable (attaque de FRY), les zones déformées plastiquement apparaissent en sombre (fig. 2). Une éprouvette fléchie lentement jusqu'au même angle de pliage que celui où s'est amorcée la rupture fragile, sectionnée et attaquée de même façon, montre les mêmes zones, mais plus clairement (fig. 3); elles sont

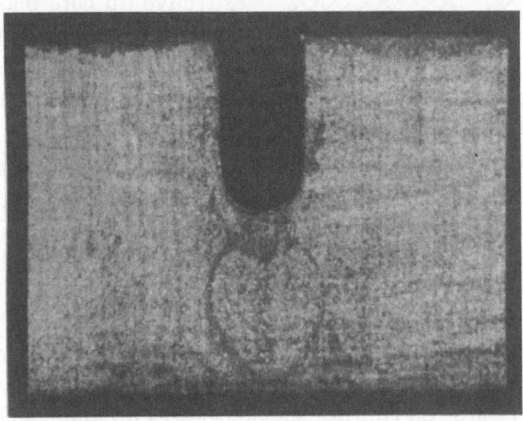

Fig. 3

schématisées sur la fig. 4: une zone de première déformation comprend une sorte de «tenaille» $AOBA'OB'$ avec des «branches» OB et OB' et une «tête» $DEOE'D'$; c'est à peu près la zone que l'on trouve par le calcul, en admettant la théorie rigide-plastique. Mais on observe en plus un coin déformé, $OECE'$, qui n'existe pas dans le calcul. Aussi n'avons nous pas encore pu évaluer les contraintes au point C; mais il est évident que le métal déformé DCD' a cédé plastiquement, en sorte que le coin C joue le rôle d'entaille.

Il est évident que «l'effet d'entaille» de ce coin C ne peut que diminuer si la zone plastique grandit et que CH diminue. On explique ainsi que le danger d'amorçage fragile soit maximum lorsque ce coin apparaît, et diminue ensuite lorsque la déformation augmente.

Ce serait là l'origine du saut de résilience: ou bien l'amorce fragile se produit lors de l'apparition du coin, ou elle ne se produit pas, et il faut attendre de fortes déformations pour pouvoir produire une amorce ductile[1].

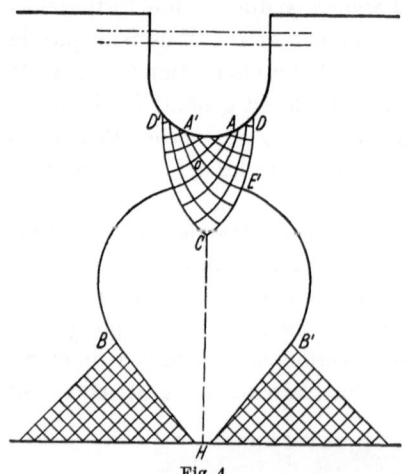

Fig. 4

Si cette interprétation est correcte, il s'en dégage un fait important: *l'amorçage fragile se produit, dans les aciers doux au moins, au contact entre une zone déformée plastiquement et une zone non déformée, ou très peu déformée* (déformation insuffisante pour donner une attaque de Fry).

On peut se demander si l'amorce se trouve du côté plastique ou du côté élastique de cette zone de contact. Mais avant de discuter ce point, voyons si cette localisation a l'air d'une loi générale; en particulier, se retrouve-t-elle dans le cas de la traction simple?

4. Essais de traction à $-196°$C. Nous avons effectué ces essais sur deux aciers doux Martin effervescents dont les compositions suivent:

$$\text{Acier } A \quad \frac{C}{0,075} \quad \frac{Mn}{0,22} ,$$
$$\text{Acier } B \quad \frac{C}{0,04} \quad \frac{Mn}{0,10} .$$

Dans tous les cas, nous avons observé sur les courbes enregistrées la chute de tension correspondant au passage de la limite élastique supérieure à l'inférieure. La rupture survient après un allongement qui varie beaucoup d'une éprouvette à l'autre, et qui varie entre 1% et 22%, cette dernière valeur étant située au-delà du début de striction, qui se produit assez tôt à ces basses temperatures, et se trouve marquée par un début de chute de la force de traction, sensible sur certaines courbes de l'acier B. Cependant, dans tous les cas, la rupture est brillante, à grains.

Ces courbes sont très plates, avec, vers 4 à 6% d'allongement, une petite incurvation vers le bas, caractéristique de la fin du palier; on l'observe à toutes les températures où celui-ci existe.

---

[1] L'élévation de temperature qui résulte de ces déformations contribue à augmenter le saut.

La forte dispersion des valeurs de l'allongement de rupture fait un peu penser à celle des résiliences à une température de la zone de transition (IIIᵉᵐᵉ partie). On peut se demander si leur répartition suit une distribution normale. Pour celà, nous avons tracé les *courbes de fréquences cumulées* correspondant aux deux aciers essayés (fig. 5). On voit que la distribution semble bimodale pour l'acier B et même trimodale pour l'acier A. Dans ce dernier cas, le 3ème mode correspond à des ruptures survenant après début de striction, faible il est vrai ; elles sont malgré cela à grain.

On remarque qu'il y a deux zones d'allongements peu probables, qui séparent trois domaines où se produisent les ruptures :

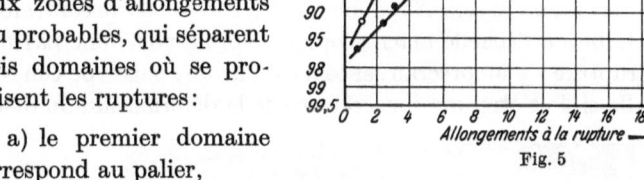

Fig. 5

a) le premier domaine correspond au palier,

b) le second, à une zone entre la fin du palier et le début de striction,

c) le troisième (pour l'acier A), à une zone qui suit de peu le début de striction.

Examinons d'un peu plus près ce qui se passe dans ces différents domaines :

a) Dans le palier, la déformation n'est pas homogène, mais se localise dans une ou plusieurs bandes de PIOBERT-LÜDERS. Lorsqu'il n'y en a qu'une, elle progresse d'un bout de l'éprouvette à l'autre, séparée de la zone non encore déformée par un sillon ou un pliage ; ce sillon produit une concentration de contrainte locale.

Lorsqu'il y a plusieurs bandes de LÜDERS, elles sont souvent en forme de coins et s'enchevêtrent, ce qui produit également en certains points des concentrations de contrainte.

Nous avons pris des diagrammes de rayons X sur le métal de part et d'autre de la rupture, mais à une petite distance de la surface de rupture proprement dite, qui comporte une couche très écrouie ; on peut ainsi apprécier l'état d'écrouissage du métal avant propagation de la rupture. On constate qu'il est très différent d'un côté et de l'autre, ce qui semble bien prouver que la rupture se fait à la limite entre une zone déformée (bande de PIOBERT-LÜDERS) et une zone non déformée.

Nous remarquerons que la première bande de Piobert-Lüders qui se forme constitue, avant qu'elle ait traversé toute l'éprouvette, un coin aigu, qui doit créer de fortes concentrations de contrainte. Ces concentrations doivent pouvoir amorcer la rupture; dans un tel cas, on pourrait avoir l'illusion, sur l'enregistrement photographique, d'une rupture brutale précédant la limite élastique, alors qu'elle suivrait en réalité les premières traces de déformation plastique. Nous avons observé ce cas exceptionnellement dans une éprouvette d'un autre acier doux; la rupture s'était produite précisément pour une charge égale à celle de la limite élastique d'autres éprouvettes identiques étirées dans les mêmes conditions. D'ailleurs des essais de Low [3] semblent bien prouver que, dans les aciers doux au moins, la rupture suit toujours le passage à la limite élastique supérieure.

b) Dans ce domaine, la contrainte moyenne est à peine supérieure à celle du domaine a). Or la déformation y est plus homogène (comme nous l'avons vérifié aux rayons X); donc les concentrations de contraintes absentes ou faibles à l'échelle macroscopique. On ne voit donc pas bien comment la rupture s'y amorcerait, si ce n'est par une baisse de cohésion du métal. Celle-ci doit être une conséquence de la déformation du métal.

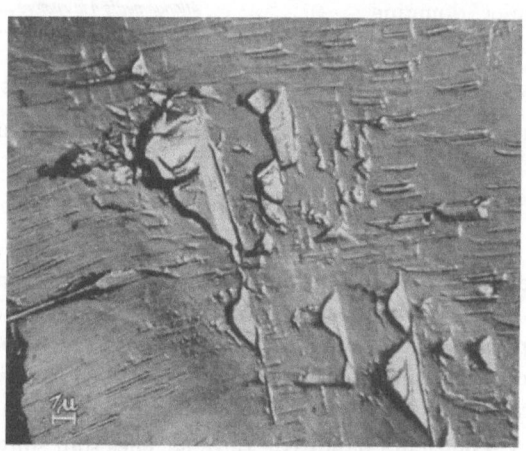

Fig. 6

Or une observation attentive de coupes de ces éprouvettes, après nickelage des surfaces de rupture, montre que le trajet de la rupture emprunte fréquemment des joints intergranulaires alors que ce cas est assez rare dans le domaine a) et absent dans le domaine c). L'amorce aussi pourrait être intergranulaire; et en tous cas les ruptures dans ce domaine semblent attribuables à une *diminution de cohésion*, manifeste surtout dans les joints ou les couches avoisinantes, et dues à la déformation.

Il se trouve que, sur un acier un peu différent (acier au Cr—Ni sensible à la fragilité de revenu), nous avons pu étudier au microscope électronique

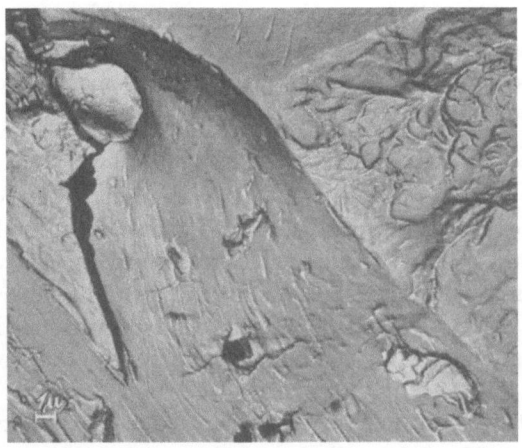

Fig. 7

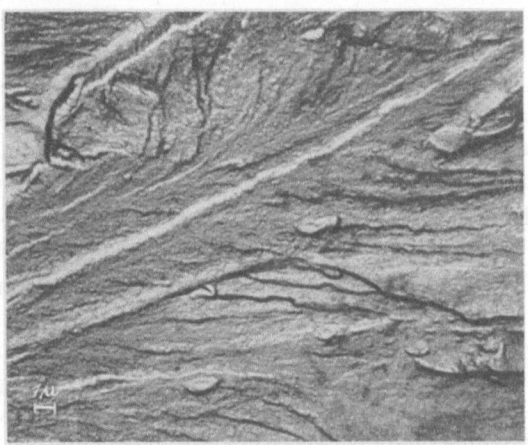

Fig. 8

des surfaces de rupture intergranulaires. Les fig. 6, 7 et 8 donnent des exemples des répliques de telles surfaces. La fig. 6 montre des pointements de métal en relief sur la surface des joints: ce sont des fragments d'un grain, restés accrochés sur l'autre; la fig. 7 montre un phénomène ana-logue, avec des fragments beaucoup plus importants. Ces deux figures sont prises sur l'état fragilisé; la fig. 8 au contraire provient de l'état tenace, où la rupture à basse température est transcristalline; on y voit une facette de clivage, dont l'aspect est tout différent des ruptures inter-granulaires qui précèdent.

Le fait que des lambeaux d'un cristal restent accrochés à l'autre prouvent que la cohésion du joint n'est pas diminuée également partout, et que souvent la rupture préfère suivre de petits bouts de chemins trans-cristallins, sans doute le long de plans de glissements déconsolidés par des empilements de dislocations.

5. Discussion des résultats et conclusions. Nos essais montrent que les modes de rupture des aciers sont très variés, surtout en ce qui concerne le rôle de la déformation préalable et des joints intergranulaires.

On admet à l'heure actuelle que les très fortes concentrations de ten-sion nécessaires à la création de microclivages, à l'échelle des cristaux, sont dues à des empilements de dislocations butant contre les joints. Ces microclivages jouent le rôle d'amorces si certaines conditions sont réali-sées [3].

Ce mécanisme peut jouer à divers stades de la déformation, que nous allons classer par valeurs croissantes:

1er stade. Les dislocations pourraient être celles créées par la *micro-déformation* qui précède la limite élastique supérieure. Un tel mécanisme ne pourrait jouer que dans des cas de fragilité extrême; nous ne l'avons jamais observé, puisque dans tous les cas les déformations atteignaient, au moins en certains points, des valeurs correspondant au palier des courbes de traction. *Ce stade est donc hypothétique.* On pourrait, d'une façon un peu schématique, dire que le joint cède ou craque: c'est-à-dire que, si les grains voisins d'un empilement de microdéformation peuvent céder plastiquement, le palier s'amorcera; sinon, la contrainte montera jusqu'à ce qu'on atteigne la cohésion du métal, et un clivage s'amorcera.

2ème stade. Expérimentalement, dans tous les cas de grande fragilité, nous avons vu qu'il y avait des zones plastiques, à déformation au moins égale à celle du palier. Nous avons montré qu'il existait un mode de rup-ture, dont l'amorçage se fait au contact entre zones plastiques et zones élastiques (en réalité microdéformées). Il est difficile de dire si l'amorce est du côté plastique ou du côté élastique; ou juste entre les deux, ce qui paraît possible vu les fortes accumulations de dislocations qui se pro-duisent à cet endroit. Des conditions favorables à l'amorçage sont égale-ment réunies du côté élastique où peuvent se superposer de fortes con-centrations de tension à échelle macroscopique (échelle des zones plasti-ques ou lignes de Piobert-Lüders) et les concentrations à échelle du grain dues à la microdéformation. De toutes façons, le point important est que dans ces zones, les concentrations de tension peuvent s'étendre sur un domaine assez grand pour que les dimensions des fissures restent toujours hypercritiques. Ce n'est pas le cas lorsque la microdéformation seule joue, comme dans le 1er stade, et que la concentration de tension est localisée dans une zone si petite (un empilement) que la fissure peut bien s'y amorcer, mais que dès qu'elle en sort, elle tombe dans une zone de contrainte faible, où elle se trouve inférieure aux dimensions critiques correspondantes.

3ème stade. Dans le cas de la traction et pour des allongements moyens, nous avons mis en évidence un troisième stade (4. domaine b). Là, le métal casse parce qu'il est moins solide; cette baisse de cohésion semble se faire sentir surtout aux joints; elle est en relation sans doute avec une abondante production de lacunes réticulaires ou d'empilements de dislocations, surtout près des joints.

Dans ce stade, la rupture s'amorce parce qu'il y a de fortes contraintes locales (empilements), et que, du fait de la déconsolidation, le rayon critique pour la croissance spontanée des fissures diminue.

D'après ce que l'on sait du comportement du métal à la traction [4], les allongements de ce stade sont proches de ceux du point de transition des courbes de traction; celui-ci correspond à un changement dans la répartition des dislocations, avec possibilités de diffusion accrues. La température a une forte influence sur ce point de transition, ce qui suggèrerait un moyen pour vérifier la corrélation entre transition et «domaine b)». Malheureusement, on ne peut obtenir de rupture fragile sur éprouvettes lisses à des températures moins basses, ce qui rend cette étude impossible.

4ème stade. Toujours pour la traction à basse température, en rencontre un nouveau stade de rupture au-delà du maximum de l'effort de traction, c'est-à-dire vers le début de striction.

Les éprouvettes cassées dans ce stade sont peu nombreuses; aussi n'avons nous pas pu mettre en évidence si ce stade est en continuité avec les ruptures fibreuses que l'on rencontre dans les strictions plus accusées. Sur une section micrographique, nous avons cependant observé en un point de la surface de rupture des grains qui semblaient fortement déformés, comme dans une rupture par arrachement; il pourrait donc y avoir continuité entre ce quatrième stade (où le clivage prédomine) et le stade de rupture fibreuse (par arrachements sur portions de plans de glissement déconsolidés) si on admet que les proportions de surface de rupture empruntant, soit des plans de clivage (100), soit des plans de glissement varient (en sens inverse) de façon continue; ou encore que, dès le début de ce quatrième stade, les facettes de clivage sont en réalité des plans de glissement «déconsolidés».

Revenons aux essais de résilience, pour nous demander à quel stade correspondent les déformations avant rupture dans les cas fragiles et dans les cas ductiles.

Dans le cas ductile, les ruptures, dans la zone d'amorce au moins, sont surtout *fibreuses*, c'est-à-dire se produisent par arrachement dans du métal fortement déformé, avec parfois des zones «à nerfs» produites par glissement. Nous sommes donc ici dans des déformations correspondant au quatrième stade, ou au-delà.

Pour les ruptures fragiles, le parallélisme est évident, elles se produisent surtout dans le 2ème stade. Mais l'aspect intergranulaire de certaines ruptures, que nous avons signalées plus haut (III, 2°) peut faire penser que dans certains cas on arrive au 3ème stade; dans ce cas, on ne sait pas très bien où est l'amorce, au coin C (fig.4), ou dans une zone plus déformée, plus près du fond d'entaille. La détermination du mécanisme exact de l'amorçage est donc plus difficile dans le cas des essais de résilience.

On voit que les essais de traction à basse température permettent une meilleure discrimination des modes de rupture, du fait que la contrainte y est homogène, et varie peu en fonction de l'allongement, les courbes de traction étant très plates. L'essai de résilience, trop brutal, escamote certains modes de rupture, ou les brouille avec les modes voisins.

Mais dans tous les cas, on voit l'influence considérable des joints intergranulaires: ce sont eux qui produisent les premiers empilements de dislocations, qui commandent à l'apparition du palier, et qui subissent sans doute une déconsolidation par la déformation. Ce n'est que pour de fortes déformations que leur action propre s'estompe.

Du fait de cette forte influence des joints, toute l'analyse que nous venons de faire n'a de valeur que pour des métaux à grain relativement fin. Lorsque le grain devient grand par rapport aux dimensions critiques des fissures, les mécanismes de rupture peuvent changer.

Bibliographie

[1] Cornu-Thenard, A.: Rev. Métall. **17**, 661 (1920).
[2] Ulmo, J., F. Bastenaire et R. Borione: C.R. Acad. Sci., Paris **237**, 59 (1953); Rev. Métall. **50**, 868 (1953). — Y. Morillon, J. Plateau et R. Borione: C.R. 27è Congrès de Chimie Industrielle, Bruxelles 1954.
[3] Low, J. R.: Trans. Amer. Soc. Met. **46A**, 163 (1954).
[4] Crussard, C.: Rev. Métall. **50**, 697 (1953).

Wave Propagation in Anelastic Materials[1]

By **E. H. Lee**, Providence (Rhode Island)

With 11 figures

1. Plastic wave propagation. The theory of the propagation of plastic waves developed independently by TAYLOR [1] and KÁRMÁN [2] considers an invariant stress-strain relation which is independent of the rate of strain. Experimental measurements are in broadly satisfactory agreement with this theory for many metals, although reproducible discrepancies are observed which are thought to be due to the influence of the rate of strain in modifying the stress-strain relation [3]. Attempts to include a strain-rate influence, such as that by MALVERN [4], have exhibited a marked improvement in predicting certain experimental observations, but have appeared to fail in the case of others. It is shown below that the major discrepancy with experiment which was considered to dictate against a theory of this type appears not to be valid.

One of the types of stress-strain relation considered by MALVERN is of the form

$$\left.\begin{array}{ll}
\dot{\varepsilon} = \dfrac{\dot{\sigma}}{E} + k\left[\sigma - f(\varepsilon)\right], & \sigma \geq f(\varepsilon), \\[2mm]
\dot{\varepsilon} = \dfrac{\dot{\sigma}}{E} & \sigma \leq f(\varepsilon)
\end{array}\right\} \tag{1}$$

where ε is the strain, σ the stress, E YOUNG's modulus, k a viscosity type constant, $\sigma = f(\varepsilon)$ the static stress-strain relation, and a dot represents differentiation with respect to time. This relation is considered here since the significance of the individual terms, and its relation with the static stress-strain curve are clearly evident, although MALVERN also considered more general relations. The first term on the right hand side represents the rate of change of the elastic component of strain, and the other term that of the plastic strain component. The rate of increase of the plastic strain is thus considered to be proportional to the over-stress, that is the amount by which the stress exceeds the static stress for the same strain. This seems a plausible way in which to introduce a strain rate effect since

[1] This work was sponsored by the Office of Naval Research, Navy Department, Washington, D. C., and permission to publish is gratefully acknowledged.

in the limit of small strain rate the dynamic stress-strain curve approaches the static one, and it is the stress above the static value which generates the rate of increase of plastic strain according to a viscosity law. The elastic component of strain is considered to be independent of the strain rate.

On physical grounds it might be considered more satisfactory to define the over-stress as the excess stress beyond that required to cause slow plastic flow at the same plastic strain, since it is the permanent plastic strain which determines the current state of the material. However the mathematical analysis is simpler if total strain is used as in (1), and the difference is negligible in most cases since the gradient of the stress-strain curve in the plastic region is small compared with the elastic modulus. Moreover, writing

$$\varepsilon = \varepsilon_p + \frac{\sigma}{E} \qquad (2)$$

where ε_p is the plastic component of strain, transformation from one form to the other can be effected. In the case of linear work-hardening discussed below, this change only modifies the magnitude of constants appearing in the differential relationship.

As shown by MALVERN [4], and in contrast to the predictions of the rate independent theory, wave fronts of increments of plastic strain are always propagated with the elastic wave velocity if (1) is obeyed. This is in agreement with experiments on the propagation of stress pulses in a wire already stressed statically [5], which differ markedly from the prediction of the rate independent theory. Because of the influence of the viscous behaviour in (1), different stress histories for the same maximum stress commonly lead to different plastic strain magnitudes, and as detailed below this leads to a discrepancy with the reported plateau of uniform strain adjacent to the impact end of a specimen subjected to constant velocity impact [3]. This disparity has been a major deterrent to the use of relations of the type (1), and in fact to my knowledge no law has been suggested which includes a strain rate influence, and which predicts such a plateau of constant strain.

This difficulty was pointed out in the discussion of MALVERN's paper [6]. The pattern of wave propagation for constant velocity impact deduced by MALVERN was similar to that obtained from the rate independent theory, so that for simplicity the latter is used for illustration in fig. 1 without loss to the qualitative argument. On the right hand side is shown the pattern of wave fronts in the (x, t) plane where x is the coordinate along the wire and t the time. Constant velocity impact for duration t_0 occurs, and thereafter the stress at the impact end is reduced to zero. KÁRMÁN [2] showed that the stress and strain are constant on straight lines through the origin, and that behind the line of maximum

strain OD a region OCD of constant stress and strain exists. Unloading
occurs across the line CD. The resulting stress histories at three sections
O, A and B are shown in fig. 1. It is clear that if the impact duration is
sufficiently short for an appreciable strain rate influence to occur, so that
the static relation does not effectively govern the whole process, the viscosity type relation (1) will give different final strains at the three positions. Over the same duration of plastic flow the overstress will have been

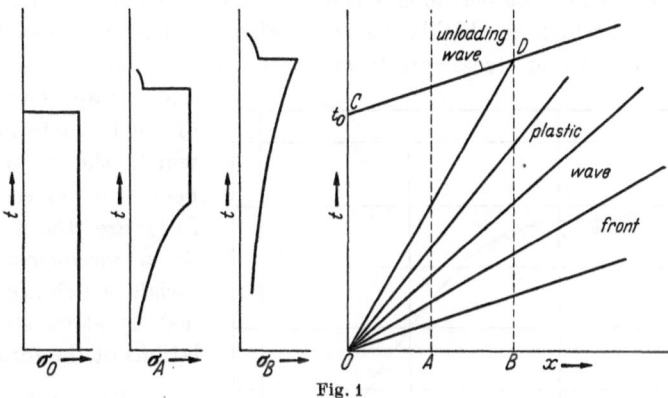

Fig. 1

greatest for the end section O, and so the deduced strain will be greater
there. However, the extent OB of the back of the plastic wave is the
length over which a plateau of constant strain has been observed experimentally [3]. The same difficulty will arise with a generalization of (1)
which involves a non-linear viscosity law.

I recently had the opportunity of discussing this difficulty with Professor WOOD, at the California Institute of Technology, and we reviewed
the original measurements of the work described in [3] to check the accuracy with which the plateau was determined. Displacements were
measured and the strain determined as the derivative of these measurements. A graphical method had been used because the consistency of the
results had not been considered sufficient for more refined analysis. A
more detailed study by differencing the displacement measurements gives
the strain distribution shown in fig. 2 for the case of specimen 8 [3] which
was a copper wire of length 80″ impacted at a speed of 109 ft./sec. The
displacement measurements were commenced 1.9″ from the impacted
end of the wire, so that the increasing strain at the end may increase
above the value shown in fig. 2. Other cases gave similar distributions.
It must be borne in mind that the consistency of these measurements
may not justify the adoption of the smooth curve drawn in fig. 2 as the
detailed strain distribution for this case, but it does appear that a marked
deviation from a plateau of uniform strain is indicated. This gives new

9*

impetus to the possibility of representing metal behaviour at high strain rates with laws of the type (1), which calls for careful reassessment of the situation. Moreover it indicates the need for developing the theory of plastic deformation with a stress-strain law of the type (1), to aid in the interpretation of experimental work. It is shown below thas recent work on the theory of wave propagation in visco-elastic materials is closely related to this problem.

2. Waves in visco-elastic materials. We now consider materials, such as plastics, which are highly rate dependent, so that it is essential to include strain and stress rate terms in the relationship between stress and strain. The group of such materials for which the theory has been developed most fully are the so-called linear visco-elastic materials which are governed by stress-strain relations of the form

$$P(\sigma) = Q(\varepsilon) \qquad (3)$$

where P and Q are linear operators in the time derivate $D = \partial/\partial t$, of the form $\sum_{0}^{n} a_r D^r$. The behaviour of certain plastics can be adequately represented by such laws.

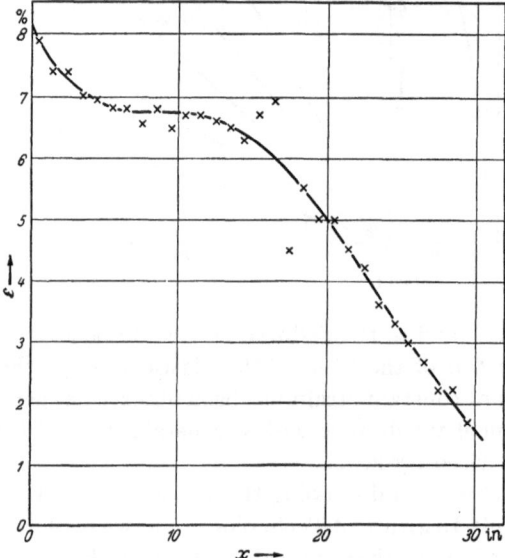

Fig. 2. Specimen No. 8. Copper wire $V_0 = 109$ ft./sec. (due to DUWEZ, WOOD and CLARK ref. [3])

For the purpose of visualizing the response of such materials to stress, it is convenient to study the so-called visco-elastic models which represent the law (3) by a physical system of springs and dashpots. Figs. 3 to 9 show a series of such models. The mechanical model is considered to be loaded by a force of magnitude σ, and the extension of the model gives the strain ε.

Qualitatively three basic types of response to stress are exhibited by visco-elastic materials: instantaneous elastic response, delayed elastic response, and viscous flow. The first is represented by the spring, fig. 3. Delayed elastic response is exhibited by the VOIGT unit, fig. 6, since when the stress is applied, it is at first carried entirely by the dashpot, and as this deforms the stress is transferred to the spring. Thus there is a delayed response to the stress, the strain increasing asymptotically to a steady value. The deformation is, however, elastic, for on removal of the stress, delayed recovery of the strain takes place. In contrast viscous flow represented by the dashpot, fig. 4, is permanent. The simplest model which includes all three types of behaviour is the four element model shown in fig. 9, the simpler models exhibiting only one or two of these phenomena.

In analysing the propagation of longitudinal waves down a rod of one of these materials, the stress-strain law must be supplemented by the equation of motion

$$\sigma_x = \varrho \, u_{tt} \tag{4}$$

where ϱ is the density, u the displacement and the subscript represents differentiation with respect to the corresponding variable. The theory can be considered within the framework of infinitesimal strain, or if LAGRANGE coordinates are used and the stress-strain laws apply to nominal values, the theory applies for finite strain magnitudes.

Fig. 3   Fig. 4   Fig. 5

Fig. 6   Fig. 7

Fig. 8   Fig. 9

In addition we need the relation between displacement and strain:

$$\varepsilon = u_x \tag{5}$$

(4) and (5) in conjunction with the stress-strain law determine the equation of wave propagation, which can be obtained by eliminating all but one dependent variable. Such equations have been obtained for all the materials represented in figs. 3 to 9, and solutions evaluated for the stress and velocity distributions in certain impact problems [7—11].

3. Plastic waves with linear work-hardening. In the case of linear work-hardening, $f(\varepsilon)$ in (1) becomes

$$f(\varepsilon) = Y + E_1\left(\varepsilon - \frac{Y}{E}\right) \tag{6}$$

where Y is the initial static yield stress, E_1 the coefficient of work-hardening, and E Young's modulus. Equation (1) then becomes

$$\dot{\varepsilon} + kE_1 - kE_1\frac{Y}{E} = \frac{\dot{\sigma}}{E} + k\sigma - kY, \quad \sigma \geq Y + E_1\left(\varepsilon - \frac{Y}{E}\right). \tag{7}$$

The constant terms can be eliminated by considering the additional stress and strain components superposed on the state at the commencement of plastic flow. For example in the case of constant velocity tensile impact on an initially unstressed semi-infinite rod, an elastic stress wave of constant magnitude Y is propagated down the rod with elastic wave speed c, producing uniform strain Y/E behind it and a uniform material velocity of magnitude $Y/\varrho c$. Thus we consider superposed on this an impact with velocity $v_0' = v_0 - Y/\varrho c$, producing stress $\sigma' = \sigma - Y$ and strain $\varepsilon' = \varepsilon - Y/E$. Equation (7) then becomes:

$$\dot{\varepsilon}' + kE_1\varepsilon' = \frac{\dot{\sigma}'}{E} + k\sigma', \quad \sigma' \geq E_1\varepsilon'. \tag{8}$$

With appropriate choice of constants, this differential relation is identical with the law for a three element visco-elastic body with two elastic elements, fig. 7. Thus the solution given in [8] and [11] can be utilized for the work-hardening plastic wave case. The solution only applies to the plastic wave problem if the inequality (8) is satisfied, and when this is violated the elastic relation

$$\dot{\varepsilon} = \frac{\dot{\sigma}}{E}$$

must be used. In applying the visco-elastic solution we shall retain the notation used in fig. 7 and will use bars over the material constants in (7) and (8).

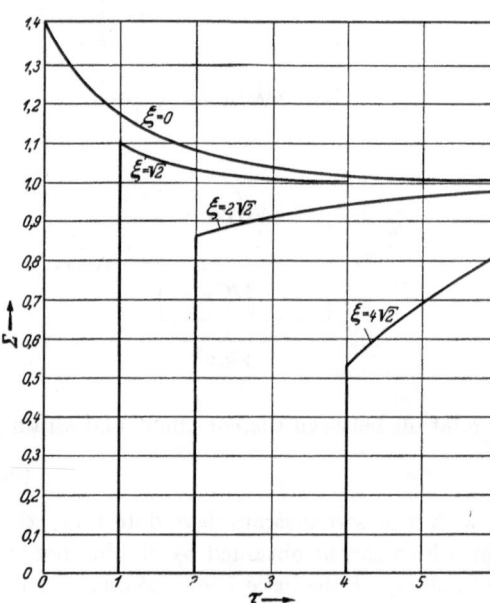

Fig. 10. Three parameter (2 elastic, 1 viscous) stress variations in constant velocity impact

Then the wave propagation equations are identical for the two problems if:

$$E = \bar{E}_1, \quad E' = \bar{E} - \bar{E}_1. \tag{9}$$

The particular visco-elastic problem evaluated in [8] and [11] was for $k = E/E' = 1$, so that $\bar{E} = 2\bar{E}_1$, and the plastic wave speed is $1/\sqrt{2}$ of the elastic wave speed. This is a much larger factor than would normally occur in a plastic wave problem, but the general form of the solution is not modified by this ratio. Fig. 10 shows the stress variation at several sections, and fig. 11 the stress distribution at several times after impact for a constant velocity impact using dimensionless units, $\xi = \sqrt{\varrho E} \mu x$,

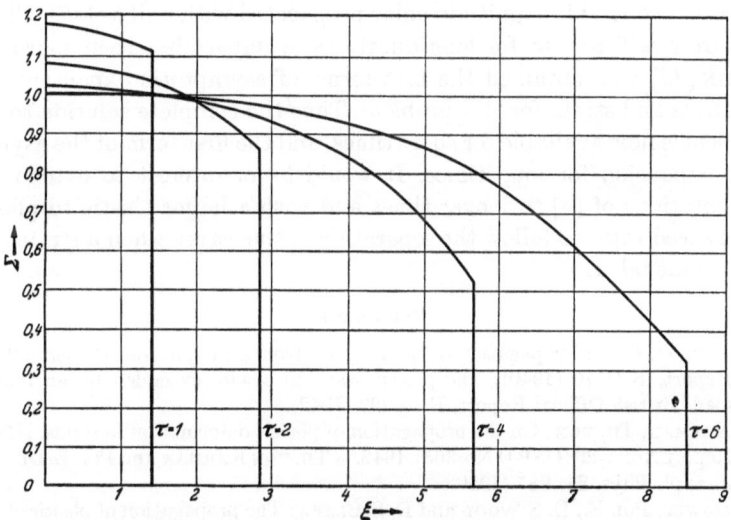

Fig. 11. Three parameter (2 elastic, 1 viscous) stress distributions in constant velocity impact

$\tau = E\mu t$, $\Sigma = \sigma/\sqrt{E\varrho}\, v_0$. The solution exhibits many of the features obtained by MALVERN [4] by numerical characteristic integration for a material with a curved work-hardening law. The initial rise in stress at the point of impact is greater than at other sections, and for all sections the dimensionless stress finally approaches unity. This is equal to the value given by the rate independent plastic-wave theory.

Because of the peak stress at the point of impact and the rapid reduction immediately following it, the end section is more likely to violate the inequality (8) than internal sections. The solution [11] enables this inequality to be checked, giving:

$$(\sigma' - E\,\varepsilon')_{x=0} = \frac{\sqrt{\varrho E}\, v_0'}{\sqrt{k\,(1+k)}}\, e^{-\frac{(1+k)\,\tau}{2\,k\,(k+1)}}\, I_0\left[\frac{\tau}{2\,k\,(k+1)}\right] > 0$$

which is always valid since $I_0 \left[\dfrac{\tau}{2\,k\,(k+1)} \right]$ remains positive for all τ. Thus no cessation of plastic flow occurs while the impact is maintained in contrast to Malvern's [4] solution with a different stress-strain curve. The present result is in agreement with the comments of Rubin [12] on this matter.

For the ratio 2 of elastic to plastic modulus in this example, the stress decreases at the impact end by a factor $1/\sqrt{2}$ which will be much less than for most metals. However the closeness of the wave speeds masks the separation of the elastic and plastic wave fronts. Since $\sum$ in fig. 11 when interpreted for the plastic wave problem represents the excess stress above the yield stress Y, the rate independent theory would give a stress of constant magnitude unity propagated with unit velocity. That this front will appear for long durations of impact has been shown by Rubin [12] who obtained the first terms of asymptotic expansions for the stress and strain for this problem. Thus the complete solution to the problem is now available for short times, and the first term of the asymptotic expansion for long times. It would be of interest to extend the computations of [8] to longer times and with a larger elastic to plastic wave speed ratio to follow the separation of the waves when a strain rate term is included.

References

[1] Taylor, G. I.: Propagation of earth waves from an explosion. British Official Report, R.C. 70 (1940); The plastic wave in a wire extended by an impact load. British Official Report, R.C. 329 (1942).

[2] Kármán, Th. von: On the propagation of plastic deformation in solids. NDRC Report No. A-29 (OSRD No. 365), 1942. — Th. von Kármán and Pol E. Duwez: J. appl. Phys. **21**, 987 (1950).

[3] Duwez, Pol E., D. S. Wood and D. S. Clark: The propagation of plastic strain in tension. NDRC Report A 99 (OSRD 931), 1942. — Pol E. Duwez, D. S. Wood, D. S. Clark and J. V. Charyk: The effect of stopped impact and reflection on the propagation of plastic strain in tension. NDRC Report A-108 (OSRD 988), 1942. — J. E. Johnson, D. S. Wood and D. S. Clark: J. appl. Mechan. **20**, 523 (1953).

[4] Malvern, L. E.: Quart. appl. Math. 8, 405 (1951). — J. appl. Mechan. Trans. ASME **73**, 203 (1951).

[5] Sternglass, E.J. and D.A.Stuart: J.appl.Mechan. **20**, 427 (1953). — C. Riparbelli: Proc. 1st Midwestern Conf. on Solid Mech. 1953, p. 148.

[6] Lee, E. H.: Discussion of Malvern [4], J. appl. Mechan. **18**, 428 (1951).

[7] Morrison, J. A. and E. H. Lee: A comparison of the propagation of longitudinal waves in rods of visco-elastic material. Report PA-TR/11, Dec. 1954.

[8] Morrison, J. A. and E. H. Lee: J. Polymer Sci. **19**, 93 (1956)

[9] Lee, E. H. and I. Kanter: J. appl. Phys. **24**, 1115 (1953).

[10] Glauz, R. D. and E. H. Lee: J. appl. Phys. **25**, 947 (1954).

[11] Morrison, J. A.: Wave propagation in rods of Voigt material and visco-elastic materials with three-parameter models, to appear in Quart. appl. Math.

[12] Rubin, R. J.: J. appl. Phys. **25**, 528 (1954).

La structure feuilletée des joints dans les métaux

Par **F. Teissier du Cros**, Paris

Avec 8 figures

1. Introduction. La métallurgie produit les métaux usuels à l'état de *polycristaux* (agrégats de grains). Les propriétés mécaniques d'une éprouvette polycristalline diffèrent nettement de celles d'un cristal unique de même métal. Lorsque pendant la déformation plastique, on observe des lignes de glissement, celles-ci forment plusieurs faisceaux, de directions différentes, qui apparaissent à la surface de chaque grain, dès le début de la déformation. Certains faisceaux s'arrêtent à la limite du grain, d'autres n'atteignent pas celle-ci, d'autres paraissent se prolonger au-delà. Les régions de la périphérie sont parfois le siège de glissements localisés, plus intenses. Bref, le grain subit de la part de ses voisins, des contraintes qui rendent sa *déformation complexe et hétérogène*.

Sur chaque surface de jonction, ou joint, qui réunit le grain à l'un des cristaux voisins, il est assujetti à des *conditions aux limites*; ces conditions ne sont pas encore bien connues, mais il est certain qu'elles dépendent de l'orientation mutuelle des cristaux, de celle du joint vis-à-vis d'eux, et enfin d'un caractère inhérent au joint, qu'on peut appeler sa *structure interne*.

Cette notion importante se dégage des études théoriques de J. M. BURGERS [*1*], F. C. FRANK [*4*], N. F. MOTT [*7*], de W. SHOCKLEY et W. T. READ [*10*]. Les expériences de JAOUL[1] ont mis en évidence l'intervention des joints intergranulaires dans le mécanisme de la déformation. Le rôle de la structure interne mérite d'être approfondi.

Le sujet est complexe: la figure formée par un joint plan, où se réunissent deux cristaux, dépend en général de cinq paramètres angulaires. On décrira ici un type de joint particulier, dépendant de quatre paramètres au plus, dont l'existence paraît vraisemblable dans l'aluminium et les métaux du même système cristallin, lorsqu'ils ont été *recuits*.

Les *dislocations cristallines* vont se présenter sous leurs deux aspects habituels: cinématique lorsqu'elles produisent en se déplaçant, le glissement interne du cristal; statique lorsque leur présence est invoquée pour

---

[1] Voir le rapport page 13 du présent compte rendu.

expliquer le changement d'orientation entre un grain et l'autre.
F. C. Frank [5] a montré qu'un groupe de dislocations mobiles, bloqué
auprès d'un joint, peut déclencher un nouveau glissement dans le grain
situé au delà, par une action à distance, sans continuité du plan de
glissement. Le joint que l'on va décrire par contre, laisse simplement
passer certaines dislocations du premier grain dans le suivant. Il ne cause
pas d'hétérogénéité spéciale.

On sait que la théorie des joints repose sur deux propositions qui ont
été énoncées en 1950, au Symposium de Pittsburgh:

1. possibilité de se représenter la désorientation de deux grains, de
positions quelconques, comme étant l'effet de dislocations distribuées sur
la surface du joint séparateur (F. C. Frank);

2. équivalence du joint plan de très petite désorientation à un *système
de dislocations* bien défini (Shockley et Read).

Double intérêt: ce système rend compte de la structure géométrique
du joint; de plus, il explique ses propriétés énergétiques. Cependant, les
cas qui relèvent de cette belle théorie ne sont pas très fréquents. Les désorientations que l'on observe entre les grains juxtaposés dans un métal
ordinaire, ne sont pas très petites; ce sont le plus souvent des *désorientations moyennes*, comprises entre 10 et 25 degrés.

La structure que l'on va décrire ne convient qu'à des joints à désorientation moyenne.

2. La structure feuilletée. Une portion de cristal possède la *structure
feuilletée* si elle renferme des surfaces cristallographiques parallèles, où
séjournent des dislocations au repos, de lignes parallèles, susceptibles de
glisser le long desdites surfaces. Les portions de cristal comprises entre
les surfaces, appelées feuillets, ont en général une certaine courbure, leur
épaisseur est microscopique. Etant dépourvus eux-mêmes de dislocations
internes, les feuillets sont le siège d'une déformation élastique, qui peut
être intense. La division géométrique en feuillets n'implique aucune discontinuité matérielle.

Le cristal *fléchi*, le cristal *polygonisé*, décrits par R. W. Cahn [2] et
par A. H. Cottrell [3], sont des exemples bien connus de structure feuilletée. On peut calculer l'énergie des dislocations à l'équilibre en adaptant
le modèle de Peierls au cas de surfaces de glissement courbes, les feuillets étant assimilés à des lamelles élastiques[1]. Enfin la structure feuilletée
peut se trouver réalisée dans le voisinage immédiat des joints. C'est ce
qu'on se propose de montrer ici.

a) Joint B_1. Les figures se rapportent à un joint du *type de* Burgers [1], à désorientation moyenne, dont l'angle est $2Z$.

---

[1] Travail restant à publier de l'auteur de cette communication.

Les angles Z (demi désorientation) qui peuvent être atteints dans le joint B_1 à l'équilibre, forment une *suite discrète*: Cot $Z = 2\,M a b^{-1}$, M désignant le *nombre entier* d'intervalles a contenus dans l'épaisseur d'un feuillet (voir fig. 1). La disposition des dislocations est celle d'un *rideau de* BURGERS[1].

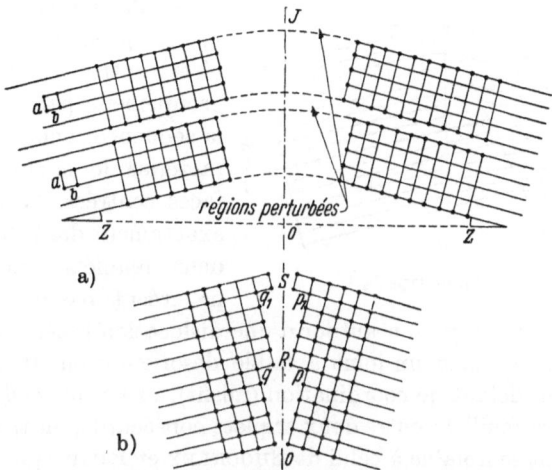

Fig. 1. Joint du type B_1. a) Région de raccordement de deux cristaux symétriques par rapport au plan OJ du joint. b) Disposition des gradins qui apparaîtraient sur les deux faces du joint sectionné. La distance verticale des paires de gradins est le double de l'épaisseur des feuillets

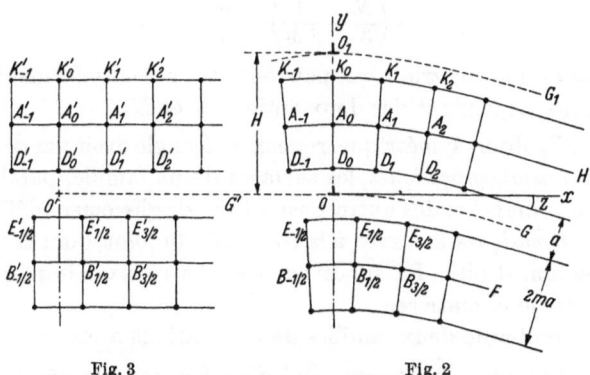

Fig. 3   Fig. 2

Fig. 2. Disposition présumée des atomes à l'équilibre, de part et d'autre des surfaces de glissement telles que OG. L'entier M est ici égal à $2m + 1$

Fig. 3. Numérotage des atomes dans les feuillets

[1] Les deux grains pourraient être soudés sans apport de matière, en réduisant les saillies au contact et rapprochant les creux. Adoptant cette manière de voir, VAN DER MERWE a calculé l'état d'équilibre du joint comme dans un problème d'indentation en élasticité plans. Cette conception, différente de la nôtre, est adaptée au cas de très petites désorientations. Voir Proc. phys. Soc., Lond., Ser. A **63**, 616 (1950).

b) Joint[1] B_2. Joint non symétrique. L'axe Oz est perpendiculaire au plan de la fig. 4. La direction du plan de coordonnées yOz bissecte le dièdre formé par les plans asymptotiques aux surfaces de glissement $G'OG$; le plan YOz du joint fait un angle ε avec le précédent. Le grain

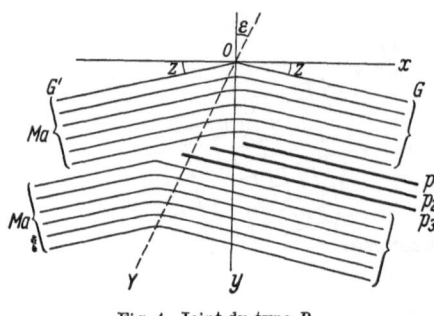

Fig. 4. Joint du type B_2

représenté à droite renferme des demi plans cristallographiques p_1, p_2, p_3 supplémentaires, insérés entre les feuillets. Le groupement par trois respecte la séquence d'empilage qui caractérise le cristal cubique à faces centrées. Trouvant place exactement dans l'intervalle de deux feuillets, ces demi-plans ne créent pas de contrainte à distance, comme il y en a autour d'une dislocation isolée, ou d'une dislocation insérée dans un joint à faible désorientation. Ils introduisent seulement un défaut de coordination (misfit), au voisinage de leur arête.

S'il y a N feuillets entre deux triplets consécutifs, on trouve, par un raisonnement semblable à celui de Shockley et Read, que

$$2\,MN \sin Z \sin \varepsilon = 3 \cos(Z + \varepsilon)$$

ou approximativement, pour ε petit,

$$\varepsilon \left(\frac{N}{3} + \frac{1}{2\,M} \right) = \frac{a}{b}\,.$$

Les angles d'asymétrie ε qui peuvent être atteints forment donc une suite discrète, dépendant des deux entiers M et N.

c) Joint F_2. Joint symétrique, rencontré à angle droit par *deux familles de surfaces cristallographiques*, les surfaces d'une famille, parallèles entre elles, rencontrant celles de l'autre. Ces surfaces de glissement délimitent des *barreaux prismatiques* incurvés à la traversée du joint, dont la fig. 5 représente la section droite. La subdivision en barreaux n'implique aucune discontinuité de la matière.

Le joint renferme deux familles de dislocations arêtes ayant le même vecteur de Burgers $\vec{B}$: chaque côté d'un barreau est décalé de $\vec{B}$ par rapport au côté du barreau voisin, quand on compte un nombre d'atomes égal dans le sens de leur longueur.

d) Joint F_4. Joint non symétrique dérivé du précédent, comme B_2 de B_1: on insère des triplets de plans cristallographiques semblables à ceux de la fig. 4, entre les feuillets d'une famille, à l'intervalle de N feuil-

---

[1] L'indice indique le nombre de paramètres dont dépend la configuration du joint. Ici ce sont les deux entiers M et N.

lets, et des triplets semblables dans l'autre famille à l'intervalle de N' feuillets. Ce joint dépend de quatre paramètres qui sont

— les entiers M, M' fixant l'épaisseur des feuillets, ou les dimensions du parallèlogramme (fig. 5), ou encore les deux composantes Z, Z' de la désorientation

— les entiers N, N', dont dépend l'asymétrie du joint.

3. Propriétés des joints à structure feuilletée. a) Les *joints symétriques feuilletés* seraient très *mobiles* dans un métal pur. On connaît la mobilité des rideaux de BURGERS; la mobilité du joint B_1 s'explique de la même manière, la translation dx du plan OJ produisant une translation $dy = 2\,\mathrm{tg}\,Z\,dx$ du grain G_2 par rapport à G_1. De

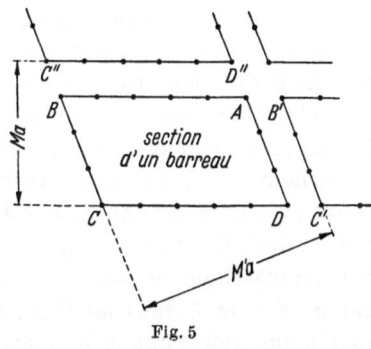

Fig. 5

même, une translation dx du plan du joint F_2, qui entraîne les deux familles de dislocations solidairement, communique à G_2 une translation, qui a une composante suivant Oy et une suivant Oz. L'application d'une contrainte de cisaillement à G_2, dans la direction de la résultante, produirait la translation sous un effort minime[1].

b) *Un joint feuilleté est franchi aisément par une dislocation*, si son plan de glissement passe d'un grain à l'autre suivant une surface cristallographique continue. Il y a une famille de tels plans qui traversent les joints B_1 et B_2, les vecteurs de BURGERS possibles sont au nombre de 3.

Il y a deux familles de plans traversant F_2 et F_4, avec un vecteur de BURGERS unique. Noter que dans tous les cas, les plans cristallographiques qui se correspondent ont des traces parallèles sur le plan du joint[2].

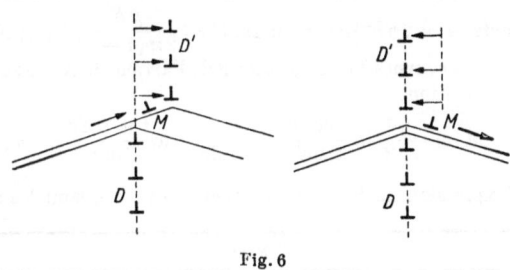

Fig. 6

La mobilité du joint symétrique lui permet de se remettre à l'alignement aussitôt après le passage d'une dislocation M isolée (fig. 6). Si cette dernière est suivie d'autres, elles rencontrent une résistance égale (tandis

---

[1] Dans les joints B_2, F_4, il est vraisemblable que les demi-plans intercalaires peuvent glisser comme les dislocations équivalentes; on y reviendra plus loin.

[2] S'il y a trois familles de plans se correspondant de cette manière, il y en a une infinité; c'est le cas particulier d'un joint mâclé.

que les dislocations successives rencontrent une résistance croissante lorsqu'elles franchissent un joint asymétrique).

L'énergie w à dépenser, pour amener M de l'infini jusque dans le plan médian OJ du joint feuilleté a été évaluée, et on l'a trouvée indépendante de l'ordonnée du plan de glissement. Par exemple dans le cas de feuillets d'épaisseur 4 a (désorientation $2\ Z = 14°$), avec un coefficient de Poisson $v = 0,35$, on a $w = 0,10\ \mu b^2$ par cm de ligne environ. L'énergie w décroît lorsque la désorientation augmente. Ces propriétés diffèrent sensiblement de celles d'un joint à faible désorientation. L'énergie d'activation pour le franchissement par une dislocation y varie beaucoup selon le plan de glissement, et sa valeur moyenne, indépendante de la désorientation, serait $w' = 0,16\ \mu b^2$ dans le cas ci-dessus[1]. La contrainte appliquée σ_i, de l'ordre de 0,01 μ par cm², qui existe dans les grains au début de la déformation plastique, ne saurait faire passer le joint à une dislocation isolée, car on a trouvé dans les feuillets un cisaillement antagoniste dont le maximum, indépendant de Z, atteint 0,13 μ environ. Mais quand il s'est formé une file d'une quinzaine de dislocations en attente, la première pourra franchir le joint sous la poussée de σ_i, puis l'arrivée d'une nouvelle dislocation en queue fera passer la deuxième ... etc.

c) La structure feuilletée serait à notre avis, *stable* dans les joints à désorientation moyenne de l'aluminium. On évalue l'énergie de formation W du joint feuilleté, en se servant des hypothèses suivantes:

---

[1] Le rideau de Burgers étant formé de dislocations ($+ y$) fixes, alignées sur l'axe des y à l'intervalle h, l'énergie potentielle élastique $U(x)$ d'une dislocation ($+ y$) mobile, d'abscisse x, d'ordonnée y const. a été calculée par A. Seeger. La plus grande ascension de $U(x)$ entre $- \infty$ et $+ \infty$ est une fonction sommable du paramètre $v = 2\pi h^{-1} y$: asc. max. $U = \dfrac{\mu b^2}{2\pi(1-v)} F(v)$; $F(v) = F(2\pi - v) = F(2\pi + v)$.

Les courbes 1 et 2 de la fig. 3 de l'article de A. Seeger (référ. in fine) représentant les fonctions

$$M(v) = \frac{1}{2} \ln \frac{\cos v}{2 \sin^2 v} + \frac{1}{2 \sin v} \ln \frac{1 + \sin v}{\cos v}; \quad m(v) = -\frac{1}{2} \ln 2(1 - \cos v);$$

l'expression de $F(v)$, et sa valeur moyenne, sont les suivantes:

$F(v)$		valeur moyenne de F
$0 \leq v \leq \dfrac{\pi}{3}$	$M(v)$	0,96
$\dfrac{\pi}{3} \leq v \leq \dfrac{\pi}{2}$	$M(v) - m(v)$	0,27
$\dfrac{\pi}{2} \leq v \leq \pi$	$- m(v)$	0,58

La moyenne de F sur l'intervalle $(0, 2\pi)$ est 0,66; ce qui conduit à l'énergie d'activation w' indiquée ci-dessus.

a) *potentiel* pour la composante de la force qui s'exerce entre deux atomes tels que D_1 et $E_{1/2}$ (fig. 3), dans la direction tangente à $G'OG$. Le potentiel V est représenté, en fonction du décalage u, sur la fig. 7.

b) *conservation des sections planes* dans les lames fléchies.

Nous ne pouvons exposer ici le calcul de W. D'autre part l'énergie W_0, relative à la formation, dans l'aluminium, d'un joint de type non spécifié, doit être donnée approximativement par une formule du genre de SHOCKLEY et READ, comme il a été constaté dans d'autres métaux. Nous

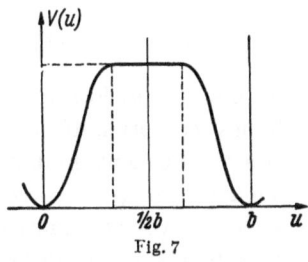

Fig. 7

avons pris $v = 0,35$ et ajusté le maximum de W_0 à la désorientation de $25°$[1].

Le tableau ci-dessous résume les résultats. L'unité d'énergie de joint est $10^{-3} \mu b$.

Epaisseur d'une lame (unité a)	2	3	4	5	6
Désorientation 2 Z (degrés)	24	17,4	14	11,6	10
W	38,4	43,7	46,6	48,8	50,6
W_0	53	51	48	44	41

Quoique approximatifs, ces chiffres suggèrent que la structure feuilletée pourrait être stable dans les joints où la désorientation dépasse une douzaine de degrés.

On constate d'autre part qu'une légère *asymétrie* du joint feuilletée modifie peu son énergie. Supposons la maille cubique, de côté b; le raisonnement consiste à transformer le joint B_1 en plusieurs étapes, jusqu'à la configuration dissymétrique B_2 qui nous intéresse:

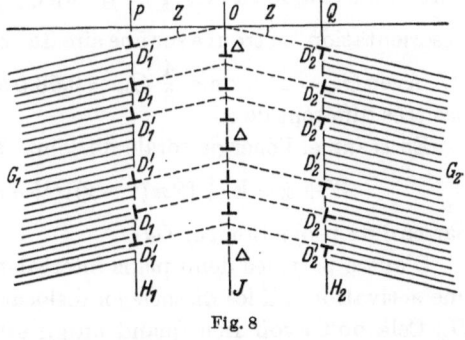

Fig. 8

1. Isoler la portion perturbée qui se trouve entre deux plans PH_1, QH_2, à la distance d de part et d'autre du plan médian OJ du joint B_1.

2. Produire l'asymétrie en faisant tourner les grains G_1 et G_2 d'un même angle par rapport au solide compris entre PH_1 et QH_2. Il apparaît une distribution de dislocations D_1, D'_1, sur PH_1 et une semblable D_2, D'_2, sur QH_2 (fig. 8).

---

[1] La formule employée est donc $W_0 = 4,27\, Z(1 - \ln. 0,08\, Z)$.

Si l'on néglige des termes en ε^2, le nombre de dislocations par cm. le long de ces plans est $\varepsilon b^{-1} \sin Z$ pour D_1, D_2; et $\varepsilon b^{-1} \cos Z$ pour D'_1, D'_2.

L'énergie d'une de ces distributions par cm² s'écrit d'après Shockley et Read

$$U = |\varepsilon| \left(A + \ln \frac{1}{|\varepsilon|} \right) (E_0 + E'_0)^1; \quad E_0 = \frac{\mu \, b \sin Z}{4\pi (1-\nu)}, \quad E' = \frac{\mu \, b \cos Z}{4\pi (1-\nu)}.$$

3. Lier les dislocations Δ, qui siègent dans les feuillets, aux files d'atomes respectifs, déplacer les plans PH_1, QH_2, de manière que leurs dislocations se meuvent dans les surfaces de glissement lorsque d diminue. Le mouvement quasi horizontal d'une paire de dislocations comme D'_1, D'_2 et le mouvement quasi vertical de D_1, D_2, n'exigent aucun travail à cause de la symétrie des contraintes des feuillets vis-à-vis du plan OJ. Finalement les paires comme D'_1, D'_2 s'annihilent en libérant l'énergie qui correspond à E'_0, et produisant un glissement relatif b du matériau situé au dessus de leur surface de glissement.

4. Libérer les Δ, cesser d'imposer l'alignement aux D_1, D_2, pour atteindre la configuration la plus stable du joint B_2.

On voit que l'énergie de la transformation, ou *énergie d'asymétrie* α, est au plus égale à celle qui restait emmagasinée à la fin de la troisième étape: $\alpha \leqslant 2 |\varepsilon| \left(A + \ln \frac{1}{|\varepsilon|} \right) E_0$, par cm² de plan OJ. Dans l'état stable du joint B_2 représenté sur la figure 4 (cristal cubique à faces centrées), D_1 et D_2 sont rassemblées à l'arête des demi-plans intercalaires p_1, p_2, p_3; le terme en $\ln \frac{1}{|\varepsilon|}$ n'existe plus, il reste $2 |\varepsilon| A E_0$.

C'est une énergie comparable à celle d'un défaut de coordination régnant sur la section droite de chaque plan intercalaire. L'énergie unitaire étant évaluée à $[4\pi (1-\nu)]^{-1} b \mu \Theta_m$ comme dans un joint à grande désorientation, on trouve sur une aire $3a$: $\alpha' = \frac{4}{3} \Theta_m |\varepsilon| E_0$, par cm² de plan OJ. Avec $\Theta_m = 25°$, on a $\frac{4}{3} \Theta_m = 0,58$ (rad.), et $2A = 0,34$; ainsi α' n'est pas très différent de α.

En résumé, l'énergie totale du joint B_2 serait

$$W + \alpha = W + [2\pi (1-\nu)]^{-1} b \mu \sin Z \cdot |\varepsilon| (1 + \ln \Theta_m)$$

par cm² de projection sur OJ.

D'autre part, les demi-plans intercalaires peuvent se mouvoir, après une activation qui les dissocie en dislocations mobiles, semblables à D_1, D_2. Celà doit avoir lieu quand un glissement se transmet à travers le joint B_2 parallèlement à ses feuillets. La dérive des plans intercalaires permet à l'asymétrie ε de changer progressivement, même s'il passe un grand nombre de dislocations sur le même plan de glissement (continuité des bandes).

---

[1] Si Θ_m est la désorientation qui rend W_0 maximum, on sait que $A = 1 + \ln \Theta_m$.

4. La formation de la structure feuilletée pendant le recuit. Le feuilletage lui-même n'est guère observable, mais l'aptitude du joint à la transmission des glissements entre un grain et l'autre peut être constatée. Il sera seulement question de *l'aluminium raffiné*.

Une éprouvette de traction polycristalline est recuite jusqu'à la grosseur de grain désirée, et polie électrolytiquement. Après un allongement plastique d'environ 1%, les lignes de glissement sont visibles au microscope ordinaire. On voit fréquemment un faisceau de lignes parallèles qui s'étend sur deux grains voisins, toutes les lignes faisant un coude à la traversée du joint intermédiaire; c'est la disposition en *chevron* signalée par LACOMBE et BEAUJARD [8]. Elle provient, selon G. J. OGILVIE [9] de la propagation du glissement plastique d'un grain à l'autre. Cet auteur a observé que:

a) La *déviation* des lignes à la traversée du joint est, dans 42% des cas, inférieure à 10°, et dans 36% des cas, comprise entre 10° et 25°.

b) Chaque ligne brisée est la trace de deux plans de glissement octaédraux qui ont leur *intersection D* dans le plan du joint; *D est une direction cristallographique* $\langle 110 \rangle$, $\langle 112 \rangle$, *ou* $\langle 123 \rangle$ *dans l'un et l'autre grains*.

c) *Le plan du joint* n'a pas une orientation remarquable par rapport aux grains.

d) *La formation de chevrons est d'autant plus fréquente que l'éprouvette a été recuite plus longtemps.*

Les observations b) et c) peuvent être expliquées en admettant que le joint possède la structure feuilletée (type B_2, ou F_4) et que les glissements se propagent à travers lui suivant les feuillets (une seule famille dans le cas F_4); du reste elles sont également conciliables avec le mécanisme de FRANK. Par contre, en ce qui concerne d), on ne voit pas bien comment un recuit préalable favoriserait ce mécanisme. Mais en principe, le recuit peut remplacer la structure peu ordonnée des joints du métal laminé, par une autre d'énergie libre plus basse.

Les joints d'orientation convenable *acquerraient au cours du recuit, la structure feuilletée* et ce changement les rendrait beaucoup plus aptes à laisser passer des glissements. Dans le cas d'une petite désorientation, c'est le schéma de SHOCKLEY et READ qui apparaît, avec un ou deux rideaux de dislocations arêtes dans le plan du joint. Le nouvel arrangement des dislocations peut s'établir à l'occasion d'un mouvement d'ensemble du joint. Appelons J le plan initial, A_l les plans octaédraux du premier grain et B_m ceux du second, C_{lm} l'intersection de A_l et B_m (l, $m = 1, 2, 3, 4$). Il est facile de voir que le plan d'un joint F_4, renfermant C_{lm} et $C_{l'm'}$ avec $l \neq l'$ et $m \neq m'$, a 72 positions possibles. Il suffit donc d'une petite rotation pour que J atteigne un tel plan. La réalisation des conditions

b) d'Ogilvie (direction cristallographique de *D*) est favorisée par l'existence d'une *texture* dans le métal utilisé[1]. Il est clair qu'un joint dont le plan fait un très petit angle avec une direction C_{lm} est représentable par le schéma B_2, augmenté de quelques dislocations étrangères, ce qui ne modifie pas sensiblement sa perméabilité vis-à-vis des dislocations qui glissent parallèlement au plan A_l puis au plan B_m. La même remarque s'applique à un joint proche de F_4.

L'auteur tient à remercier M. Lacombe, Directeur du Centre de Recherches métallurgiques de l'Ecole des Mines, qui lui a suggéré cette étude; M.Jaoul qui lui a facilité la tâche de la manière la plus obligeante et lui a permis d'examiner la formation des chevrons, enfin, M. Jacques Friedel, qui lui a fait part de plusieurs observations utiles.

Bibliographie

[1] Burgers, J. M.: Proc. nederl. Akad. Wetenschap. **42**, 293 (1939).
[2] Cahn, R. W.: J. Inst. Met. **76**, 121 (1949).
[3] Cottrell, A. H.: Dislocations and plastic flow in crystals. London 1953.
[4] Frank, F. C.: (Resultant content of dislocations in a boundary). Plastic deformation of crystalline solids, p. 150. Pittsburgh symposium 1950.
[5] Frank, F. C.: (Initiation of slip in a grain near a boundary). Plastic deformation of crystalline solids, p. 95. Pittsburgh symposium 1950.
[6] Friedel, Cullity et Crussard: Acta Métall. **1**, 79 (1953).
[7] Mott, N. F.: Proc. phys. Soc., Lond. **60**, 391 (1948).
[8] Lacombe et Beaujard: J. Inst. Met. **74**, 1 (1948).
[9] Ogilvie, G. J.: J. Inst. Met. **1469**, 491 (1953).
[10] Read, W. T. et W. Shockley: Plastic deformation of crystalline solids, p.124. Pittsburgh symposium 1950.
[11] A. Seeger: Naturwissenschaften, **22**, 526 (1951).

---

[1] Parmi les paires de grains voisins, une forte proportion ont, d'après Friedel, Cullity et Crussard, leurs réseaux presque symétriques par rapport à un plan *PM* (pseudo-mâcle), lequel est souvent très proche d'un plan cristallographique de bas indice dans chacun des deux grains. Il existe quatre droites C_{lm} voisines de *PM* et la probabilité qu'elles forment un très petit angle avec une direction de bas indice dans les deux réseaux est beaucoup plus grande que pour une C_{lm} quelconque, entre grains orientés au hasard. Le cas où *D* a des dénominations différentes dans les deux grains est laissé de côté, il demanderait une étude spéciale.

The Theory of Piecewise Linear Isotropic Plasticity[1]

By **P. G. Hodge** jr., Brooklyn (N.Y.)

With 14 figures

1. Introduction. Two qualities are required of any theoretical description of a physical phenomenon. It must be a sufficiently simple theory to be mathematically tractable and to yield predictions, and it must be sufficiently inclusive so that these predictions are of some value. In the case of structures subjected to sufficiently small strains, the classical theory of elasticity satisfies both of these requirements in a wide variety of applications. However, the theory of plasticity is in no such fortunate state. On the one hand, experimental evidence [1, 2] indicates that even the most elaborate theories yet proposed do not show any consistent relation with reality, while on the other hand, comparatively few practically important theoretical solutions have been obtained with even the simplest of theories.

Under these circumstances, there are three important lines of investigation to be pursued. Carefully designed experiments are needed to test theoretical results and assist in formulating theories. New mathematical (including numerical) techniques are needed to make predictions based upon increasingly complex theories. Finally, new theoretical studies are needed to make solutions easier and more useful. In the present paper we shall be concerned entirely with the third of these approaches.

The theory with which we shall deal is not by an means a new theory, but rather a new viewpoint. Although the name "piecewise linear" is perhaps original, some of the concepts date back to the beginnings of plasticity theory. This subject will be taken up in more detail in the next section where we will give a brief historical survey leading up to the present paper.

The remainder of the paper will be devoted to the particular physical problem of a circular cylindrical shell (fig. 1) subject to a uniform exterior pressure p. All of the results can, of course, be presented in more general terms, but it is hoped that the concepts involved will be more clear in

    [1] The results presented in this paper were obtained in the course of research conducted under Contract Nonr—267(00), Project Nr 360—001 with the Office of Naval Research.

10*

reference to specific problem. Further, we shall attempt to keep the mathematics to a minimum in the present exposition, and refer to source materials for more explicit detail.

As previously mentioned, section 2 will give a historical introduction to the problem. We will also define the principal ingredients of any theory

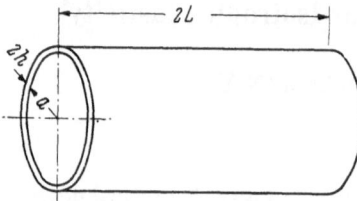

Fig. 1. Dimensions of shell

of plasticity and particularize to the title theory. In section 3 we will restate the theory in terms of a cylindrical shell. Section 4 will deal directly with the simplest case of the theory, the so-called rigid-perfectly plastic material, in which elastic strains are neglected and there is no strain-hardening. In section 5 we will consider the general theory where both elastic strains and strain-hardening are considered simultaneously. Section 6 will then take up the intermediate theories involving elastic strains or strain hardening, but not both.

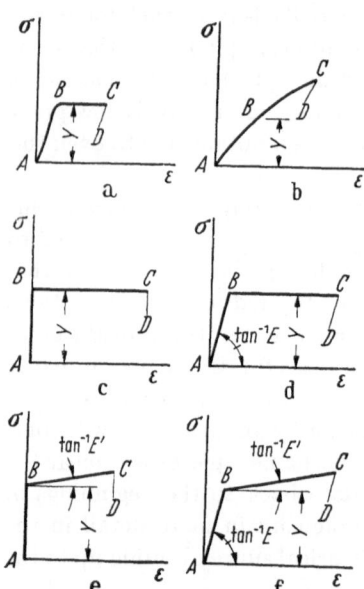

Fig. 2. Stress strain curves a) with sharp yield point, b) with conventional yield point, c) Rigid-perfectly plastic, d) elastic-perfectly plastic, e) rigid-strain hardening, f) elastic-strain hardening

Since specified solutions will be obtained for the problems to be considered they may be regarded as solved. However, for more complex physical problems, complete solutions may not be readily obtainable. For this reason it will be valuable to consider some approximate methods of solution. For a perfectly plastic material, the well known theorems of limit analysis provide a powerful tool for extracting certain important information without first obtaining a complete solution. However, these are not applicable to a strain hardening material. In section 7 we will present two minimum principles which may be used to advantage in this case.

Finally, the paper will conclude with a summary of the results and a discussion of their limitations.

2. History of the basic theory. A primary ingredient of any theory of plasticity is the yield condition. For a material under simple tension, this consists simply of the value of the yield stress Y. For some materials, such as mild steel, this may be sharply defined (fig. 2a) while for other materials, such as aluminium, it may be conventional (fig. 2 b). However, once it has been defined, the material is said to be elastic and is assumed to follow HOOKE's law so long as the stress σ remains less then Y, and otherwise to be plastic.

For multiaxial states of stress, it is necessary to specify some relation between the stresses, rather than just an single stress. The earliest proposed yield condition was due to TRESCA [3] and stated that the maximum shearing stress could not exceed the yield stress in pure shear. In the particular case where one of the principal stresses, σ_3, is zero, this may be written

$$f(\sigma_1, \sigma_2) = \max[\,|\sigma_1|,\ |\sigma_2|,\ |\sigma_1 - \sigma_2|\,] - Y \leqq 0, \qquad (2.1)$$

where Y is the tensile yield stress. The level curves of this function are shown in fig. 3 a. In particular, the region inside the curve $f = 0$ represents elastic behavior.

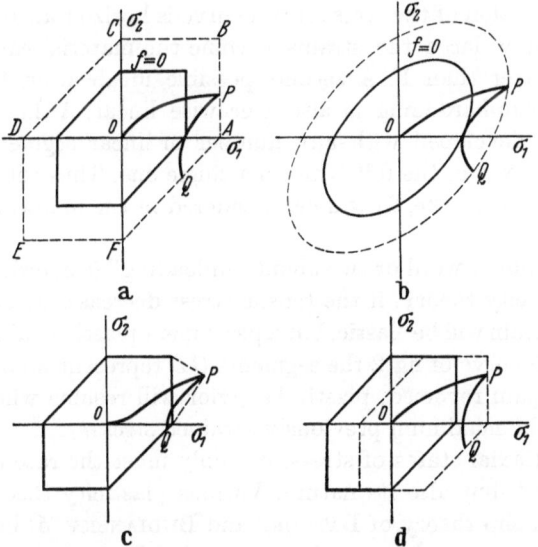

Fig. 3. Level curves of yield functions ———— original yield condition – – – hardened yield condition ———— loading path a) TRESCA, with isotropic hardening, b) MISES, with isotropic hardening, c) TRESCA, with slip theory hardening, d) TRESCA, with PRAGER hardening

In 1913 MISES [4] suggested that the octahedredal shearing stress was critical rather than the maximum shearing stress, and that this quantity could not exceed the yield stress in pure shear. Thus, for plane stress,

$$f(\sigma_1, \sigma_2) = \sigma_1^2 - \sigma_1 \sigma_2 + \sigma_2^2 - Y^2. \qquad (2.2)$$

The level curves of this function are shown in fig. 3 b, and the region inside the curve $f = 0$ again represents elastic behavior.

The MISES condition is obviously non-linear, since the stresses in equation (2.2) are raised to the second power. On the other hand, if we know which of the six faces in fig. 3 a is applicable, the TRESCA condition is certainly linear. However, since the linear coefficients are different for each of the six faces, the term "piecewise linear" may be used to describe the entire TRESCA condition. As we shall see, such a piecewise linear yield condition possesses some, but not all, of the advantages of a linear theory. The remainder of the paper will be concerned only with piecewise linear yield conditions.

A second ingredient in any theory of plasticity is the rate of hardening after the yield stress has been reached. For a material in simple tension this may be expressed as $d\varepsilon/d\sigma$. In general, for real materials (figs. 2 a, 2 b) this will be a varying function of the stress σ. However, various idealizations are permissible. For a perfectly plastic material, $d\varepsilon/d\sigma$ is assumed to be infinite once the yield stress has been reached. Thus, in figs. 2 c, 2 d, the plastic portion of the stress strain curve is horizontal. It follows that no matter how large the strains become the material cannot support a stress greater than Y. A second possible idealization is to assume that the strain-hardening is also piecewise linear. Although this concept can be developed with any number of linear segments, we shall restrict ourselves in the following to a single one. Thus the stress strain curves shown in figs. 2 e, 2 f will be considered in the following, as well as figs. 2 c, 2 d.

At this point a word or two about "unloading" is in order. According to any plasticity theory, if the tensile stress decreases at any time, the changes in strain will be elastic, but a permanent plastic strain will remain. Thus in each curve of fig. 2 the segments CD represent unloading. If the loading is again reversed, plastic behavior will resume when the stress is equal to the maximum previously attained stress.

For multiaxial states of stress, not only must the rate of hardening be considered, but also its nature. Various plasticity theories, such as the so-called slip theory of BATDORF and BUDIANSKY [5] have assumed that the yield curve changes in shape as the hardening progresses (fig. 3 c). Recently, PRAGER [6] has proposed a theory in which the yield curve maintains its size and orientation but changes its center (fig. 3 d). The present "isotropic" theory assumes that the yield curve maintains its shape and orientation. Thus the current yield curve is always one of the level curves of the yield function (fig. 3 a).

Whenever the state of stress changes in such a fashion that the stress point leaves the yield curve (path PQ in figs. 3), unloading ensues and the resulting changes in strain are purely elastic. Plastic behavior will resume when the stress point returns to some point on the hardened yield curve (point Q in figs. 3).

We shall consider here only the so-called "flow theories" of plasticity, so that the next ingredient of a plasticity theoryis the flow law. Historically, the first flow laws were proposed by MISES [4], PRANDTL [7], and REUSS [8]. The PRANDTL-REUSS law states that the total strain rate $\dot{\varepsilon}$ can be decomposed into an elastic and a plastic part,

$$\dot{\varepsilon} = \dot{\varepsilon}_e + \dot{\varepsilon}_p, \tag{2.3}$$

where the elastic part is given by the differentiated form of HOOKE's law. The plastic strains are incompressible and their rate of change is proportional to the existing state of stress deviation[1]. Thus

$$(\dot{\varepsilon}_1)_p = \dot{\lambda} (\sigma_1)_{\text{dev}}, \text{ etc.,} \tag{2.4}$$

where the dot indicates differentiation with respect to time. For a perfectly plastic material, λ is an undetermined positive function, while for a strain-hardening material it is a known function which describes the hardening.

Up until 1928, practically all theoretical research in plasticity theory was done on the basis of the stress strain laws (2.4). A few solutions were obtained using them in conjunction with the TRESCA yield condition, but for the most part problems were more easily formulated and solutions more easily interpreted when the MISES yield condition was adopted. In 1928, MISES [9] proposed the theory of the plastic potential. According to this theory, a certain strain-rate vector was required to be normal to and outward directed from the yield curve, so that yield condition and flow law were intimately connected. In terms of principle stresses, the strain-rate vector has the components $\dot{\varepsilon}_1$, $\dot{\varepsilon}_2$, $\dot{\varepsilon}_3$ and it is easily shown that if this vector is given by equation (2.4), then it is in fact, normal to the MISES yield curve. Later, HILL [10] and DRUCKER [11, 12] gave strong theoretical arguments in favor of the plastic potential flow law.

Let us look at this law as it would be applied to the TRESCA yield condition in a problem of plane stress. The component $\dot{\varepsilon}_3$ can be regarded as defined by the incompressibility condition,

$$\dot{\varepsilon}_3 = - (\dot{\varepsilon}_1 + \dot{\varepsilon}_2), \tag{2.5}$$

so that it is sufficient to discuss the problem entirely in terms of $\dot{\varepsilon}_1$ and $\dot{\varepsilon}_2$. In other words, the vector with components $\dot{\varepsilon}_1$, $\dot{\varepsilon}_2$ must be normal to the hexagon in fig. 3 a. It follows immediately that on side AB, $\dot{\varepsilon}_2 = 0$, $\dot{\varepsilon}_1 \geqq 0$, while on side FA, $\dot{\varepsilon}_1 = - \dot{\varepsilon}_2 \geqq 0$, with similar expressions for the remaining sides.

However, at the corner A, a difficulty is encountered since the normal is not defined. This problem is partially resolved by referring back to

---

[1] The stress deviation is defined as the difference between the total stress and the mean normal stress: $(\sigma_1)_{\text{dev}} = \sigma_1 - (\sigma_1 + \sigma_2 + \sigma_3)/3$, etc.

DRUCKER's proof [12] of orthogonality based on certain thermodynamic assumptions. The actual result proved by DRUCKER may be stated in the form "every stress vector drawn from a point on the yield curve and lying inside or on the yield curve must intersect the strain-rate vector at an angle not less than 90°". For a point on AB, for example, this leaves the unique normal to the curve as the only permissible direction.

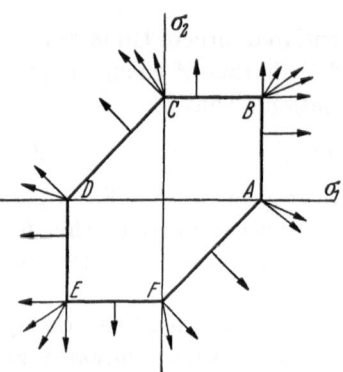

However, at the corner A any vector between the limiting normals on AB and DA is permissible. Typical strain rate vectors are shown in fig. 4.

It thus appears that a piecewise linear yield curve and associated plastic potential flow law possess a certain undesirable uniqueness. If the strain rate is given normal to a side, the stress is not unique, while if the stress is given at a corner the strain rate is not unique. Despite this fact, DRUCKER,

Fig. 4. TRESCA yield condition and strain rate vectors

PRAGER, and GREENBERG [13] showed that the rate of dissipation of energy was unique and that the theorems of limit analysis were valid. Then in 1953 KOITER [14] showed that all known uniqueness in the large and variational principles were generally valid for piecewise linear yield conditions. About the same time PRAGER [15] solved a particular problem which involved strain-hardening at the corner of the yield curve. SANDERS [16, 17] has also obtained some general results which hold at the corners of the yield function. Finally, the present author [18, 19]

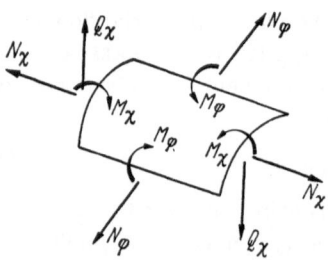

has obtained explicit formulations for the flow law at the corner for an isotropic strain-hardening material.

3. Theory of a cylindrical shell. The spirit of thin shell theory is to express the problem in terms of certain generalized stresses and strains (stress and strain resultants) rather than three dimensional stress and strain components. For a cylindrical shell under radially symmetric loading, the only stress resul-

Fig. 5. Stress resultants for cylindrical shell

tants which do not vanish due to symmetry are the axial and circumferential direct stresses N_x and N_φ, the corresponding moments M_x and M_φ, and the axial shear Q_x (fig. 5). Since by hypothesis, there is no end load, $N_x = 0$. Further, PRAGER [20] has shown that only those generalized stresses which do work need to be considered. To determine these we must first define the corresponding generalized strains. It will be assumed that the deformations are all small, so that the geometric part of conventional elastic analysis is applicable. Thus, if w represents the inward radial displacement, the extensions and curvatures of the middle surface are [21]

$$e_x = \frac{d\,u}{d\,x}\,, \qquad e_\varphi = -\frac{w}{a}\,, \tag{3.1}$$

$$\varkappa_x = -\frac{d^2\,w}{d\,x^2}\,, \qquad \varkappa_\varphi = 0\,. \tag{3.2}$$

Further, the deformation due to shear is neglected. Therefore, regarding only those quantities which do work we are left with the two generalized stresses N_φ and M_x and the corresponding generalized strains e_φ and $\varkappa_x$.

In the following, it will prove convenient to introduce dimensionless variables defined by

$$\left.\begin{aligned}
n &= \frac{N_\varphi}{2Yh}\,, & m &= \frac{M_x}{Yh^2}\,, \\[2mm]
W &= \frac{w}{a}\,, & c &= \frac{L}{\sqrt{ah}}\,, & y &= \frac{x}{L}\,, \\[2mm]
e &= e_x = -W\,, & \varkappa &= \left(\frac{h}{2}\right)\varkappa_x = -\frac{W''}{2c^2}\,,
\end{aligned}\right\} \tag{3.3}$$

where primes indicate differention with respect to y.

If generalized stresses are to be used, it is necessary to express the yield condition in terms of n and m, i.e., to construct the yield curve in an n, m space. DRUCKER [22], ONAT [23], and HODGE [24] have shown that the solid curve in fig. 6 corresponds to the TRESCA yield condition. However, despite the fact that this is based on a piecewise linear yield condition, it is obviously non-linear in terms of n and m. Therefore, if piecewise linear theory is to be used, the curve must be approximated by straight line segments. One obvious such approximation is the hexagon shown as a dotted curve in fig. 6, and DRUCKER [22] and HODGE [24, 25] have solved various problems on this basis.

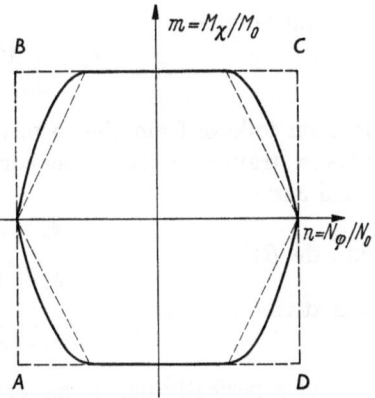

Fig. 6. Yield condition for shell
——— TRESCA condition   — · — hexagon condition   — — — — square condition

However, in order to keep to present analysis as simple as possible we shall here consider the square approximation shown as a dashed curve in fig. 6. Thus the original elastic domain is defined by

$$-1 < n < 1\,, \quad -1 < m < 1\,, \tag{3.4}$$

and the initial yield curve is

$$n = 1, \quad m = 1, \quad n = -1, \quad m = -1. \tag{3.5}$$

The piecewise linear curves shown in fig. 6 may be interpreted in either of two ways. On the one hand, the shell material may be assumed to satisfy the TRESCA yield condition and the analysis based on a piecewise linear curve is an approximate solution. Equally, the piecewise linear curve can be thought of as the true one for the shell, and the shell material then satisfies some approximation to the TRESCA yield condition. We shall here adopt this latter viewpoint and henceforth deal only with the generalized stresses and strains.

In addition to satisfying the yield condition, the stresses must be in equilibrium. Due to symmetry considerations only axial moments and radial forces must be considered. If the shear resultant is eliminated between these two equations, the eqilibrium equation is obtained in the form

$$\frac{m''}{2c^2} + n + P = 0, \tag{3.6}$$

where the dimensionless pressure P is defined by

$$P = \frac{pa}{2Yh}. \tag{3.7}$$

Next, we must consider the flow rule. First, we observe that in terms of the generalized stresses and strains the specific power of deformation per unit volume is

$$\dot{\Delta} = \frac{(N_\varphi \dot{e}_\varphi + M_x \dot{\varkappa}_x)}{2h} = Y(n\dot{e} + m\dot{\varkappa}). \tag{3.8}$$

It then follows from the results of PRAGER [20] that the appropriate plastic strain-rate vector has components $\dot{e}$, $\dot{\varkappa}$. Thus, on side CD of the yield curve

$$\dot{e}_p = \dot{\lambda}, \quad \dot{\varkappa}_p = 0, \tag{3.9a}$$

on side BC,

$$\dot{e}_p = 0, \quad \dot{\varkappa}_p = \dot{\mu}, \tag{3.9b}$$

and at the corner C,

$$\dot{e}_p = \dot{\lambda}, \quad \dot{\varkappa}_p = \dot{\mu}. \tag{3.9c}$$

For a perfectly plastic material λ and $\dot{\mu}$ ust be non-negative, but are otherwise arbitrary. For a strain-hardening material they may be expressed in terms of the dimensionless plastic specific power

$$\dot{D}_p = \frac{\dot{\Delta}_p}{Y} = n\dot{e}_p + m\dot{\varkappa}_p. \tag{3.10}$$

To this end, it is convenient to define the "stress intensity" Q by

$$Q = \max(n, m, -n, -m). \tag{3.11}$$

For the isotropic material here considered, Q is a measure of the size of that level curve of the yield condition which defines further yielding. The following relations between λ, μ and $\dot{D}_p$ are then easily obtained.

$$\left.\begin{array}{ll} \text{on } CD, \ Q = n, & \dot{\lambda} = \dfrac{\dot{D}_p}{Q}, \\[2ex] \text{on } BC, \ Q = m, & \dot{\mu} = \dfrac{\dot{D}_p}{Q}, \\[2ex] \text{at } C, \quad Q = n = m, & \dot{\lambda} + \dot{\mu} = \dfrac{\dot{D}_p}{Q}. \end{array}\right\} \tag{3.12}$$

Thus, prescription of the specific power completely determines the flow law except for the necessary lack of uniqueness at the corners. This formulation at a corner was first used in an example by PRAGER [15]. Later HODGE [18, 19] generalized it and showed that the same result was obtained if the corner were regarded as the limiting case of a smooth curve.

Finally, the flow law can be expressed in terms of the results of a tension test. In accordance with our definition of piecewise linear plasticity it will be assumed that an experimentally observed curve has been approximated by a curve such as fig. 2f. The plastic power for simple tension can be shown [18] to be given by

$$\frac{\dot{D}_p}{Q} = F\dot{Q}, \tag{3.13}$$

where

$$F = \frac{Y}{E'} - \frac{Y}{E}. \tag{3.14}$$

Substitution of equation (3.14) into (3.13) and thence in (3.9) will then give the flow law in terms of the constant F.

4. Rigid-perfectly plastic material. If elastic strains and strain hardening are both to be neglected, then the entire flow law is given by equations (3.9) where λ and μ are arbitrary. Therefore, considering also equations (3.3) the flow law on CD is

$$\dot{e} = -\dot{W} = \dot{\lambda} \geq 0, \quad \dot{\varkappa} = -\frac{\dot{W}''}{2\,c^2} = 0$$

with similar results for the other regimes[1]. The complete result is shown in table 1.

The use of table 1 is best illustrated by means of examples. To this end, let us consider first a shell which is simply supported at either end and subject to a uniform exterior pressure p. Due to symmetry we need consider only half the shell, $0 \leq y \leq 1$.

---

[1] The term "plastic regime" was introduced by HOPKINS and PRAGER [26] to denote either a side or a corner as desired.

Table 1. *Stress and strain rates for a rigid-perfectly plastic material*

Regime	n	m	$\dot{W}$	$\dfrac{\dot{W}''}{2\,c^2}$
AB	-1	$-1 < m < 1$	$+$	0
B	-1	1	$+$	$-$
BC	$-1 < n < 1$	1	0	$-$
C	1	1	$-$	$-$
CD	1	$-1 < m < 1$	$-$	0
D	1	-1	$-$	$+$
DA	$-1 < n < 1$	-1	0	$+$
A	-1	-1	$+$	$+$

Physically it is reasonable to expect the hoop stress to be everywhere compressive, so that $n < 0$. Therefore, the side CD and corners C and D of the yield curve (fig. 6) need not be considered. If part of the shell lies on side BC, then the velocity function $\dot{W}$ must satisfy $\dot{W} = 0$, $\dot{W}'' < 0$ which is obviously absurd. Since a similar argument holds for DA, it remains only to consider side AB and corners A and B.

As a first hypothesis we shall assume that the entire stress profile is on side AB, i.e., the shell stresses are everywhere in regime AB. Then, it follows from table 1, that $\dot{W}$ must be a linear function of y which vanishes at $y = 0$. Thus

$$W = \dot{W}_0\, y. \tag{4.1}$$

Now, symmetry would ordinarily demand that $\dot{W}'$ vanish at $y = 1$ and hence that $\dot{W}_0 = 0$. This, in turn implies no displacement at all and hence the shell would be rigid regardless of the pressure, a result which is obviously absurd.

The resolution of the above paradox lies in the idealizations introduced by assuming a rigid-perfectly plastic material. As a consequence of these idealizations, certain continuities which are normally associated with elasticity theory no longer exist. In particular, slope need not be continuous and hence W' need not vanish. A circumference of the shell where this occurs is called a "hinge circle" [26]. However, some attention must be given to the interpretation of this discontinuity. For a real material of which we have a reasonable idealization, there will evidently be a narrow region of rapid change of $\dot{W}'$. This, in turn, corresponds to a very large average absolute value of $\dot{W}''$, while $\dot{W}$ itself remains finite. Therefore, in the limit as this region tends to zero, the ratio $\dot{W}''/\dot{W}$ tends to infinity, while $\dot{W}$ remains finite and non-zero. With reference to table 1 and regard for the signs involved, this can happen only at the corner B of the yield curve. Therefore, we can ignore the boundary condition on $\dot{W}$ at $y = 1$ only at the expense of introducing the condition

$$m(1) = +1. \tag{4.2}$$

Under the hypothesis that the entire stress profile be in regime $A\,B$, $n = -1$ and the equilibrium equation (3.6) can be solved for m:

$$m = -c^2 (P - 1) y^2 + A\, y + B. \tag{4.3}$$

At the simple support $m = 0$ and at the center of symmety the shear which is proportional to m' must vanish. Thus

$$m = c^2 (P - 1)(2\,y - y^2). \tag{4.4}$$

This solution satisfies the original conditions on m but it still remains to satisfy (4.2). Evidently this will be valid only if

$$P = P_0 = 1 + \frac{1}{c^2}. \tag{4.5}$$

The above solution is typical of rigid-perfectly plastic problems. A unique value of the pressure is obtained, but the velocity is determined only to with an arbitrary constant multiplier. This multiplier is a necessary consequence of the horizontal stress strain curve where any positive strain can correspond to the yield stress.

The load P_0 may be interpreted as follows. For $P < P_0$, part of the shell may or may not be plastic, but there remains a statically determinate core of rigid material so that all motion is prohibited. For $P = P_0$ this core of rigid material is no longer determinate (for the shell problem it vanishes, in other problems part of the material may move as a rigid body) and quasi-static motion satisfying the equilibrium equations may take place. For $P > P_0$, no equilibrium configuration exists and inertia terms must be accounted for. The load P_0 is referred to as the "yield load" or "collapse load".

As a second example, we consider a shell which is simply supported at $y = 0$ and clamped at $y = 2$. As before, we assume that the entire shell is in regime $A\,B$ so that the velocity is linear. Obviously no single linear function will suffice so that we consider

$$\dot{W} = \begin{cases} \dot{W}_0 \dfrac{y}{\eta} & (0 < y < \eta), \\[2mm] \dot{W}_0 \dfrac{(2 - y)}{(2 - \eta)} & (\eta < y < 2), \end{cases} \tag{4.6a}$$

where η is to be determined. Since m and m' must both be continuous, and m must satisfy a second order equation, m must be analytic. At $y = 0$, no moment is transmitted, while at $y = \eta$ and at the clamped end there are hinge circles. Consideration of signs shows that m must satisfy

$$m(0) = 0, \quad m(\eta) = +1, \quad m(2) = -1.$$

Therefore,

$$m = -\left(\frac{y}{2}\right) + c^2 (P - 1)(2\,y - y^2), \tag{4.6b}$$

and P and η are related by

$$P = 1 + \frac{1}{2\,c^2}\,\frac{2+\eta}{\eta\,(2-\eta)}\,. \qquad (4.6\,\text{c})$$

At first glance it might seem that equations (4.6) define a solution for any value of η between zero and two, and hence for a considerable range of P. However, the solution cannot be accepted unless the stress profile actually lies on the finite segment $A\,B$. In other words, in addition to satisfying the equation $n = -1$, the stresses must satisfy the inequality $-1 \leq m \leq 1$. Now, since $m = +1$ at the interior point $y = \eta$, m will exceed one somewhere in the vicinity of η unless $m'(\eta) = 0$. This additional equation then serves to determine

$$\left.\begin{aligned}
\eta &= 2\left(\sqrt{2}-1\right) = 0.828\,, \\[2mm]
P_0 &= 1 + \frac{(3 + 2\sqrt{2})}{(4\,c^2)} = 1 + \frac{1.457}{c^2}\,.
\end{aligned}\right\} \qquad (4.7)$$

Similar computations are easily carried out for other common end conditions and the following results are obtained for the collapse load:

$$\left.\begin{array}{lll}
\text{free-free} & P_0 = 1\,, & \text{(a)} \\[2mm]
\text{simple-free} & P_0 = 1\,, & \text{(b)} \\[2mm]
\text{clamped-free} & P_0 = 1 + \dfrac{1}{4\,c^2}\,, & \text{(c)} \\[2mm]
\text{simple-simple} & P_0 = 1 + \dfrac{1}{c^2}\,, & \text{(d)} \\[2mm]
\text{simple-clamped} & P_0 = 1 + \dfrac{1.457}{c^2}\,, & \text{(e)} \\[2mm]
\text{clamped-clamped} & P_0 = 1 + \dfrac{2}{c^2}\,. & \text{(f)}
\end{array}\right\} \qquad (4.8)$$

The resulting collapse loads are shown as functions of the parameter c in fig. 7.

5. Elastic-strain hardening material. In the general case where we take account of both the elastic deformations and the strain hardening, equations (3.9) must be supplemented by the elastic strain rates. These latter may be written [21]

$$\dot{e}_e = \alpha\,\dot{n}\,, \quad \dot{\varkappa}_e = \beta\,\dot{m}\,, \qquad (5.1)$$

where

$$\alpha = \frac{Y}{E}\,, \quad \beta = \left(\frac{3}{4}\right)\left(\frac{Y}{E}\right)(1-\nu^2)\,. \qquad (5.2)$$

Further λ and $\dot{\mu}$ are now known functions of the stress intensity Q as indicated by equations (3.12) and (3.14).

We shall illustrate the techniques used in piecewise linear plasticity by means of a simple example. Although the method is easily generalized, it must be admitted that for more complex examples the computations may become laborious.

Let us consider, then, an unsupported shell subjected to uniform external pressure p and to internal ring loads $\lambda p L$. We assume and later verify that each element of the shell is either always elastic as the load is slowly increased, or else is elastic up to a certain load and thenceforth is plastic on regime AB. In the first case, the complete flow law is given by equations (5.1) which are trivially integrated with respect to time, the initial condition being zero stress and zero strain. Thus

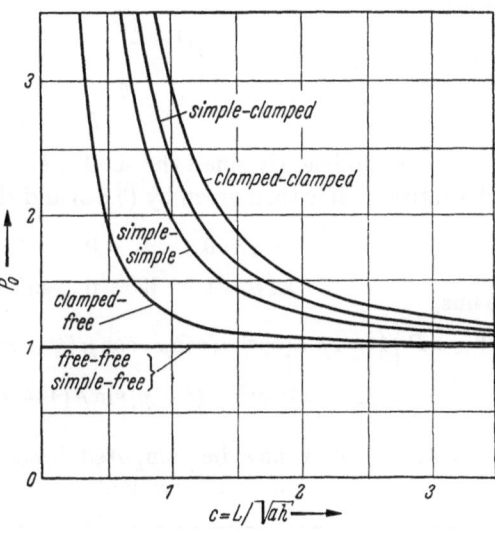

Fig. 7.
Collapse load as a function of c for various end supports

$$e = -W = \alpha n, \\ \varkappa = -\frac{W''}{2c^2} = \beta m. \quad \Big\} \quad (5.3)$$

For the second type of particle equations (5.3) hold as long as $n > -1$. For $n \leq -1$, it follows from a generalization of (3.9) that

$$\dot{e} = \alpha \dot{n} - \lambda, \quad \dot{\varkappa} = \beta \dot{m}. \quad (5.4)$$

Further, it follows from (3.14) and the generalization of (3.12) that

$$\lambda = \frac{\dot{D}_p}{Q} = F\dot{Q} = -F\dot{n}. \quad (5.5)$$

Therefore the flow law on AB can be written

$$\dot{e} = (\alpha + F)\dot{n}, \quad \dot{\varkappa} = \beta \dot{m}. \quad (5.6)$$

Finally, equations (5.6) may be integrated with respect to time, using the initial condition that (5.3) must hold at $n = -1$. The resulting stress strain law then becomes

$$e = -W = (\alpha + F)n + F, \quad \varkappa = -\frac{W''}{2c^2} = \beta m. \quad (5.7)$$

Equations (5.3) for the elastic particle or (5.7) for the plastic particle are trivially solved for the stresses n and m in terms of the displacement W.

If we then substitute the results into the equilibrium equation (3.6), we obtain in each case a fourth order linear differential equation with constant coefficients for W:

$$\text{elastic}: \quad W'''' + 4\mu^4 W = 4\alpha\mu^4 P, \tag{5.8 a}$$

$$\text{plastic}: \quad W'''' + 4\mu_1^4 W = 4(\alpha + F)\mu_1^4 \left(P - \frac{F}{\alpha + F}\right), \tag{5.8 b}$$

where

$$\left. \begin{aligned} \mu^4 &= \frac{\beta c^4}{\alpha}, \\ \mu_1^4 &= \frac{\beta c^4}{\alpha + F} = \frac{\alpha\mu^4}{\alpha + F}. \end{aligned} \right\} \tag{5.8 c}$$

For P sufficiently small the shell is entirely elastic. The solution for the entire shell is then given by (5.8a) and the boundary conditions

$$\left. \begin{aligned} \text{at } y = 0, \quad & m = 0, \quad m' = 2c^2\lambda P, \\ \text{at } y = 1, \quad & \dot{W}' = 0, \quad m' = 0. \end{aligned} \right\} \tag{5.9}$$

Thus,

$$W = \alpha P \Big\{ 1 - 4\mu\lambda\left[\cosh\mu\cos\mu\cosh\mu(1-y)\cos\mu(1-y) \right.$$
$$\left. + \sinh\mu\sin\mu\sinh\mu(1-y)\sin\mu(1-y)\right]\frac{1}{\sinh 2\mu + \sin 2\mu} \Big\}. \tag{5.10}$$

The stresses may now be computed from equations (5.3) so that the complete elastic solution is available.

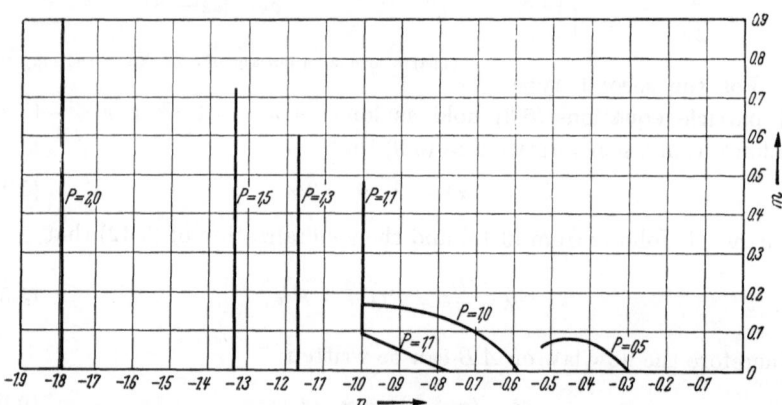

Fig. 8. Stress profiles of elastic-strain hardening shell under uniform pressure and ring loads for selected values of P

In order to present the remainder of the analysis in specific terms, we shall assume particular values for the parameters of the problem as follows:

$$\left. \begin{aligned} c &= 2.201, \quad \nu = 0.3, \quad \lambda = 0.1, \\ \frac{Y}{E} &= 1.75 \cdot 10^{-3}, \quad \frac{E}{E'} = 400. \end{aligned} \right\} \tag{5.11a}$$

It then follows from equations (5.2), (3.14), and (5.8 c), that

$$\left.\begin{aligned} \alpha &= 1.75 \cdot 10^{-3}, \ \beta = 1.193 \cdot 10^{-3}, \ F = 3.99\,\alpha\,, \\ \mu &= 2.000\,, \qquad \mu_1 = 0.4472\,. \end{aligned}\right\} \qquad (5.11\,\text{b})$$

In fig. 8 we have plotted the "stress profile" of the shell for $P = 0.5$ and for the particular values (5.11). This is the image in the stress plane of the state of stress throughout the shell. Since the stress profile is entirely within the initial yield square the shell is, indeed, elastic for this value of P.

Since n and m are each linear in P for the elastic solution, the stress profile will expand uniformly out from the origin as P increases, until the end $y = 1$ touches the side $A\,B$ at a load P_1. This load is determined by the condition $n(1) = -1$, and for the values (5.11) turns out to be

$$P_1 = 0.955\,. \qquad (5.12)$$

For P somewhat greater than P_1 the shell will be elastic for $0 < y < \eta$ and plastic on $A\,B$ for $\eta < y < 1$. Solving each of equations (5.8) for the appropriate range, we obtain a solution involving eight arbitrary constants, as well as the unknown value of η. To determine these constants, we have the four boundary conditions (5.9) together with conditions at the interface $y = \eta$. Four additional equations are provided by the conditions

W, W', n, m all continuous.   (5.13)

The final necessary relation is obtained by choosing η so that the continuous value of n is equal to the initial yield value

$$n(\eta) = -1\,. \qquad (5.14)$$

When the general solution of equations (5.8) is substituted into conditions (5.9), (5.13) and (5.14), a set of nine equations for nine unknowns is determined. All eight of the arbitrary constants appear in

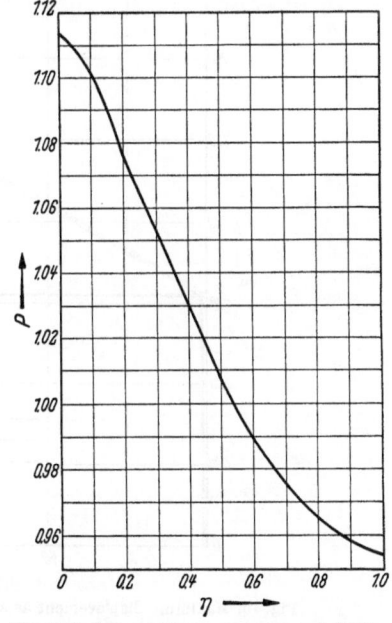

Fig. 9. Relation between pressure and elastic-plastic interface for shell under uniform pressure and ring loads

linear form, but the interface value η is highly non-linear. This difficulty is easily overcome by a change in viewpoint. Since P occurs linearly in the equations, we fix a series of values of η and determine the corre-

sponding P as well as the eight constants in each case. A curve showing
the relation between P and η is then constructed so that the solution for
any given P is readily obtained (fig. 9). Details of the calculations have
been omitted but may be found in [27].

If P continues to increase, it is evident from fig. 9 that η decreases
and eventually reaches zero for

$$P = P_2 = 1.114 . \tag{5.15}$$

In other words, for $P > P_2$ the entire stress profile (fig. 8) is outside of
the initial elastic regime and hence is in regime $A\,B$. Thus equation (5.8 b)
applies over the entire shell. The four constants are determined from the
boundary conditions (5.9), and the solution may be written

$$W = -F + (F + \alpha) P \left\{ 1 - 4\mu_1 \lambda \left[\cosh \mu_1 \cos \mu_1 \cosh \mu_1 (1-y) \cos \mu_1 (1-y) \right. \right.$$

$$\left. \left. + \sinh \mu_1 \sin \mu_1 \sinh \mu_1 (1-y) \sin \mu_1 (1-y) \right] \frac{1}{\sinh 2\mu_1 + \sin 2\mu_1} \right\} \tag{5.16}$$

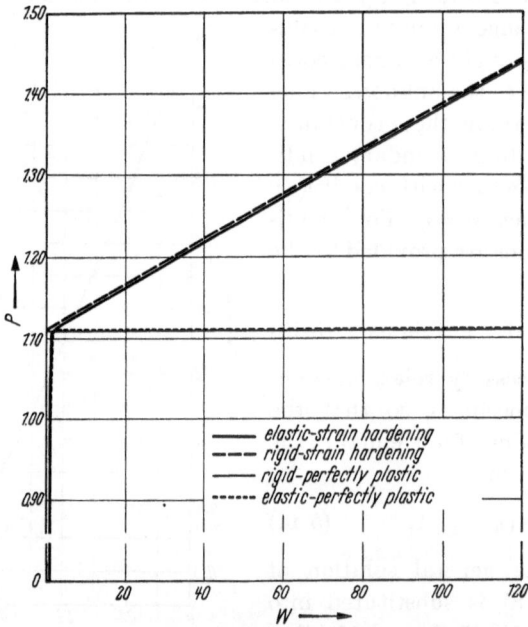

Fig. 10. Maximum displacement as a function of load for various materials

The stresses are now found from equations (5.7) and it is readily verified
that this solution is valid for all $P > P_2$. The maximum displacement
and maximum compressive hoop stress are shown as functions of the
load by the heavy curves in figs. 10 and 11, respectively. Details of the
calculations may again be found in [27].

6. Elastic-perfectly plastic and rigid-strain hardening materials. The treatments in the two preceding sections represent the two extreme cases of considering neither or both of elastic strains and strain hardening. In some applications it may be desirable to consider one but not the other. Naturally the resulting solution will, in general, be more complex that of section 4 and simpler than that of section 5.

The techniques of solution are similar to those of the previous two sections and will not be discussed here. Rather, let us consider qualitatively the relations between the various materials, regarding the elastic-strain hardening material as real and the others as approximations to it. We look first at the maximum displacement at the center of the shell as the load increases. For the real material, this is shown by the heavy curve in

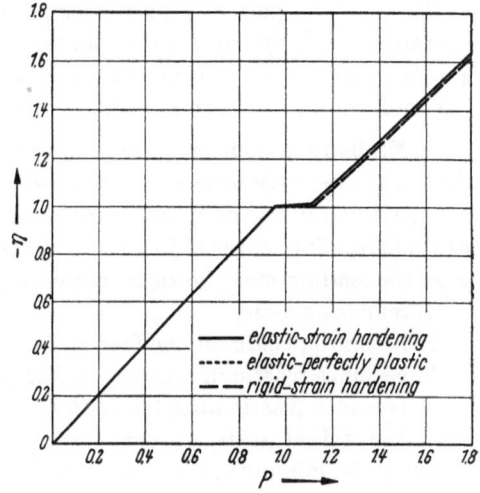

Fig. 11. Maximum hoop stress as a function of load for various materials

fig. 10. When $P < P_1$ the displacement is linear, for $P_1 < P < P_2$ the displacement increases more and more rapidly with load, and for $P > P_2$ a small increment produces an extremely large increase in displacement.

The rigid-perfectly plastic material was discussed in section 4. The load displacement curve for the present example (light curve, fig. 10) consists of two perpendicular straight line segments. For $P < P_0$, $W = 0$, while for $P = P_0$, W can assume any finite value. Despite its crude simplicity, the curve is a reasonable idealization of the exact one. In particular, the collapse load P_0 is close to the load P_2 at which the real deformations begin to run way ($P_0 = 1.111$, $P_2 = 1.114$).

The rigid-perfectly plastic material cannot withstand a load greater than P_0 so that it is obviouslyan inappropriate idealization for large loads. A compromise here is a rigid-strain hardening material. The load displacement curve for such a material is given as a dashedcurve in fig. 10 (details may be found in [27]). It is evidently a very good approximation to the real material for large loads. In particular, the load P_s at which deformations startis close to P_2.

If primary interest is in the range of loads between P_1 and P_2, then neither of the above approximations is useful, but we may consider an elastic-perfectly plastic material. In this case, the maximum allowable load is given by the same value P_0 as if elastic strains were neglected. The resulting curve taken from [27] is given as dotted in fig. 10. To the scale used, or even to a much larger one, the difference is impercetible until P approaches P_0.

Fig. 11 shows the maximum hoop stress as a function of load. Here again, the rigid-perfectly plastic material gives a reasonably accurate overall picture, the rigid-strain hardening comes closer for large loads, and the elastic-perfectly plastic analysis is certainly justified for loads less then P_0. Further comparisons of the solutions may be found in [27].

7. Minimum principles. If the loading on the shell is suitably restricted, then it has recently been shown [18, 19] that the principles of minimum potential energy and minimum complementary energy apply to an elastic-strain hardening material just as in the case of elasticity. For this to be true, the loading must be such that each particle of the shell either

1. remains elastic,
2. becomes plastic on one face and stays there,
3. becomes plastic in a corner and stays there, or
4. becomes plastic on a face and then moves to an adjacent corner.

Both the actual state of stress and all approximate states used in applying the minimum principles must satisfy these four conditions.

The principles are well known in elasticity but they will be stated here in the specific form applicable to cylindrical shells. The first principle states that among all displacement fields which satisfy all displace-

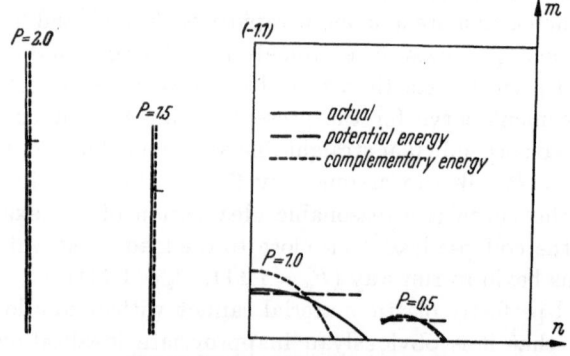

Fig. 12. Stress profiles at various loads

ment boundary conditions, the actual field minimizes the dimensionless potential energy

$$I = \int_0^1 (U - PW)\, dy, \tag{7.1}$$

where

$$U = \int (n\, de + m\, d\varkappa). \tag{7.2}$$

The second principle states that among all states of equilibrated stresses satisfying the stress boundary conditions, the actual state minimizes the complementary energy

$$I_c = \int\limits_0^1 U_c \, dy, \qquad\qquad (7.3)$$

where

$$U_c = \int (e \, dn + \varkappa \, dm). \qquad\qquad (7.4)$$

We will illustrate these principles with reference to the shell previously considered[1] in section 5. A suitable velocity field is

$$W = A + B(1-y)^2, \qquad\qquad (7.5)$$

where A and B are to be determined so as to minimize I. In order to account for the ring loads, equation (7.1) must be rewritten in the form

$$I = \int\limits_0^1 (U - PW) \, dy + \lambda PW(0). \qquad\qquad (7.6)$$

We first assume that the entire shell is elastic and use the stress-strain law (5.3) in order to evaluate (7.2), the result being

$$2U = \alpha n^2 + \beta m^2. \qquad (7.7)$$

Equations (5.3), (7.7), (7.6) and (7.5) can now be used to obtain I as a quadratic function of A and B. Next A and B are found by minimizing I, and hence a complete "solution" has been found. The first dashed curves in fig. 12 shows the resulting stress profile for $P = 0.5$, while the dashed curves in figs. 13 and 14 show the maximum displacement and hoop stress, respectively, as functions of load.

This solution continues to be valid until $n(1) = -1$, the corresponding value of P being

$$P^* = 0.955 \qquad (7.8)$$

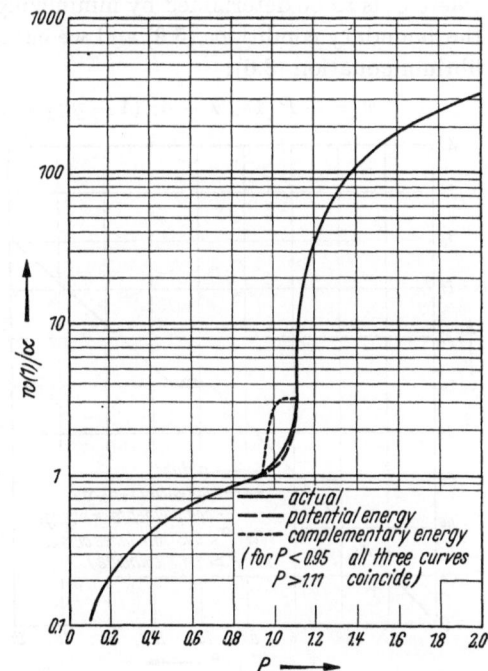

Fig. 13. Maximum displacement as a function of load

---

[1] This same example was treated in [18]. Here we shall give only the highlights of the solution and the interested reader is referred to [18] for more detail.

for the particular shell given by (5.11). For greater values of P, the shell will be elastic for $0 < y < \eta$ and plastic on face $A\,B$ for $\eta < y < 1$, where η will be determined so as to make the stresses continuous. In the plastic region U must be evaluated from (5.7), the result being

$$2\,U = \alpha\,n^2 + \beta\,m^2 + F\,(n^2 - 1)\,. \tag{7.9}$$

Equation (7.7) is still valid for $0 < y < \eta$, so that I can again be evaluated and minimized and a complete solution obtained. Pertinent results are shown by the dashed curves in figs. 12—14.

This solution remains valid so long as $P^* < P < P^{**}$ where

$$P^{**} = 1.116\,. \tag{7.10}$$

For still greater P the entire shell is plastic and equations (7.9) and (5.7) must be used throughout. Pertinent results are again shown by dashed curves in figs. 12—14.

The application of the principle of minimum complementary energy is carried out in similar fashion starting with the bending moment distribution:

$$m = c^2\,\lambda\,(2\,y - y^2)\,[P\,(1 - y)^2 + C\,(2\,y - y^2)]\,, \tag{7.11}$$

where C is to be determined by minimizing I_c. Obviously (7.11) satisfies the boundary conditions (5.9) and we choose n so as to satisfy the equilibrium equation (3.6):

$$n = -\,P\,[1 + \lambda - 6\,\lambda\,(1 - y)^2] + 2\,C\,\lambda\,[1 - 3\,(1 - y)^2]\,. \tag{7.12}$$

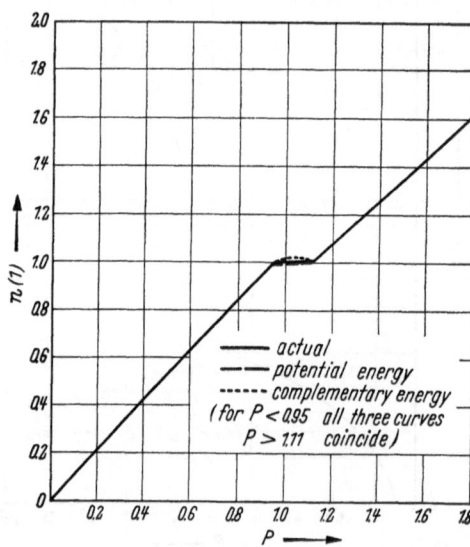

The procedure is now quite similar to the application of minimum potential energy. It is found that the predictions are elastic for $0 < P < P_c^*$, elastic-plastic for $P_c^* < P < P_c^{**}$ and fully plastic for $P_c^{**} < P$ where

$$\left.\begin{aligned} P_c^* &= 0.952, \\ P_c^{**} &= 1.115. \end{aligned}\right\} \tag{7.13}$$

Pertinent results are shown by the dotted curves in figs. 12—14. Also shown are the exact solution data from section 5. The significance of the curves will be commented on in the concluding section of the paper.

Fig. 14. Maximum compressive stress as a function of load

8. Conclusion and limitations. In the preceding paper we have described a theory of plasticity which endeavors to strike a reasonable compromise between the opposing demands of physical reality and mathematical simplicity. Following a basic description of the elements to be used in the theory and of the physical problem to be illustrated, the development took place in two parts. First the theory itself was expounded and illustrated both for the general case and for various specific cases. Second, we considered techniques for approximating the solution for a strain hardening material.

The ultimate evaluation of any theory must rest upon a detailed comparison with experiment. Lacking this, we can still give a general indication of some of the limitations. With regards to the existence of an initial yield function, there seems to be general agreement. However, not only is there lack of agreement as to the precise form of this function, but there are certain indications [28, 29] that some properties of the solution may be seriously affected by slight changes in the yield condition. On the other hand, it is a consequence of the theorems of limit analysis that any lower bound on the collapse load based upon an approximate yield surface which lies wholly within the actual surface is a lower bound on the actual collapse load, while any upper bound based upon an approximate yield surface wholly without the actual surface is an actual upper bound. Thus small changes in the yield surface produce predictable changes in the collapse load and in general these are small. Obviously further investigation in this direction is needed. In particular, the effect of approximations on the displacement solution is unknown.

However, the major point of modern disagreement is not with the initial yield condition or with the flow law, but rather with the nature of hardening. If the loading is prescribed in such a way that the strain rate vector has a constant direction, then all three of the theories illustrated in fig. 3 yield the same predictions. However, if the strain rate vector has a variable direction the theories will all differ, particularly if the loading involves more than one face. The case of unloading and reloading in the opposite direction is particularly significant. In this case PRAGER's theory is probably the most realistic since it takes account of the well known BAUSCHINGER effect.

Certainly some critically designed experiments would be most welcome in this connection. In view of the fact that the piecewise linear theory is the most easily manipulated, such experiments should endeavor not only to decide the truth or falsity of the various theories, but also conditions under which the predictions of one theory may reasonably be applied to the solutions of another.

Returning to the strain rate vector, HANDELMAN and WARNER [30] have shown that the case where this has a constant direction is the only one where a flow law without singularities can be integrated. For the piecewise linear theory here presented, more generality is allowed, since the flow law can be integrated so long as the particle remains in a particular plastic regime. As a matter of fact, the flow law can be "piecewise integrated" with respect to any number of regimes. This process eliminates the time dependence but introduces additional unknowns in the form of the stress intensity at which the transfer takes place. These remarks apply to both the general and special theories of sections 4, 5 and 6. For the minimum principles of section 7, the loading must be further restricted as indicated.

At this point, let us assume that we are agreed upon the applicability of the general piecewise linear theory of plasticity as presented in section 5, but that the resulting mathematical problem is intractable. We can obtain an approximate solution either by using one of the simpler theories or by using a minimum principle. On the basis of the single example investigated, it appears that either method can be used to advantage. For loads such that the shell is partially plastic, the elastic-perfectly plastic analysis gives results very close to the exact, as do the energy approximations. In particular, as might be expected, the potential energy approach gives a better estimate of the displacement (fig. 13), and the complementary energy predicts the stresses more accurately (figs. 12, 14).

If the load is such that the shell is fully plastic the result is even more startling. In this case the rigid-strain hardening material is certainly a reasonable one while both energy approximations coincide with the actual solution to the scale used.

In conclusion, then, the theory of piecewise linear plasticity represents an attempt to provide a theory which is both mathematically tractable and physically realistic. Compromises have been made in both directions, but it is hoped that the net result may be of some value to the engineer who must find some solution to problems of inelastic behavior.

References

[1] PHILLIPS, ARIS: J. appl. Mechan. 19, 496 (1952).
[2] STOCKTON, F. D. and D. C. DRUCKER: J. Colloid. Sc. 5, 239 (1950).
[3] TRESCA, H.: Mém. pres. par div. sav. 18, 733 (1868).
[4] MISES, R. VON: Göttinger Nachr. math.-phys. Kl., S. 582 (1913).
[5] BATDORF, S. B. and BERNARD BUDIANSKY: A mathematical theory of plasticity based on the concept of slip, NACA TN 1871, 1949.
[6] PRAGER, W.: The theory of plasticity — a survey of recent achievements. James Clayton Lecture, Inst. Mech. Engrs. London 1955.
[7] PRANDTL, L.: Proc. 1st Internat. Congr. appl. Mechan., Delft 1924, p. 43.
[8] REUSS, E.: Z. angew. Math. Mech. 10, 266 (1930).
[9] MISES, R. VON: Z. angew. Math. Mech. 8, 161 (1928).
[10] HILL, R.: Phil. Mag. 40, H. 7, 971 (1949).
[11] DRUCKER, D. C.: Quart. appl. Math. 7, 411 (1950).
[12] DRUCKER, D. C.: Proc. 1st U. S. Nat. Congr. Appl. Mech., Chicago 1951, p. 487.

[13] DRUCKER, D. C., W. PRAGER, and H. J. GREENBERG: Quart. appl. Math. 9, 381 (1952).

[14] KOITER, W. T.: Quart. appl. Math. 11, 350 (1953).

[15] PRAGER, W.: J. appl. Mechan. 20, 317 (1953).

[16] SANDERS JR., J. L.: Proc. 2nd. U. S. Nat. Congr. Appl. Mech., Ann Arbor 1954, p. 455.

[17] SANDERS JR., J. L.: GDAM Report No. A 11–106, Brown University, Providence, R. I., 1954.

[18] HODGE JR., P. G.: PIBAL Report No. 297, Polytechnic Inst. of Brooklyn, Brooklyn, New York 1955.

[19] HODGE JR., P. G.: PIBAL Report No. 298, Polytechnic Inst. of Brooklyn, Brooklyn, New York 1955.

[20] PRAGER, W.: Proc. 8th Inter. Congr. Appl. Mech., Istanbul 1952, p. 65.

[21] TIMOSHENKO, S.: Theory of plates and shells. New York 1940.

[22] DRUCKER, D. C.: Proc. 1st Midw. Conf. Solid Mech., Urbana 1953, p. 158.

[23] ONAT, E. T.: Quart. appl. Math. 13, 63 (1955).

[24] HODGE JR., P. G.: J. appl. Mechan. 21, 336 (1954).

[25] HODGE JR., P. G.: J. appl. Mechan. 23, 73 (1956).

[26] HOPKINS, H. G. and W. PRAGER: J. Mech. Phys. Solids 2, 1 (1953).

[27] HODGE JR., P. G. and F. ROMANO: J. Mech Phys. Solids 4 (1956).

[28] HODGE JR., P. G.: J. Math. Phys. 29, 38 (1950).

[29] HILL, R.: J. appl. Mechan. 21, 93 (1954).

[30] HANDELMAN, G. H. and W. H. WARNER: J. Math. Phys. 33, 157 (1954).

Contribution to Discussion of the Paper by P. G. Hodge jr.

By **H. G. Hopkins,** Sevenoaks (Kent)

In recent years there has been a definite trend towards a preference for the TRESCA yield criterion, as opposed to that of VON MISES, for a material under combined stress, in the theoretical treatment of problems in plasticity. An essential feature of the TRESCA yield criterion is the occurrence of singular points, i.e. points on the yield surface where there is no unique normal. It is therefore of considerable interest that some recent work of J. F. W. BISHOP has shown that the occurrence of such points may be justified on the basis of the physics of single metal crystals. In many problems in plasticity, the adoption of a piecewise linear yield criterion, such as for example that of TRESCA, leads to simplifications in the mathematical analysis. Indeed, it is well-known that such simplifications can be very considerable, and in fact some problems that have been solved in this way would otherwise be almost intractable. Now the yield condition is not always expressed directly in terms of stresses. It is often convenient to introduce an approximate yield criterion expressed in terms of generalised stresses. This procedure is usually adopted in discussing problems involving engineering structural elements such as plates and shells. Even if the original yield criterion is piecewise linear, this is not necessarily true of the corresponding yield criterion expressed in terms of generalised stresses. In Professor HODGE's theory, it is assumed in such cases that this latter yield criterion is approximated by one that is piecewise linear. Thus the yield criterion is now taken to be represented by a polyhedron in generalised stress space. Although the errors introduced by this approximation may always be reduced through refinements in the yield polyhedron, and may in some cases be estimated, the procedure has the obvious disadvantage of introducing an *ad hoc* yield criterion with no direct interpretation except in terms of a hypothetical material. On the other hand, the procedure has the same striking advantages that occur through the adoption of the Tresca yield criterion in place of that of von Mises. This situation is well illustrated in a wide variety of problems that have been discussed in published work. In his paper, Professor Hodge has given a very clear discussion of this approach within a very general framework, and has amply illustrated the general situation through detailed study of a particular problem. The term "piecewise linear theory" concisely expresses the essential idea in this approach.

Contribution to Discussion of the Paper by P. G. Hodge jr.

By E. H. Lee, Providence (Rhode Island)

Professor Hodge has clearly brought out the value of the use of the Tresca yield condition and associated flow rule in obtaining solutions of boundary value problems of plastic flow. In terms of Lode's variables, this flow criterion appears to differ considerably from the experimental results of Taylor and Quinney: Phil. Trans. Roy. Soc., Lond., Ser. A **230**, 323 (1931), and for this reason in the past the Mises flow law has been preferred in theoretical work. However, some recent solutions based on the Tresca criterion have been checked experimentally (J. Foulkes and E. T. Onat, Tests of behaviour of circular plates under transverse load, Brown University Report DA—3172/3, May 1955) and have been found to give remarkable agreement with experiment. In the problem of a loaded circular plate, the characteristics change from radial lines to logarithmic spirals at a certain radius and such a change was observed in markings on the plate surface. It seems therefore that solutions obtained by this method are more satisfactory than had been anticipated. The reason for this may be due to concentration of the corresponding points in the Lode diagram in the region of the diagonal in problems of non-homogeneous stress and strain, due to the freedom of the strain rate vector at the corners of the Tresca yield hexagon. It would seem worthwhile to look into this, for as Professor Hodge has shown, this technique offers an extremely powerful means of solution of plastic flow problems, which will be the more valuable when the basis for its accuracy is better understood.

Nueva teoría de la plasticidad

Por M.Velasco de Pando, Madrid

La teoría clásica de la plasticidad, en la que sobresalen los nombres de SAINT VENANT, LEVY, REUSS, VON MISES, etc. admite que durante los procesos plásticos se cumple una condición de fluencia, es decir, que una función de las tensiones permanece constante. Los cuerpos que cumplen esta condición se llaman «plásticos perfectos». La condición de fluencia que generalmente se admite hoy, es la siguiente:

$$(\sigma_1 - \sigma_2)^2 + (\sigma_2 - \sigma_3)^2 + (\sigma_3 - \sigma_1)^2 = \text{const},$$

llamada criterio de VON MISES.

Se deduce de lo expuesto que la teoría clásica de la plasticidad solo se aplica a los plásticos perfectos. Pero los metales, aun los metales dúctiles, los cuerpos más importantes para las aplicaciones no son plásticos perfectos. Todas las experiencas que podemos realizar conducen al resultado de observar un aumento del discriminante de plasticidad durante los procesos plásticos monótonamente crecientes; así ocurre por ejemplo, en la probeta hueca estirada y en el tubo de pared estrecha sometido a torsión; incluso ocurre también en las probetas macizas de tamaño normal, si bien en éstas el fenómeno es más complejo por la existencia de tensiones radiales, circunferenciales y aun tangenciales, éstas últimas fuera de la sección estringida.

Estos hechos se recogen por los autores de plasticidad, como por ejemplo, por PRAGER, diciendo que cuando hay aumento de las tensiones es que existe «ecrouissage» «hardening» es decir, endurecimiento. No parece que para estos cuerpos endurecibles se posea una teoría totalmente satisfactoria. Es cierto que PRAGER indica diversas posibilidades, pero no se ha llegado, que sepamos, a une teoría completa y desarrollada.

Por lo expuesto he creído oportuno el momento para ensayar una nueva teoría de la plasticidad de estos cuerpos endurecibles, la cual he desarrollado en varios artículos publicados en revistas técnicas españolas, en mi librito «Plasticidad» y en un folleto en inglés que he entregado a los Profesores concurrentes a este Coloquio.

Fundamento mi teoría en el siguiente postulado: «Si un proceso de plastificación se divide en tramos pequeños, hay en cada tramo una correspondencia biunívoca entre los incrementos o variaciones de las tensiones y los incrementos de las deformaciones.»

Conviene aclarar lo que entendemos por división en tramos pequeños de un proceso de plastificación. Una transformación plástica se realiza durante un cierto tiempo, de modo que en la descripción del fenómeno intervendrán las tres coordenadas x, y, z de un punto del cuerpo, supuesto identificable, que serán funciones del tiempo t.

Por tanto, las ecuaciones dependerán de cuatro variables independientes, a saber, las tres variables espaciales x, y, z y el tiempo t.

Pues bien, efectuemos un cambio de variable sobre t introduciendo un parámetro λ tal que $\lambda = f(t)$.

Este parámetro λ lo tomamos proporcional a la *acción determinante* del proceso plástico, la cual puede ser de dos clases:

1a: un corrimiento forzado de ciertos puntos de la superficie,

2a: una carga exterior variable.

Ejemplos de la primera clase tenemos en la probeta estirada y en la masa comprimida entre dos platos, en los cuales tomamos λ proporcional a la distancia entre las cabezas de la probeta o al acercamiento de los platos.

Ejemblos de la clase segunda, tenemos en la flexión de una viga con momento flexor variable en el tiempo, o un tubo con presión interior también variable.

Unas veces el parámetro λ tiene un crecimiento monótono durante el proceso de plastificación y otras puede tomar valores oscilantes.

Lo que sí admitimos es que la primera derivada $f'(t)$ es pequeña, es decir, que el proceso es lento y que la segunda derivada $f''(t)$ es despreciable, con lo cual no hay que tener en cuenta las aceleraciones y resulta una sucesión de estados de equilibrio.

Pues bien, un proceso elemental es el que corresponde a una variación pequeña $\delta\lambda$ del parámetro λ, a la cual corresponderían pequeñas variaciones δu, δv, δw de los corrimientos de un punto, de las tensiones desarrolladas $\delta\sigma_x$, $\delta\sigma_y$, $\delta\sigma_z$, $\delta\tau_{xy}$, ... y de las deformaciones $\delta\varepsilon_x = \partial\,\delta u/\partial x$, ...

El postulado fundamental se traducirá por ciertas ecuaciones entre los incrementos de las tensiones, y de las deformaciones, ecuaciones que a causa de la pequeñez de los incrementos serán lineales, y que, teniendo en cuenta el «efecto Poisson», podrían escribirse en la siguiente forma:

$$\delta\varepsilon_x = \frac{\partial\,\delta u}{\partial x} = \frac{\delta\sigma_x}{T_1} - \frac{\delta\sigma_y}{Q_2} - \frac{\delta\sigma_z}{Q_3},$$

$$\frac{\partial\,\delta v}{\partial y} = \frac{\delta\sigma_y}{T_2} - \frac{\delta\sigma_z}{Q_3} - \frac{\partial\sigma_x}{Q_1},$$

$$\vdots$$

$$\cdot\quad\cdot\quad\cdot\quad\cdot\quad\cdot\quad\cdot\quad\cdot\quad\cdot$$

$$G_1\left(\frac{\partial\,\delta u}{\partial y} + \frac{\partial\,\delta v}{\partial x}\right) = \delta\tau_{xy} \quad \text{etc.}$$

En general los parámetros T_1, T_2, ... Q_1, Q_2, ... que intervienen en estas ecuaciones serán funciones de las tensiones en el momento actual y de la historia de las deformaciones. Por tanto, no hay entre estas ecuaciones y las de LEVY-REUSS que se admiten en la plasticidad clásica, una diferencia tan sustancial puesto que los incrementos de las deformaciones dependen de las tensiones así como de los incrementos de éstas; esto último es lo que no ocurre en la plasticidad clásica.

Hay dos casos muy importantes en los que las experiencias realizadas por el Profesor STÜSSI, de Zurich, nos permiten ercribir las ecuaciones anteriores en forma aproximada. Dichas experiencias han consistido en producir estados variables de tensión en tubos de pared delgada y han conducido a la conclusión de que estableciendo un estado plástico por una gran tensión normal plastificante, la adición de otras tensiones normales o tangenciales produce deformaciones que están con las tensiones adicionales introducidas como en periodo elástico.

En otra serie de experiencias de STÜSSI se produce la plastificación por una fuerte tensión tangencial y entonces otras tensiones producen aumentos de las deformaciones como en periodo elástico.

De estas experiencias se deduce que en el caso de una fuerte tensión normal las ecuaciones que traducen nuestro postulado pueden escribirse en la siguiente forma:

$$\frac{\partial \delta u}{\partial x} = \frac{\delta \sigma_x}{T} - \frac{\delta \sigma_y}{m E} - \frac{\delta \sigma_z}{m E},$$

$$\frac{\partial \delta v}{\partial y} = \frac{\delta \sigma_y}{E} - \frac{\delta \sigma_z}{m E} - \frac{\delta \sigma_x}{Q},$$

$$\vdots$$

$$\cdot \quad \cdot \quad \cdot \quad \cdot \quad \cdot \quad \cdot \quad \cdot \quad \cdot$$

$$G_0 \left(\frac{\partial \delta u}{\partial y} + \frac{\partial \delta v}{\partial x} \right) = \delta \tau_{xy} \quad \text{etc.}$$

Los coeficientes T, Q de estas ecuaciones son en general, funciones de la tensión σ_x, pero estas funciones varían muy moderadamente y, en primera aproximación, pueden considerarse constantes.

En el caso de plastificación producida por una fucrte tensión tangencial tenemos una ecuación del tipo

$$G \left(\frac{\partial \delta u}{\partial y} + \frac{\partial \delta v}{\partial x} \right) = \delta \tau_{xy}$$

y las demás pueden tomarse como en periodo elástico.

Conviene observar que estas ccuaciones revelan una anisotropia cilíndrica, producida por la propia plastificación y aun suponiendo que el material fuese isótropo en el pcríodo elástico y antes de producirse deformaciones plásticas.

Se supone en las anteriores ecuaciones que el incremento de la tensión plastificante es del mismo signo que ésta, es decir, que la tensión plastificante crece en valor absoluto. Si, por el contrario, decrece, se aplican las curvas de descarga.

En cada tramo $\delta \lambda$ dada la superficie al empezar el tramo y las condiciones de carga, se produce un estado determinado de equilibrio, según es fácil ver por el número y aspecto de las ecuaciones de que disponemos.

Un proceso de plastificación habrá que estudiarlo por tramos, partiendo del periodo elástico o, si se quiere, simplificando éste suponiendo el cuerpo rígido hasta que comience la plastificación. De un tramo a otro varía la forma de la superficie circunstancia que influye más o menos en la marcha del fenómeno.

Para plastificaciones moderadas, puede admitirse que T y Q son constantes y que las variaciones de las coordenadas x, y, z, del punto M cualquiera son pequeñas, con lo cual se puede demostrar que

$$\sum \frac{\partial \delta u}{\partial x} = \frac{\partial u}{\partial x}$$

y se obtienen las ecuaciones siguientes entre tensiones y deformaciones:

$$\left. \begin{aligned} \frac{\partial u}{\partial x} &= \frac{\sigma_x - \sigma_E(1-h)}{T} - \frac{\sigma_y}{mE} - \frac{\sigma_z}{mE}, \\ \frac{\partial v}{\partial y} &= \frac{\sigma_y}{E} - \frac{\sigma_x - \sigma_E(1-h)}{Q} - \frac{\sigma_z}{mE}, \\ \vdots \\ \cdots \cdots \cdots \cdots \cdots \end{aligned} \right\} \quad \left(h = \frac{T}{E} \right)$$

$$G_0 \left(\frac{\partial u}{\partial y} + \frac{\partial v}{\partial x} \right) = \tau_{xy}.$$

Estas relaciones suponen que la relación tensión-dilatación se toma en forma de dos rectas, estilizando la curva experimental, y que se ha seguido para la deéformación un camino monótono determinado.

De otro modo, es sabido que no pueden en plasticidad admitirse ecuaciones determinadas entre tensiones y deformaciones.

Otros autores como HENCKY han admitido también relaciones entre tensiones y deformaciones con análogas hipótesis. Porque, insistimos, en que en el periodo plástico no pueden existir, en general, relaciones finitas entre tensiones y deformaciones.

Por el método indicado he estudiado los siguientes ejemplos:

1. De plastificaciones producidas por una fuerte tensión normal: flexión de las vigas, con o sin esfuerzo cortante; envolventes cilíndricas; cuña; cuchilla; masa plástica de gran anchura comprendida entre dos platos; probeta rectangular estirada; plastómetro de disco; probeta cilíndrica estirada; laminación de barra rectangular; trefilado.

2. Plastificación producida por una tensión tangencial; torsión de barras cilíndricas o de forma cualquiera.

En los casos de dos dimensiones las ecuaciones diferenciales obtenidas son de tipo hiperbólico, es decir, de características reales.

Un método aproximado que permite obtener muy rápidamente la marcha general de los fenómenos consiste en despreciar en las ecuaciones los términos que tienen E en denominador.

Con esto en dos dimensiones se llega a la ecuación de la cuerda vibrante ly en 3 dimensiones a la ecuación de las ondas. En ambas el papel del tiempo lo desempeña la coordenada dirigida según la dirección de la tensión plastificante.

Otro método expedito lo proporciona el empleo de las dilataciones ogarítmicas, tomando

$$\sum \frac{\partial \, \delta \, u}{\partial \, x} = \ln \left(1 + \frac{\partial \, u}{\partial \, x_0} \right).$$

Con este método puede estudiarse de una vez una plastificación profunda.

Respecto a mi teoría, la clásica, no es un caso particular sino en caso singular, pues en un diagrama funcional, el caso de la recta horizontal significa que ha desaparecido la relación funcional.

Estoy dispuesto a admitir que mi teoría tenga todavía imperfecciones pues desde 1950 en que empecé a estudiar con ella ciertos casos particulares, he visto que muchas cosas pueden perfeccionarse. Pero hay que tener en cuenta que en la teoría clásica de la plasticidad han trabajado muchos y eminentes cerebros y en la mia solo ha trabajado el mio modestísimo y que, como dice BELL, autor norteamericano de una «Historia de las matemáticas», a veces una teoría es imperfecta porque ni está muerta ni es inútil.

The Theory of Deformation of Non-hardening Rigid-Plastic Plates under Transverse Load

By **H. G. Hopkins**, Sevenoaks (Kent)

With 2 figures

1. Introduction. Only recently has the study of the bending of thin plates been made on the basis of the incremental theory of plasticity. The first approach leading to the exact solution of problems of this type was made by HOPKINS and PRAGER [1] who discussed the quasi-static yielding of a circular plate subjected to transverse load under conditions of rotational symmetry. The plate material was taken to be non-hardening rigid-plastic and to obey the yield condition and flow rule of TRESCA. The basic assumptions made by these authors were similar to those of the conventional engineering elastic theory of thin plates (see, for example, TIMOSHENKO [2]) and the effect of transverse shear strain-rate was neglected. A notable feature was that, within the framework of the theory, exact solutions were obtained to certain important problems such as, for example, those involving either uniformly-distributed or concentrated loads. The solution of similar problems for other yield conditions, in particular that of VON MISES, has been studied by HOPKINS and WANG [3], and for combined loads has been studied by DRUCKER and HOPKINS [4]. The problem of the limits of economy of material has been studied by HOPKINS and PRAGER [5]. The above studies all refer to quasi-static problems, and the extension of the fundamental analysis to dynamic problems was first made by HOPKINS and PRAGER [6], the effect of rotatory inertia being neglected. Again it was notable that exact solutions could be obtained to certain problems. Thus the complete solution was obtained for a simply-supported circular plate subjected to a uniformly-distributed load which is suddenly applied and, after a certain time, suddenly removed. The solution of certain impulsive load problems for simply-supported and built-in edge conditions has been given by WANG [7] and by WANG and HOPKINS [8], respectively. The problems treated theoretically by HOPKINS and PRAGER [1] were first investigated experimentally by HAYTHORNETHWAITE [9]; for subsequent experimental work see the references cited by PRAGER [10]. An extended summary of the greater part of the work contained in [1–9] is given by PRAGER [10].

Finally, PRAGER [11] has given a method for solving statical problems when the plate material is work-hardening rigid-plastic.

Hitherto attention has been confined almost exclusively to the case of rotational symmetry. The exceptions are the introductory analysis by PRAGER and HODGE [12] and the treatment of discontinuities by PRAGER [13] for plates made of VON MISES material, and the discussion of the minimum weight design of simply-supported convex plates by PRAGER [14] for the case of TRESCA material. In problems of the present type, and indeed fairly generally, the use of the TRESCA, rather than say the VON MISES, yield condition and associated flow rule results in striking mathematical simplifications. For example, in the solution of the problem of the uniformly-loaded, simply-supported, circular plate, PELL and PRAGER [15] adopting VON MISES material could only obtain an approximate solution through the use of the limit design theorems, whereas HOPKINS and PRAGER [1] adopting TRESCA material easily found the exact solution. The purpose of the present paper is to remove the restriction to conditions of rotational symmetry. For the reason just stated TRESCA material is adopted.

This paper will present an introduction to the general theory of the small deformations of a thin uniform plate under transverse load. The plate is made of non-hardening rigid-plastic material obeying the yield condition and flow rule of TRESCA. The objective is to parallel existing theories of certain plastic fields such as, for example, the simpler one of plane strain. The approach resembles closely that adopted in the conventional engineering elastic theory of thin plates. Thus the basic assumptions of the plastic theory are similar to those made in the elastic theory, and here the effects of transverse shear strain and rotatory inertia will be neglected. The field equations involve the stress-moments and the middle-surface curvature-rates as the generalized stresses and strain-rates. These equations are given first in terms of rectangular cartesian co-ordinates. Now the directions of principal stress-moment are likely to be mathematically-preferred. This conjecture has been proved correct for certain plastic regimes but this is not discussed however in the present paper. Therefore the governing equations are rewritten in terms of orthogonal curvilinear co-ordinates taken along these directions.

2. The field equations in cartesian co-ordinates. It is convenient to think of the plate as being horizontal. Let $O(x_1, x_2, x_2)$ be a right-handed rectangular cartesian frame of reference, the origin O being taken in the middle surface and the x_3-axis being directed vertically upwards. Let $2h$ be the plate thickness.

For simplicity, standard rectangular cartesian tensor notation is used. All tensors are either three- or two-dimensional, their components being defined with respect to the frame $O(x_1, x_2, x_3)$ or $O(x_1, x_2)$, and being distinguished through the use of suffixes i, $j = 1$, 2, 3 or $k, l = 1, 2$, respectively.

Let σ_{ij} and v_i denote the three-dimensional stress tensor and velocity vector, respectively. Then if ε_{ij} is the strain-rate tensor,

$$\varepsilon_{ij} = \frac{1}{2} (v_{j,i} + v_{i,j}). \tag{1}$$

If body forces are negligible then the stresses satisfy the equations of motion

$$\sigma_{ij,j} = \varrho\, v_{i,t} \tag{2}$$

where ϱ is the density of the material and t is the time. The material is assumed to be homogeneous and isotropic and not to work-harden. Then the yield condition is

$$f = f(\sigma_{ij}) = 0, \tag{3}$$

where the scalar f is negative or zero for stress-states below or at yield, respectively, and f only involves invariants of the stress tensor. According to the theory of the VON MISES plastic potential, the flow rule is now

$$\varepsilon_{ij} = \begin{cases} \lambda \dfrac{\partial f}{\partial \sigma_{ij}} & \text{if} \quad f = 0 \quad (\lambda \geq 0), \\ 0 & \text{if} \quad f < 0, \end{cases} \tag{4}$$

where f is regarded as a function of *nine* independent quantities σ_{ij} in forming the partial derivatives $\partial f / \partial \sigma_{ij}$.

The procedure is now to approximate (*1*) to (4) within the framework of the assumptions of thin plate theory. There are the following four main assumptions.

a) Linear elements initially normal to the middle surface preserve this normality during plastic deformation.

b) The stresses acting in planes normal to the middle surface are more significant in governing transverse displacement than are the stresses acting in planes parallel to the middle surface.

c) The yield function is that of TRESCA; the yield function is also the plastic potential.

d) The effects of transverse shear strain-rate and rotatory inertia may be neglected.

In thin plate theory attention is given only to quantities defined in the middle surface. Here, these quantities are the stress-resultant vector

$$Q_k = \int_{-h}^{h} \sigma_{3k}\, dx_3, \tag{5}$$

the stress-moment tensor

$$M_{kl} = \int_{-h}^{h} \sigma_{kl}\, x_3\, dx_3, \tag{6}$$

and the curvature-rate tensor

$$\varkappa_{kl} = - v_{,kl} \qquad [v = v_3(x_1, x_2, 0)] \tag{7}$$

where v is the middle-surface transverse velocity.

The fundamental premise is that the full three-dimensional equations (1) to (4) involving σ_{ij} and ε_{ij} may be adequately approximated by certain two-dimensional equations involving Q_k, M_{kl} and $\varkappa_{kl}$. The derivation of the final equations is straightforward (see [1], [2] and [12]) and therefore the details will be omitted.

The equations of motion are

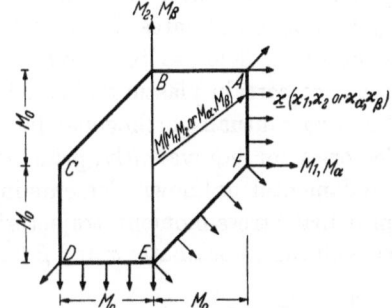

Fig. 1. TRESCA yield condition

$$Q_{k,k} + p = 2\varrho h \frac{\partial v}{\partial t}, \tag{8}$$

$$M_{kl,l} - Q_k = 0, \tag{9}$$

where p is the upwards-acting pressure applied to the plate, and hence

$$M_{kl,kl} + p = 2\varrho h \frac{\partial v}{\partial t}. \tag{10}$$

The yield condition, expressed in terms of principal bending moments M_1 and M_2, is

$$\text{max.} \left| M_1 - M_2, M_1, M_2 \right| = M_0, \tag{11}$$

M_0 being the fully-plastic bending moment $\sigma_0 h^2$ where σ_0 is the tensile yield stress. The yield condition is represented by the hexagon $ABCDEF$ of fig. 1 drawn in a plane in which M_1 and M_2 are taken as rectangular cartesian co-ordinates. Let the yield condition be written

$$f = f(M_{kl}) = 0 \tag{12}$$

where the scalar f is negative or zero for stress-states below or at yield, respectively. Then the flow rule is

$$\varkappa_{kl} = \begin{cases} \lambda \dfrac{\partial f}{\partial M_{kl}} & \text{if} \quad f = 0 \quad (\lambda \geq 0), \\ 0 & \text{if} \quad f < 0, \end{cases} \tag{13}$$

where f is regarded as a function of *four* independent quantities M_{kl} in forming the derivatives $\partial f/\partial M_{kl}$. In connection with (11) note that

$$M_1, M_2 = \frac{1}{2}(M_{11} + M_{22}) \pm \frac{1}{2}[(M_{11} - M_{22})^2 + 4M_{12}^2]^{1/2} \tag{14}$$

where M_1 and M_2 are new distinguished by the condition

$$M_1 \geq M_2. \tag{15}$$

The vertices of the yield hexagon are singular points, and at such a point f is to be replaced by

$$(1 - q)f_1 + qf_2 \quad (0 \le q \le 1) \tag{16}$$

where q is otherwise arbitrary and f_1 and f_2 are the yield functions in the immediate vicinity of the singular point.

3. The field equations in curvilinear co-ordinates. The field equations will now be alternatively expressed in terms of an orthogonal curvilinear system of co-ordinates. The directions of principal stress-moment at any point of the field are likely to be mathematically-preferred directions, at least for certain plastic regimes. In this event, attention is focussed on the two mutually-orthogonal families of principal stress-moment trajectories, or equivalently, principal curvature-rate trajectories, as the fundamental unknown elements in the problem. Let the two families of principal stress-moment trajectories be identified through orthogonal curvilinear co-ordinates α and β (see fig. 2). Here $\alpha(x_1, x_2)$ and $\beta(x_1, x_2)$, together with their first order derivatives, are supposed continuous functions such that

$$\frac{\partial \alpha}{\partial x_1} \frac{\partial \beta}{\partial x_1} + \frac{\partial \alpha}{\partial x_2} \frac{\partial \beta}{\partial x_2} = 0. \tag{17}$$

The exact nature of α and β does not need to be specified here, and these co-ordinates may be chosen to suit the particular problem in hand.

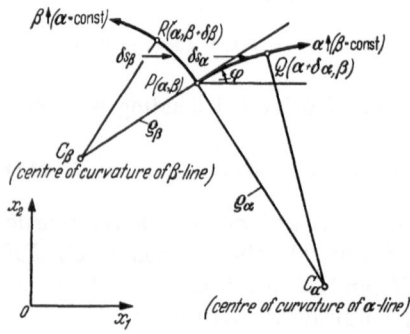

Fig. 2. Geometry of the α, β-co-ordinate system

A member of one family of principal stress-moment trajectories is defined by an equation $\beta = $ const. and is called an α-line; and similarly a member of the other family is defined by an equation $\alpha = $ const. and is called a β-line. The two families are distinguished through the choice

$$M_\alpha = M_1, \quad M_\beta = M_2 \quad (M_\alpha \ge M_\beta) \tag{18}$$

where M_α and M_β are the stress-moments acting on elements normal to the lines $\alpha = $ const. and $\beta = $ const., respectively. The directions of principal stress-moment are of course unique except in the case of stress-moment isotropy, and then one of the plastic regimes A and D must apply.

The fundamental relationships will now be developed in terms of the geometry of the α, β-co-ordinate system. In general, the α- and β-lines are unique. It is necessary only to specify positive directions along these lines. If a right-handed convention is adopted for the moving frame formed by the directions of increasing values of α, β and x_3, then there are only two ways in which this can be done. It is supposed that one of these two ways is chosen.

Let $\varphi(\alpha, \beta)$ be the anti-clockwise angle of rotation required to bring the positive x_1-direction into coincidence with the positive tangent to the α-line through the point $P(\alpha, \beta)$. Let

$$\delta s_\alpha = h_\alpha(\alpha, \beta)\,\delta\alpha\,, \quad \delta s_\beta = h_\beta(\alpha, \beta)\,\delta\beta \quad (h_\alpha, h_\beta > 0) \tag{19}$$

be the elements of length on the α- and β-lines, respectively. Let c_α, c_β and ϱ_α, ϱ_β be the (algebraic) curvatures and radii of curvature of the α- and β-lines, respectively, where

$$c_\alpha = \frac{1}{\varrho_\alpha} = -\frac{\partial\varphi}{\partial s_\alpha} = -\frac{\partial\varphi}{h_\alpha\,\partial\alpha}\,, \quad c_\beta = \frac{1}{\varrho_\beta} = \frac{\partial\varphi}{\partial s_\beta} = \frac{\partial\varphi}{h_\beta\,\partial\beta} \tag{20}$$

(see fig. 2). The orthogonality of the co-ordinate axes requires, from simple geometry, that

$$\frac{\partial h_\alpha}{\partial\beta} = h_\alpha h_\beta c_\alpha\,, \quad \frac{\partial h_\beta}{\partial\alpha} = h_\alpha h_\beta c_\beta\,. \tag{21}$$

Moreover, since

$$\frac{\partial^2\varphi}{\partial\alpha\,\partial\beta} = \frac{\partial^2\varphi}{\partial\beta\,\partial\alpha}\,,$$

it follows that c_α and c_β satisfy the Lamé condition

$$\frac{\partial c_\alpha}{\partial s_\beta} + \frac{\partial c_\beta}{\partial s_\alpha} + c_\alpha^2 + c_\beta^2 = 0\,. \tag{22}$$

Thus h_α, h_β, c_α and c_β are not all independent functions of α and β but must satisfy the three differential equations (21) and (22).

The principal stress-moments M_α and M_β and the vertical shear forces Q_α and Q_β satisfy the equations of motion

$$\frac{\partial Q_\alpha}{\partial s_\alpha} + \frac{\partial Q_\beta}{\partial s_\beta} + c_\beta Q_\alpha + c_\alpha Q_\beta + p = 2\varrho\,h\,\frac{\partial v}{\partial t}\,, \tag{23}$$

$$\left.\begin{array}{l}\dfrac{\partial M_\alpha}{\partial s_\alpha} + c_\beta(M_\alpha - M_\beta) - Q_\alpha = 0\,, \\[2mm] \dfrac{\partial M_\beta}{\partial s_\beta} + c_\alpha(M_\beta - M_\alpha) - Q_\beta = 0\,.\end{array}\right\} \tag{24}$$

These equations are most easily established from (8) and (9) through the use of moving axes theory. The quantities Q_α and Q_β may be eliminated between these equations to give

$$\frac{\partial^2 M_\alpha}{\partial s_\alpha^2} + \frac{\partial^2 M_\beta}{\partial s_\beta^2} + c_\beta\frac{\partial}{\partial s_\alpha}(2M_\alpha - M_\beta) + c_\alpha\frac{\partial}{\partial s_\beta}(2M_\beta - M_\alpha)$$

$$+ \left(\frac{\partial c_\beta}{\partial s_\alpha} + c_\beta^2 - \frac{\partial c_\alpha}{\partial s_\beta} - c_\alpha^2\right)(M_\alpha - M_\beta) + p = 2\varrho\,h\,\frac{\partial v}{\partial t}\,. \tag{25}$$

The curvature-rates along the α- and β-lines are, respectively,

$$\varkappa_\alpha = -\frac{\partial^2 v}{\partial s_\alpha^2} - c_\alpha \frac{\partial v}{\partial s_\beta}, \quad \varkappa_\beta = -\frac{\partial^2 v}{\partial s_\beta^2} - c_\beta \frac{\partial v}{\partial s_\alpha}. \tag{26}$$

Now the condition of isotropy requires $\varkappa_\alpha$ and $\varkappa_\beta$ to be principal curvature-rates, and therefore

$$\frac{\partial^2 v}{\partial s_\alpha \partial s_\beta} - c_\alpha \frac{\partial v}{\partial s_\alpha} \equiv \frac{\partial^2 v}{\partial s_\beta \partial s_\alpha} - c_\beta \frac{\partial v}{\partial s_\beta} = 0. \tag{27}$$

The yield condition and flow rule for the various plastic regimes are as follows.

(i) $\quad A: M_\alpha = M_\beta = M_0, \ \varkappa_\alpha = q\lambda, \ \varkappa_\beta = (1-q)\lambda;$

$\qquad D: M_\alpha = M_\beta = -M_0, \ \varkappa_\alpha = -(1-q)\lambda, \ \varkappa_\beta = -q\lambda;$ $\left.\right\} \left(\frac{1}{2} \leq q \leq 1\right);$

(ii) $FA: M_\alpha = M_0, \ 0 < M_\beta < M_0, \ \varkappa_\alpha = \lambda, \ \varkappa_\beta = 0;$

$\qquad DE: M_\beta = -M_0, \ -M_0 < M_\alpha < 0, \ \varkappa_\alpha = 0, \ \varkappa_\beta = -\lambda;$

(iii) $\quad F: M_\alpha = M_0, \ M_\beta = 0, \ \varkappa_\alpha = q\lambda, \ \varkappa_\beta = -(1-q)\lambda \left(\frac{1}{2} \leq q \leq 1\right);$

$\qquad E: M_\alpha = 0, \ M_\beta = -M_0, \ \varkappa_\alpha = q\lambda, \ \varkappa_\beta = -(1-q)\lambda \left(0 \leq q \leq \frac{1}{2}\right);$

(iv) $EF: 0 < M_\alpha < M_0, \ M_\alpha - M_\beta = M_0, \ \varkappa_\alpha = \frac{1}{2}\lambda, \ \varkappa_\beta = -\frac{1}{2}\lambda;$

$$\left.\right\} (\lambda \gtreqqless 0). \tag{28}$$

In these results certain plastic regimes are grouped together because they are not essentially distinct from a mathematical standpoint. Thus there are four basic types of plastic regime of which two correspond to edges of the TRESCA yield hexagon and two correspond to vertices of this hexagon.

4. Concluding remarks. The next stage in the analysis is to discuss the properties and methods of integration of the governing equations for the various plastic regimes. This further analysis is of course an essential preliminary to the solution of specific problems of interest, and the details will be given elsewhere.

References

[1] HOPKINS, H. G. and W. PRAGER: J. Mech. Phys. Solids 2, 1 (1953).
[2] TIMOSHENKO, S.: Theory of Plates and Shells. New York and London 1940.
[3] HOPKINS, H. G. and A. J. WANG: J. Mech. Phys. Solids 3, 117 (1954).
[4] DRUCKER, D. C. and H. G. HOPKINS: Proc. Sec. United States Nat. Congr. Appl. Mech., Ann Arbor, Mich. 1954 (Amer. Soc. Mech. Engrs., New York 1954), p. 517.
[5] HOPKINS, H. G. and W. PRAGER: J. appl. Mechan. 22, 372 (1955).

[6] HOPKINS, H. G. and W. PRAGER: Z. angew. Math. Phys. 5, 317 (1954).

[7] WANG, A.: J. appl. Mechan. 22, 375 (1955).

[8] WANG, A. J. and H. G. HOPKINS: J. Mech. Phys. Solids 3, 22 (1954).

[9] HAYTHORNETHWAITE, R. M.: Proc. Sec. United States Nat. Congr. Appl. Mech., Ann Arbor, Mich. 1954 (Amer. Soc. Mech. Engrs., New York 1954), p. 521.

[10] PRAGER, W.: Bull. techn. Suisse rom., No. 6, 85, March 1955.

[11] PRAGER, W.: Brown University Tech. Rep. No. A 11—123, July 1955.

[12] PRAGER, W. and P. G. HODGE JR.: Theory of Perfectly Plastic Solids, p. 252 et seq. New York and London 1951.

[13] PRAGER, W.: Proc. Sec. United States Congr. Appl. Mech., Ann Arbor, Mich. 1954 (Amer. Soc. Mech. Engrs., New York 1954), p. 21.

[14] PRAGER, W.: Brown University Tech. Rep. DA—798/18, March 1955.

[15] PELL, W. H. and W. PRAGER: Proc. First United States Congr. Appl. Mech., Chicago, Ill., 1951 (Amer. Soc. Mech. Engrs., New York 1952), p. 547.

Theory of Melting and Yield Strength[1]

By **H. Aroeste**, Pasadena (Calif.)

The theory of melting proposed by FÜRTH [1], which has been related to the rupture strength, has been criticized as yielding fortuitous results largely because rupture strengths have been associated more definitely with surface phenomena [2]. It may therefore be of interest to try to relate FÜRTH's theory or a modification thereof to the yield strength, which is less surface-dependent.

If we restrict ourselves to single crystals of high purity, the FRANK-READ mechanism for yielding presumably applies. The increase in energy, U, in forming a semicircular dislocation of radius R from an edge dislocation of width ζ may be written as

$$U = -\frac{1}{2}\pi R^2 \tau b + U_e + U_m,\qquad(1)$$

where τ is the applied shear stress, b is the BURGERS vector, and U_e an U_m represent, respectively, the increase in elastic and misfit energy. Making the simple assumption that the mixed semicircular dislocation is half edge and half screw, one obtains [3] approximately for U_e and U_m:

$$U_e \approx \frac{2\pi - 4 - \pi\sigma}{8\pi(1-\sigma)} R\mu b^2 \ln\left(\frac{4R}{\zeta} - 1\right),\qquad(2)$$

$$U_m \approx \frac{2\pi - 4 - \pi\sigma}{8\pi(1-\sigma)} R\mu b^2.\qquad(3)$$

Here μ and σ are respectively the shear modulus and POISSON's ratio of an isotropic crystal. In writing U_e we have put ε, the usual lower limit of integration, approximately equal to $\zeta/2$. The general conclusions drawn later are not particularly dependent on this choice.

The critical yield stress, τ_c, will correspond to the value of τ when $\partial U/\partial R = 0$. Thus we obtain

$$\tau_c \approx \frac{k}{\pi Rb} \ln\left(\frac{4R}{\zeta} + 1\right),\qquad(4)$$

---

[1] The author's lecture, "Dislocations and Melting of Crystals", was based on the contents of a letter published in Phys. Rev. 97, 1723 (1955), which is reproduced here.

where

$$k = \frac{2\pi - 4 - \pi\sigma}{8\mu(1 - \sigma)}\,\mu\,b^2 \,. \tag{5}$$

Equations (4) and (5) are derived for an initial edge dislocation because it may easily be shown that a screw or mixed straight dislocation will give a higher value for τ_c.

If we take for ζ the result obtained by FOREMAN, JASWON, and WOOD [4] that

$$\zeta = \frac{\mu b}{2\pi(1 - \sigma)\tau_m}\,, \tag{6}$$

where τ_m is the theoretical shear strength, and use for τ_m the lowest value thus far derived [5], i.e., about $\mu/30$, we obtain

$$\zeta \approx \frac{15\,b}{\pi(1 - \sigma)} \,. \tag{7}$$

Now, from the theory of FÜRTH, which assumes that melting is due to the break up of a block structure, we may write that

$$2R \approx \frac{6\,b\,\varLambda}{Q}\,, \tag{8}$$

where $\varLambda$ is the heat of sublimation and Q is the heat of melting. Using equations (7) and (8) in (4), we obtain

$$\tau_c \approx \frac{2\pi - 4 - \pi}{24\pi^2(1 - \sigma)\varLambda}\,\mu\,Q\ln\left[\frac{12\pi(1 - \sigma)}{15Q}\,\varLambda + 1\right]. \tag{9}$$

This formula is in principle applicable to unworked materials close to the absolute zero. The yield strengths of a few metal crystals such as Zn and Cd and of the ionic crystal NaCl have been measured at very low temperatures. One may also very roughly extrapolate the data on other crystals to absolute zero. The yield strengths derived from formula (9) are at least an order of magnitude too high. One may lessen the discrepancy a bit by using the suggestion of FISHER[1] that a single-ended source near the surface should start to operate at one half the stress needed for a double-ended source of the same length.

There are two further ways to achieve lower results. One is to assume that the crystal is not homogeneously blocked and that there are some much longer blocks which are operative. Another, which is more interesting and more likely, is to modify FÜRTH's formula to read, say,

$$2R \approx 6\,\frac{\zeta}{2}\,\frac{\varLambda}{Q}\,. \tag{10}$$

This assumes that the mechanism of melting is tied intrinsically not only

---

[1] See reference [3], p. 86.

to the block size, but also to a width between blocks, where the atoms are misfit and may be expected to enter into the mechanism first.

The author wishes to thank Professor H. S. Tsien for helpful discussion.

References

[1] Fürth, R.: Phil. Mag. 40, 1227 (1949), the other references cited there.
[2] Frenkel, J.: Kinetic Theory of Liquids, p. 101. London 1946.
[3] See for example A. Cottrell: Dislocations and Plastics Flow in Crystals, London 1953, for basic formulas.
[4] Foreman, Jaswon, and Wood: Proc. phys. Soc., Lond. Ser. A 64, 156 (1951).
[5] Mackenzie, J.: thesis University of Bristol 1949 (unpublished).

II
Nichtlineare Elastizität und Vermischtes
Non-linear Elasticity and Miscellaneous Problems

Ein nichtlineares Elastizitätsgesetz;
Aufbau und Anwendungsmöglichkeiten

Von H. Kauderer, Stuttgart

Mit 1 Abbildung

1. Die Linearisierungsprozesse der klassischen Elastizitätstheorie. Hat man für einen festen Körper zu einer vorgeschriebenen Belastung und zu bestimmten Lagerungsbedingungen die Spannungen und Verzerrungen zu berechnen, so muß sich der Zusammenhang zwischen diesen Größen an jeder Stelle des Körpers durch ein mathematisch zu formulierendes Gesetz wiedergeben lassen.

Ein solches Gesetz sollte hierbei zwei Grundforderungen, die sich zumeist widersprechen werden, optimal zu befriedigen suchen: Es soll einerseits diejenigen physikalischen Eigenschaften des Materials, auf deren Berücksichtigung man besonderen Wert legt, möglichst getreu wiedergeben, und es soll andererseits von möglichst einfacher Gestalt sein.

Der letzteren Forderung kommt zweifellos das HOOKEsche Gesetz in seiner Beschränkung auf kleine Verzerrungen, in der man es in der klassischen Elastizitätstheorie benützt, am weitesten entgegen.

Dem HOOKEschen Gesetz liegen zwei verschiedene Linearisierungsprozesse zugrunde. Der erste, welcher der „geometrische" genannt sei, weil er sich auf die Geometrie des deformierten Körpers bezieht, besteht darin, daß man sich in der Darstellung der Verzerrungen, nämlich der Dehnungen ε_x, ε_y, ε_z und der Scherungen ψ_{xy}, ψ_{yz}, ψ_{zx}, die in den Ableitungen der Verschiebungskomponenten u, v, w nach den Koordinaten x, y, z bekanntlich die nichtlineare Form

$$\varepsilon_x = \sqrt{1 + 2\frac{\partial u}{\partial x} + \left(\frac{\partial u}{\partial x}\right)^2 + \left(\frac{\partial v}{\partial x}\right)^2 + \left(\frac{\partial w}{\partial x}\right)^2} - 1$$

und zyklisch weiter sowie

$$\psi_{xy} = \frac{1}{(1+\varepsilon_x)(1+\varepsilon_y)}\left(\frac{\partial u}{\partial y} + \frac{\partial v}{\partial x} + \frac{\partial u}{\partial x}\frac{\partial u}{\partial y} + \frac{\partial v}{\partial x}\frac{\partial v}{\partial y} + \frac{\partial w}{\partial x}\frac{\partial w}{\partial y}\right)$$

und zyklisch weiter haben, auf die linearisierten Ausdrücke

$$\varepsilon_x = \frac{\partial u}{\partial x}, \quad \varepsilon_y = \frac{\partial v}{\partial y}, \quad \varepsilon_z = \frac{\partial w}{\partial z};$$

$$\psi_{xy} = \frac{\partial u}{\partial y} + \frac{\partial v}{\partial x}, \quad \psi_{yz} = \frac{\partial v}{\partial z} + \frac{\partial w}{\partial y}, \quad \psi_{zx} = \frac{\partial w}{\partial x} + \frac{\partial u}{\partial z}$$

beschränkt und außerdem die Spannungen auf den unverformten statt auf den verformten Körper bezieht. Dies hat zur Folge, daß die Verzerrungen, welche die Theorie bei der Anwendung auf endliche Verschiebungen liefert, mit Fehlern behaftet sein können, die von der Größenordnung der Quadrate der Verzerrungen sind. Praktisch kann man, wie etwa der Zugversuch zeigt, mit den linearisierten Verzerrungen bis zu Größenordnungen von einem Prozent rechnen, da man ja bei technischen Festigkeitsrechnungen einen relativen Fehler von dieser Größe noch durchaus in Kauf zu nehmen pflegt.

Der zweite Linearisierungsprozeß, welcher der „physikalische" genannt sei, ermöglicht eine zweifache Deutung. Um dies klarzustellen, seien hier in üblicher Weise die Tensoren des Spannungs- und des Verzerrungszustandes

$$T = \begin{pmatrix} \sigma_x & \tau_{xy} & \tau_{zx} \\ \tau_{xy} & \sigma_y & \tau_{yz} \\ \tau_{zx} & \tau_{yz} & \sigma_z \end{pmatrix}, \quad D = \begin{pmatrix} \varepsilon_x & \frac{1}{2}\psi_{xy} & \frac{1}{2}\psi_{zx} \\ \frac{1}{2}\psi_{xy} & \varepsilon_y & \frac{1}{2}\psi_{yz} \\ \frac{1}{2}\psi_{zx} & \frac{1}{2}\psi_{yz} & \varepsilon_z \end{pmatrix}$$

eingeführt, und diese sogleich in ihre Kugeltensoren

$$T_0 = \begin{pmatrix} \sigma_0 & 0 & 0 \\ 0 & \sigma_0 & 0 \\ 0 & 0 & \sigma_0 \end{pmatrix}, \quad D_0 = \begin{pmatrix} \varepsilon_0 & 0 & 0 \\ 0 & \varepsilon_0 & 0 \\ 0 & 0 & \varepsilon_0 \end{pmatrix}$$

mit der Mittelspannung $\sigma_0 = \frac{1}{3}(\sigma_x + \sigma_y + \sigma_z)$ und der Mitteldehnung $\varepsilon_0 = \frac{1}{3}(\varepsilon_x + \varepsilon_y + \varepsilon_z)$ und ihre Deviatoren

$$T' = \begin{pmatrix} \sigma_x - \sigma_0 & \tau_{xy} & \tau_{zx} \\ \tau_{xy} & \sigma_y - \sigma_0 & \tau_{yz} \\ \tau_{zx} & \tau_{yz} & \sigma_z - \sigma_0 \end{pmatrix}, \quad D' = \begin{pmatrix} \varepsilon_x - \varepsilon_0 & \frac{1}{2}\psi_{xy} & \frac{1}{2}\psi_{zx} \\ \frac{1}{2}\psi_{xy} & \varepsilon_y - \varepsilon_0 & \frac{1}{2}\psi_{yz} \\ \frac{1}{2}\psi_{zx} & \frac{1}{2}\psi_{yz} & \varepsilon_z - \varepsilon_0 \end{pmatrix}$$

zerlegt, so daß

$$T = T_0 + T', \quad D = D_0 + D'$$

wird. Das HOOKEsche Gesetz kann man dann mit zwei Konstanten, dem Kompressionsmodul K und dem Schubmodul G, bekanntlich auf die einfache Gestalt

$$\sigma_0 = 3 K \varepsilon_0, \quad T' = 2 G D'$$

bringen. Es stellt also die Proportionalität der beiden Kugeltensoren und der beiden Deviatoren fest.

Auf jeden Fall enthält das HOOKEsche Gesetz in dieser Form eine *erste physikalische Aussage*. Sie besteht darin, daß der durch den Kugeltensor D_0 dargestellte Anteil an der gesamten Verzerrung des Körperelements allein durch den Kugeltensoranteil T_0 des dort herrschenden Spannungszustands bestimmt wird, und ebenso der Deviatoranteil D' der Verzerrung allein durch den Deviatoranteil T' des Spannungszustands.

Beschränkt man den physikalischen Inhalt des HOOKEschen Gesetzes auf diese Aussage, so folgen die obigen Gleichungen zwangsläufig als eine Grenzform für infinitesimal kleine Verzerrungen; ihre Linearität ist dann nur eine Folge davon, daß man in Gleichungen zwischen infinitesimalen Größen Glieder höherer Potenz gegenüber linearen Gliedern streichen darf.

Das HOOKEsche Gesetz enthält jedoch sofort dann eine viel *weitergehende physikalische Aussage*, wenn man seinen Gültigkeitsbereich nicht auf den Grenzzustand infinitesimaler Verzerrungen beschränkt, sondern auch noch auf zwar kleine, aber durchaus endliche, bis zu einer bestimmten „Proportionalitätsgrenze" reichende Verzerrungen erweitert.

Inwieweit diese zweite Auffassung des HOOKEschen Gesetzes zulässig ist, hängt von der Natur des Werkstoffes ab. Es gibt ja sowohl Werkstoffe (z. B. Stahl), die eine ausgeprägte endliche Elastizitätsgrenze haben, als auch solche, wie etwa gewisse Nichteisenmetalle oder Kunststoffe, bei denen sich eine solche Grenze nicht mehr scharf definieren läßt. Bei den meisten Stoffen liegt nun aber die Proportionalitätsgrenze immer noch weit unterhalb derjenigen Grenze für die Verzerrungen, die man infolge der geometrischen Linearisierung der Verzerrungsgrößen nicht überschreiten darf, wenn der zu erwartende relative Fehler der Rechenergebnisse nicht über ein technisch zulässiges Maß hinausgehen soll. So gibt es also einen weiten Bereich von Verzerrungen, in dem die geometrische Linearisierung noch durchaus erlaubt ist, während die *weitergehende* physikalische Aussage des HOOKEschen Gesetzes nicht mehr gilt. Gerade dieser Verformungsbereich, der etwa bis zu Verzerrungen von einem Prozent reichen mag, spielt in den technischen Anwendungen eine wichtige Rolle. Es erscheint daher zweckmäßig, für diesen Bereich ein geeignetes Spannungs-Dehnungsgesetz aufzustellen. —

Es soll nun zunächst gezeigt werden (Ziff. 2), welche allgemeine Gestalt ein solches Gesetz annehmen muß, wenn man vier sinnvoll erscheinende Grundforderungen an das elastische Verhalten der festen Körper stellt. Sodann seien zwei Minimalsätze angeführt, die sich aus diesem Gesetz folgern lassen (Ziff. 3). Hierauf sollen Fragen behandelt werden, die sich mit der praktischen Anwendung des Gesetzes befassen. Zuerst experimentell (Ziff. 4), welche Versuche zur Ermittlung der im Gesetz auftretenden Werkstoffkonstanten durchzuführen sind, und sodann mathematisch (Ziff. 5), welche Rechenverfahren zu empfehlen sind, um das Gesetz auf die Lösung praktischer Aufgaben anwenden zu können. Schließlich seien noch einige typische numerische Ergebnisse angeführt, die man bei der Anwendung des Gesetzes auf einzelne Probleme der Elastomechanik erhält (Ziff. 6).

2. Aufbau des nichtlinearen Elastizitätsgesetzes. Um die allgemeine Gestalt des Verformungsgesetzes zu gewinnen [1, 2], seien nun die vier Forderungen an das elastische Verhalten des Werkstoffes aufgestellt, die auf Grund des oben Bemerkten gerechtfertigt erscheinen dürften.

Die *erste Forderung* soll darin bestehen, daß sich die auf die Volumeneinheit bezogene, an einem Element des Körpers aufzubringende Formänderungsarbeit

$$A = \int (\sigma_x\, d\varepsilon_x + \sigma_y\, d\varepsilon_y + \sigma_z\, d\varepsilon_z + \tau_{xy}\, d\psi_{xy} + \tau_{yz}\, d\psi_{yz} + \tau_{zx}\, d\psi_{zx})$$

als eine eindeutige Funktion der Komponenten des Verzerrungstensors des Elements darstellen lasse. —

Diese Forderung wird sicher dann erfüllt sein, wenn der Spannungszustand des Körpers sich allein durch den *Zustand* seiner Verzerrung eindeutig bestimmen läßt, das heißt also, wenn der Körper „vollkommen elastisch" ist im üblichen Sinn dieses Begriffs.

Möchte man das aufzustellende Gesetz auch noch auf solche nicht vollkommen elastischen Körper anwenden, bei denen der Spannungszustand nicht nur vom jeweiligen *Zustand* der Verzerrung, sondern auch von dem *Weg* abhängt, auf dem der Körper in diesen Verzerrungszustand gebracht worden ist, so hat man noch zuzulassen, daß für diesen Weg bestimmte Vorschriften gemacht werden, so etwa, daß die Verzerrung in den durch den Tensor D bestimmten Endzustand derart vor sich gehen soll, daß alle Zwischenzustände $c \cdot D$ durchlaufen werden, wobei c mit der Zeit monoton von Null bis Eins wächst. Man hat sich dann allerdings damit abzufinden, daß das Verzerrungsgesetz, das wegen dieser Eigenschaft als „Elastizitätsgesetz" bezeichnet sei, bei nicht vollkommen elastischen Körpern jeweils nur für einen erstmaligen Verzerrungsvorgang anwendbar sein wird.

Die *zweite Forderung*, die wir stellen wollen, lautet, daß der Werkstoff homogen und isotrop sei, also hinsichtlich seines elastischen Verhaltens keine ausgezeichneten Stellen oder Richtungen aufweisen soll. — Diese Forderung hat zur Folge, daß der Ausdruck für die Formänderungsarbeit A allein von den gegenüber Drehungen des Koordinatensystems unabhängigen Invarianten des Tensors D, nämlich

$$J_1 = \varepsilon_x + \varepsilon_y + \varepsilon_z, \quad J_2 = \varepsilon_x \varepsilon_y + \varepsilon_y \varepsilon_z + \varepsilon_z \varepsilon_x - \frac{1}{4}\left(\psi_{xy}^2 + \psi_{yz}^2 + \psi_{zx}^2\right),$$

$$J_3 = \begin{vmatrix} \varepsilon_x & \dfrac{1}{2}\psi_{xy} & \dfrac{1}{2}\psi_{zx} \\[2mm] \dfrac{1}{2}\psi_{xy} & \varepsilon_y & \dfrac{1}{2}\psi_{yz} \\[2mm] \dfrac{1}{2}\psi_{zx} & \dfrac{1}{2}\psi_{yz} & \varepsilon_z \end{vmatrix}$$

abhängen kann [*3*].

Die *dritte Forderung* soll zum Ausdruck bringen, daß wir die erste physikalische Aussage des HOOKEschen Gesetzes auch für das nichtlineare Elastizitätsgesetz beibehalten wollen: Denken wir uns wieder die Zerlegungen

$$D = D_0 + D', \quad T = T_0 + T'$$

vorgenommen, so soll die gestaltsgetreue Volumenänderung D_0 allein durch T_0 und die volumengetreue Gestaltsänderung D' allein durch T' bestimmt werden, und umgekehrt. —

Diese Forderung steht, speziell für den Eintritt des Fließvorganges, auch im Einklang mit der VON MISESschen Plastizitätshypothese.

Wir untersuchen nun zunächst, inwieweit diese drei ersten Forderungen schon die Gestalt beeinflussen, die der Ausdruck für die Formänderungsarbeit A annehmen wird. Bekanntlich läßt sich diese in die zwei Anteile der Volumenänderungsarbeit A_0 und der Gestaltänderungsarbeit A' additiv zerlegen, wobei

$$A_0 = 3\int \sigma_0 \, d\varepsilon_0,$$

$$A' = \int\left(\sigma_x' \, d\varepsilon_x' + \sigma_y' \, d\varepsilon_y' + \sigma_z' \, d\varepsilon_z' + \tau_{xy} \, d\psi_{xy} + \tau_{yz} \, d\psi_{yz} + \tau_{zx} \, d\psi_{zx}\right)$$

mit

$$\sigma_x' = \sigma_x - \sigma_0, \quad \sigma_y' = \sigma_y - \sigma_0, \quad \sigma_z' = \sigma_z - \sigma_0;$$

$$\varepsilon_x' = \varepsilon_x - \varepsilon_0, \quad \varepsilon_y' = \varepsilon_y - \varepsilon_0, \quad \varepsilon_z' = \varepsilon_z - \varepsilon_0$$

zu setzen ist. Es hängt somit A' nur von den Komponenten der Deviatoren T' und D' des Spannungs- und des Verzerrungszustandes ab, und A_0 nur von den beiden Skalaren σ_0 und ε_0 der Kugeltensoren T_0 und D_0. Nach der dritten Forderung muß sich also A_0 allein in ε_0 und, wiederum wegen der geforderten Isotropie, A' allein in den zwei Invarianten

$$J_2' = \varepsilon_x' \varepsilon_y' + \varepsilon_y' \varepsilon_z' + \varepsilon_z' \varepsilon_x' - \frac{1}{4} (\psi_{xy}^2 + \psi_{yz}^2 + \psi_{zx}^2),$$

$$J_3' = \begin{vmatrix} \varepsilon_x' & \dfrac{1}{2} \psi_{xy} & \dfrac{1}{2} \psi_{zx} \\[2mm] \dfrac{1}{2} \psi_{xy} & \varepsilon_y' & \dfrac{1}{2} \psi_{yz} \\[2mm] \dfrac{1}{2} \psi_{zx} & \dfrac{1}{2} \psi_{yz} & \varepsilon_z' \end{vmatrix}$$

von D' ausdrücken lassen. Führt man noch an Stelle von J_2' besser das „Schermaß" ψ_0 durch

$$\psi_0^2 = \frac{4}{3} \left[\varepsilon_x'^2 + \varepsilon_y'^2 + \varepsilon_z'^2 + \frac{1}{2} (\psi_{xy}^2 + \psi_{yz}^2 + \psi_{zx}^2) \right] = -\frac{8}{3} J_2'$$

ein, so muß A in die zwei Summanden

$$A = A_0 (\varepsilon_0) + A' (\psi_0^2, J_3')$$

zerlegbar sein.

Man kann nun zeigen, daß A' tatsächlich nicht von der Invarianten J_3' abhängen kann. Hierzu wollen wir annehmen, der Verzerrungstensor $D = D_0 + D'$ werde infinitesimal um $dD = dD_0 + dD'$ geändert. Dann wird von den Spannungen am Elementarquader hierbei die Arbeit

$$dA = dA_0 + dA'$$

geleistet. Nun ist einerseits, durch formale Rechnung gebildet,

$$dA = \frac{\partial A}{\partial \varepsilon_x} d\varepsilon_x + \frac{\partial A}{\partial \varepsilon_y} d\varepsilon_y + \frac{\partial A}{\partial \varepsilon_z} d\varepsilon_z + \frac{\partial A}{\partial \psi_{xy}} d\psi_{xy} + \frac{\partial A}{\partial \psi_{yz}} d\psi_{yz} + \frac{\partial A}{\partial \psi_{zx}} d\psi_{zx},$$

$$dA_0 = \frac{dA_0}{d\varepsilon_0} d\varepsilon_0,$$

$$dA' = \frac{\partial A'}{\partial \varepsilon_x'} d\varepsilon_x' + \frac{\partial A'}{\partial \varepsilon_y'} d\varepsilon_y' + \frac{\partial A'}{\partial \varepsilon_z'} d\varepsilon_z' + \frac{\partial A'}{\partial \psi_{xy}} d\psi_{xy} + \frac{\partial A'}{\partial \psi_{yz}} d\psi_{yz} + \frac{dA'}{d\psi_{zx}} d\psi_{zx}.$$

Andererseits ist aber, wie man am Elementarquader nachweist, die infinitesimale Zunahme der Arbeitsbeträge

$$dA = \sigma_x d\varepsilon_x + \sigma_y d\varepsilon_y + \sigma_z d\varepsilon_z + \tau_{xy} d\psi_{xy} + \tau_{yz} d\psi_{yz} + \tau_{zx} d\psi_{zx},$$

$$dA_0 = 3\sigma_0 d\varepsilon_0,$$

$$dA' = \sigma_x' d\varepsilon_x' + \sigma_y' d\varepsilon_y' + \sigma_z' d\varepsilon_z' + \tau_{xy} d\psi_{xy} + \tau_{yz} d\psi_{yz} + \tau_{zx} d\psi_{zx}.$$

Der Vergleich entsprechender Ausdrücke, die für beliebige Differentiale der Verzerrungsgrößen übereinstimmen sollen, liefert

$$\frac{\partial A}{\partial \varepsilon_x} = \sigma_x, \dots, \dots; \qquad \frac{\partial A}{\partial \psi_{xy}} = \tau_{xy}, \dots, \dots; \qquad \frac{dA}{d\varepsilon_0} = 3\sigma_0;$$

$$\frac{\partial A'}{\partial \varepsilon_x'} = \sigma_x', \dots, \dots; \qquad \frac{\partial A'}{\partial \psi_{xy}} = \tau_{xy}, \dots, \dots$$

So werden wegen

$$\frac{\partial \psi_0^2}{\partial \varepsilon_x'} = \frac{8}{3}\,\varepsilon_x', \ldots, \ldots; \qquad \frac{\partial J_3'}{\partial \varepsilon_x'} = \varepsilon_y'\,\varepsilon_z' - \frac{1}{4}\,\psi_{yz}^2$$

die Spannungen

$$\sigma_x' = \frac{\partial A'}{\partial \varepsilon_x'} = \frac{\partial A'}{\partial \psi_0^2}\,\frac{\partial \psi_0^2}{\partial \varepsilon_x'} + \frac{\partial A'}{\partial J_3'}\,\frac{\partial J_3'}{\partial \varepsilon_x'} = \frac{8}{3}\,\varepsilon_x'\,\frac{\partial A'}{\partial \psi_0^2} + \left(\varepsilon_y'\,\varepsilon_z' - \frac{1}{4}\,\psi_{yz}^2\right)\frac{\partial A'}{\partial J_3'}$$

und zyklisch weiter.

Addiert man die Gleichungen für σ_x', σ_y' und σ_z', so erhält man wegen

$$\sigma_x' + \sigma_y' + \sigma_z' = 0$$

links den Wert Null. Rechts wird, weil ebenso

$$\varepsilon_x' + \varepsilon_y' + \varepsilon_z' = 0$$

ist, nur die Summe der zweiten Glieder übrig bleiben; das heißt, es muß

$$0 = \left[\left(\varepsilon_y'\,\varepsilon_z' - \frac{1}{4}\,\psi_{yz}^2\right) + \left(\varepsilon_z'\,\varepsilon_x' - \frac{1}{4}\,\psi_{zx}^2\right) + \left(\varepsilon_x'\,\varepsilon_y' - \frac{1}{4}\,\psi_{xy}^2\right)\right]\frac{\partial A'}{\partial J_3'}$$

werden. Der Ausdruck in der eckigen Klammer ist aber gleich $J_2' = -\dfrac{3}{8}\,\psi_0^2$.

Da nun ψ_0^2 nicht identisch gleich Null sein kann für alle Verzerrungs-zustände, so muß $\partial A'/\partial J_3' = 0$ werden; das heißt, A' kann nicht von J_3' abhängen. Dann wird aber

$$\sigma_x' = \frac{8}{3}\,\frac{dA'}{d\psi_0^2}\,\varepsilon_x', \qquad \sigma_y' = \frac{8}{3}\,\frac{dA'}{d\psi_0^2}\,\varepsilon_y', \qquad \sigma_z' = \frac{8}{3}\,\frac{dA'}{d\psi_0^2}\,\varepsilon_z'$$

und, da

$$\frac{\partial \psi_0^2}{\partial \psi_{xy}} = \frac{4}{3}\,\psi_{xy}$$

und zyklisch weiter ist,

$$\tau_{xy} = \frac{4}{3}\,\frac{dA'}{d\psi_0^2}\,\psi_{xy}, \qquad \tau_{yz} = \frac{4}{3}\,\frac{dA'}{d\psi_0^2}\,\psi_{yz}, \qquad \tau_{zx} = \frac{4}{3}\,\frac{dA'}{d\psi_0^2}\,\psi_{zx}.$$

Die Beziehungen zwischen Spannungen und Verzerrungen lassen sich daher jetzt in der folgenden Form zusammenfassen

$$\sigma_0 = \frac{1}{3}\,\frac{dA_0}{d\varepsilon_0}, \qquad T' = \frac{8}{3}\,\frac{dA'}{d\psi_0^2}\,D'.$$

Die beiden Funktionen $dA/d\varepsilon_0$ und $dA'/d\psi_0^2$ bestimmen somit vollständig das elastische Verhalten des Körpers.

Wir stellen nunmehr noch die *vierte Forderung*, die jedoch nur formaler Natur ist: Das aufzustellende Elastizitätsgesetz soll speziell für infinitesimal kleine Verzerrungen die Gestalt des HOOKEschen Gesetzes annehmen, nämlich

$$\sigma_0 = 3\,K\,\varepsilon_0, \qquad T' = 2\,G\,D'.$$

Um diese Forderung zu erfüllen, setzen wir die Ableitungen von A_0 und A' speziell in der Form

$$\frac{dA_0}{d\varepsilon_0} = 9K\,\varepsilon_0\,\varkappa(\varepsilon_0)\,, \quad \frac{dA'}{d\psi_0^2} = \frac{3}{4}\,G\,\gamma\,(\psi_0^2)$$

an, wobei wir zwei neue Funktionen einführen, nämlich die „Dehnungsfunktion" $\varkappa(\varepsilon_0)$ und die „Scherungsfunktion" $\gamma(\psi_0^2)$, die beide die Eigenschaft besitzen sollen, daß

$$\lim_{\varepsilon_0=0} \varkappa(\varepsilon_0) = 1\,, \quad \lim_{\psi_0^2=0} \gamma\,(\psi_0^2) = 1$$

wird. Das Elastizitätsgesetz lautet dann

$$\sigma_0 = 3K\varkappa(\varepsilon_0)\,\varepsilon_0\,, \quad T' = 2G\gamma\,(\psi_0^2)\,D'.$$

Die Anwendung des Gesetzes wird besonders bequem, wenn es gelingt, die beiden Funktionen durch Potenzreihen mit den Entwicklungskoeffizienten $\varkappa_i$ und γ_{2j} darzustellen:

$$\varkappa(\varepsilon_0) = 1 + \varkappa_1\,\varepsilon_0 + \varkappa_2\,\varepsilon_0^2 + \cdots\,, \quad \gamma\,(\psi_0^2) = 1 + \gamma_2\,\psi_0^2 + \gamma_4\,\psi_0^4 + \cdots.$$

Selbstverständlich läßt sich das Elastizitätsgesetz auch für die einzelnen Komponenten des Spannungstensors T anschreiben. Man erhält dann

$$\sigma_x = 3K\varkappa(\varepsilon_0)\,\varepsilon_0 + 2G\gamma\,(\psi_0^2)\,(\varepsilon_x - \varepsilon_0)\,,$$
$$\sigma_y = 3K\varkappa(\varepsilon_0)\,\varepsilon_0 + 2G\gamma\,(\psi_0^2)\,(\varepsilon_y - \varepsilon_0)\,,$$
$$\sigma_z = 3K\varkappa(\varepsilon_0)\,\varepsilon_0 + 2G\gamma\,(\psi_0^2)\,(\varepsilon_z - \varepsilon_0)\,;$$
$$\tau_{xy} = G\gamma\,(\psi_0^2)\,\psi_{xy}\,, \quad \tau_{yz} = G\gamma\,(\psi_0^2)\,\psi_{yz}\,, \quad \tau_{zx} = G\gamma\,(\psi_0^2)\,\psi_{zx}.$$

Bildet man in der Gleichung für den Deviator T' beiderseits die zweite Invariante, so ergibt sich mit der invarianten Spannungsgröße

$$\tau_0 = \sqrt{\frac{2}{3}}\,\sqrt{\frac{1}{3}\,(\sigma_x^2 + \sigma_y^2 + \sigma_z^2 - \sigma_x\sigma_y - \sigma_y\sigma_z - \sigma_z\sigma_x) + \tau_{xy}^2 + \tau_{yz}^2 + \tau_{zx}^2}\,,$$

die zur zweiten Invarianten des Spannungsdeviators proportional ist, die Beziehung

$$\tau_0 = G\gamma\,(\psi_0^2)\,\psi_0.$$

Um nun auch umgekehrt die Verzerrungen durch die Spannungen auszudrücken, führt man am besten zuerst die dimensionslosen Spannungsgrößen

$$s_0 = \sigma_0/3K\,, \quad t_0 = \tau_0/G$$

ein und bildet hiermit die Umkehrungen der Beziehungen $\sigma_0 = 3K\varkappa(\varepsilon_0)\,\varepsilon_0$ und $\tau_0 = G\gamma\,(\psi_0^2)\,\psi_0$ in der Form

$$\varepsilon_0 = k\,(s_0)\,s_0 \quad \text{und} \quad \psi_0 = g\,(t_0^2)\,t_0$$

mit den beiden Funktionen $k(s_0)$ und $g(t_0^2)$, die als ,,*Kompressions-*'' und als ,,*Schubfunktion*'' bezeichnet seien, und die man selbst wieder durch Reihenentwicklungen nach Potenzen von s_0 bzw. von t_0^2 darstellen kann:

$$k(s_0) = 1 + k_1 s_0 + k_2 s_0^2 + \cdots, \quad g(t_0^2) = 1 + g_2 t_0^2 + g_4 t_0^4 + \cdots,$$

wobei sich die Koeffizienten k_i und g_{2j} aus den Koeffizienten der Funktionen $\varkappa(\varepsilon_0)$ und $\gamma(\psi_0^2)$ nach den Gesetzen der Reihenumkehrung berechnen lassen. Mit diesen Funktionen nimmt das Elastizitätsgesetz die folgende Gestalt an

$$\varepsilon_x = \frac{1}{3K} k(s_0) \sigma_0 + \frac{1}{2G} g(t_0^2)(\sigma_x - \sigma_0),$$

$$\varepsilon_y = \frac{1}{3K} k(s_0) \sigma_0 + \frac{1}{2G} g(t_0^2)(\sigma_y - \sigma_0),$$

$$\varepsilon_z = \frac{1}{3K} k(s_0) \sigma_0 + \frac{1}{2G} g(t_0^2)(\sigma_z - \sigma_0);$$

$$\psi_{xy} = \frac{1}{G} g(t_0^2) \tau_{xy}, \quad \psi_{yz} = \frac{1}{G} g(t_0^2) \tau_{yz}, \quad \psi_{zx} = \frac{1}{G} g(t_0^2) \tau_{zx}.$$

Es sei hier nochmals darauf hingewiesen, daß das Elastizitätsgesetz ein Näherungsgesetz sein wird, bei dem infolge der geometrischen Linearisierung die in ihm vorkommenden Verzerrungsgrößen mit Fehlern von der Größenordnung η^2 behaftet sein können, wenn η der Maximalwert der Verzerrungen ist, für die wir das Gesetz noch anwenden wollen. Es hätte deshalb keinen Sinn, von den Reihenentwicklungen der Dehnungs- und der Scherungsfunktion überhaupt Gebrauch zu machen und nicht kurzerhand die vier Funktionen $\varkappa(\varepsilon_0)$, $\gamma(\psi_0^2)$, $k(s_0)$ und $g(t_0^2)$ gleich Eins zu setzen, wenn die Koeffizienten $\varkappa_i$, γ_{2j} und k_i, g_{2j} in der Nähe von Eins liegen würden oder gar noch kleiner wären, da dann der Fehler, den man mit der Vernachlässigung dieser Koeffizienten beginge, von der gleichen Größenordnung wäre wie der ohnehin schon durch die geometrische Linearisierung verursachte. Nun ist aber z.B. für Kupfer der Wert von $\gamma_2 = -7{,}27 \cdot 10^6$, so daß man das Glied $\gamma_2 \psi_0^2$ in der Entwicklung von $\gamma(\psi_0^2)$ bei Problemen, bei denen Verzerrungen von der Größenordnung $0{,}1\%$ vorkommen, durchaus zu berücksichtigen hat, während man es natürlich bei Problemen mit sehr kleinen Verzerrungen, etwa von der Größenordnung 10^{-5}, vernachlässigen kann und mit dem HOOKEschen Gesetz rechnen darf.

3. Minimalsätze. Wie für das HOOKEsche Gesetz, so lassen sich auch für das nichtlineare Elastizitätsgesetz Minimalsätze für die Formänderungsarbeit aufstellen [4]. Hierbei ist es allerdings zweckmäßig, zwei Begriffe auseinanderzuhalten, die man beim HOOKEschen Gesetz nicht zu unterscheiden braucht. Es ist dies einerseits die Formänderungsarbeit je Volumeneinheit A und andererseits eine Größe, die ebenfalls die Dimension einer Arbeit je Volumeneinheit besitzt und durch das folgende Integral definiert ist

$$\overline{A} = \int (\varepsilon_x \, d\sigma_x + \varepsilon_y \, d\sigma_y + \varepsilon_z \, d\sigma_z + \psi_{xy} \, d\tau_{xy} + \psi_{yz} \, d\tau_{yz} + \psi_{zx} \, d\tau_{zx}) \, .$$

Sie sei als die „konjugierte Formänderungsarbeit" je Volumeneinheit bezeichnet.

Denkt man sich in A die Spannungen nach dem Elastizitätsgesetz in den Verzerrungen ausgedrückt, so werden die partiellen Ableitungen von A nach den Verzerrungskomponenten gleich den entsprechenden Spannungskomponenten. Für $\overline{A}$ läßt sich nun ganz entsprechend zeigen: Drückt man dort die Verzerrungskomponenten durch die Spannungskomponenten aus, so liefern die partiellen Ableitungen von $\overline{A}$ nach den Spannungskomponenten wieder die entsprechenden Verzerrungskomponenten.

Es gibt nun, wie beim Hookeschen Gesetz, zwei Minimalsätze über die Formänderungsarbeit [5]. Der eine, der sich auf die Variation der Verschiebungen bezieht, ist genau gleichlautend wie beim Hookeschen Gesetz. Der andere, der sich auf die Variation der Spannungen bezieht (das sog. Castiglianosche Prinzip), läßt sich hingegen erst dann entsprechend wie beim Hookeschen Gesetz formulieren, wenn man in dem Ausdruck für die sog. „Ergänzungsarbeit" an Stelle von A die konjugierte Formänderungsarbeit $\overline{A}$ benützt. Man hat — als Voraussetzung für das Auftreten eines wirklichen Minimums — lediglich zu fordern, daß die Funktionen $\sigma_0 = 3 K \varkappa (\varepsilon_0) \varepsilon_0$ und $\tau_0 = G \gamma (\psi_0^2) \psi_0$ monoton wachsen, was im Bereich der kleinen Verzerrungen für die meisten Stoffe zutrifft.

4. Grundversuche zur Bestimmung der Konstanten. Über die Frage, wie man vorzugehen hat, um die Konstanten G und K sowie die Koeffizienten der Dehnungs- und der Scherungsfunktion zu bestimmen, sei hier nur einiges Grundsätzliche bemerkt.

An sich würden zwei Grundversuche genügen, erstens ein solcher, bei dem der Probekörper einem hydrostatischen Spannungszustand unterworfen wird (und zwar auf Druck und Zug), und ein weiterer, bei dem die Beanspruchung aus einer reinen Gestaltsänderung ohne Volumenänderung besteht. Während sich ein Versuch der letzteren Art verhältnismäßig leicht durchführen läßt, etwa indem man ein dünnwandiges Rohr mit Kreisquerschnitt auf Torsion beansprucht, und sich ebenso ein hydrostatischer Druckversuch ausführen läßt, so stößt man bei einem hydrostatischen Zugversuch auf Schwierigkeiten. Man kann sie umgehen, indem man dafür einen einachsigen Zugversuch in der üblichen Form ausführt und dann auf rechnerischem Weg aus den Ergebnissen dieses und etwa des Torsionsversuches auf die Funktion $\varkappa (\varepsilon_0)$ und die Koeffizienten $\varkappa_i$ sowie auf die Konstante K schließt. Auch ein einachsiger Zug-

Druck-Versuch genügt schon, vorausgesetzt, daß man außer der Längs-auch noch die Querdehnung sehr genau bestimmen kann. Hier liegen allerdings einige bis jetzt noch nicht ganz befriedigend gelöste experimentelle Probleme.

Die folgenden Zahlwerte wurden aus Versuchsergebnissen errechnet, die von Roš und EICHINGER [6] an gezogenen und tordierten Stäben erzielt worden sind.

Man erhält so z. B. für Aluminiumbronze [7]

$$K = 1{,}35 \cdot 10^6 \, \text{kg/cm}^2, \quad k(s_0) = 1 + 38{,}0 \cdot 10^6 s_0^2 + 120{,}4 \cdot 10^{12} s_0^4,$$
$$G = 0{,}477 \cdot 10^6 \, \text{kg/cm}^2, \quad g(t_0^2) = 1 + 0{,}040 \cdot 10^6 t_0^2 + 0{,}195 \cdot 10^{12} t_0^4;$$

für Kupfer [8]

$$K = 1{,}33 \cdot 10^6 \, \text{kg/cm}^2, \quad \varkappa(\varepsilon_0) = 1,$$
$$G = 0{,}470 \cdot 10^6 \, \text{kg/cm}^2, \quad \gamma(\psi_0^2) = 1 - 7{,}26 \cdot 10^6 \psi_0^2.$$

5. Rechenverfahren zur Anwendung des Gesetzes. Möchte man unter Verwendung dieses Gesetzes die gleichen Aufgaben lösen, die schon mit dem HOOKEschen Gesetz behandelt worden sind, so wird man zuerst Umschau halten nach solchen Aufgaben, für die es eine exakte Lösung in geschlossener Form gibt. Hierbei erweist sich, wie zu erwarten war, die Gruppe dieser Aufgaben als viel kleiner als bei der Anwendung des HOOKEschen Gesetzes. Sie enthält im wesentlichen nur die Probleme des gleichförmigen Spannungszustandes und außerdem das Torsionsproblem für Stäbe mit kreis- oder kreisringförmigem Querschnitt. Bei diesem zeigt es sich, daß die Verzerrungen im Stabquerschnitt den gleichen linearen Verlauf über den Radius nehmen können wie beim HOOKEschen Gesetz, wobei dann natürlich die Schubspannungen nicht mehr linear von innen nach außen anwachsen können. Das Problem der gleichförmigen Biegung eines prismatischen Stabes durch freie Endmomente läßt hingegen schon keine exakte Lösung in geschlossener Form mehr zu.

Um nun zu weiteren Ergebnissen zu gelangen, bieten sich verschiedene Verfahren dar, die man auch sonst bei nichtlinearen Differentialgleichungen anzuwenden pflegt.

Als erstes sei das Verfahren der Reihenentwicklung der Verschiebungs-oder der Spannungskomponenten nach Potenzen der Koordinaten genannt, das allerdings trotz seiner grundsätzlich sehr allgemeinen Verwendbarkeit meistens auf so komplizierte Bestimmungsgleichungen für die Entwicklungskoeffizienten führt, daß sich seine Benützung nur in sehr wenigen Fällen, wie etwa solchen, bei denen die Verschiebungen nur von einer einzigen Ortskoordinate abhängen, empfiehlt. Insbesondere dürfen hierbei die Dehnungs- und die Scherungsfunktion keine komplizierte Gestalt haben. Die einfachste, auch physikalisch sinnvolle Annahme ist

$$\varkappa(\varepsilon_0) \equiv 1, \quad \gamma(\psi_0^2) \equiv 1 + \gamma_2 \psi_0^2$$

oder auch ein entsprechender Ansatz für die Kompressions- und die Schubfunktion:

$$k\,(s_0) \equiv 1, \quad g\,(t_0^2) \equiv 1 + g_2\,t_0^2\,.$$

Übrigens scheint der Ansatz $\varkappa\,(\varepsilon_0) \equiv 1$ oder $k\,(s_0) \equiv 1$ bei vielen Stoffen sehr weitgehend richtig zu sein; an der Nichtlinearität ist im allgemeinen in viel stärkerem Maß die Gestaltsänderung schuld als die Volumenänderung.

Bessere Erfolge erzielt man mit dem Verfahren der Störungsrechnung. Auch sie wendet man meistens an auf eine Kompressions- und eine Schubfunktion der angegebenen einfachen Gestalt. Ist die Lösung beim Hookeschen Gesetz bekannt, so entwickelt man die gesuchte Lösung nach Potenzen von γ_2 oder von g_2 als Parameter. Die Funktionen der Ortskoordinaten, die als Koeffizienten der Potenzen von γ_2 oder von g_2 auftreten, hat man dann als Lösungen linearer, nichthomogener Differentialgleichungen zu bestimmen, die allerdings meistens schon beim zweiten Glied sehr kompliziert werden, so daß man nur selten über das erste Störglied hinausgehen wird.

Ein sehr fruchtbares Verfahren besteht in der Anwendung der direkten Methoden der Variationsrechnung von RITZ oder GALERKIN auf die Minimalsätze über die Formänderungsarbeit. Man setzt beim Verfahren von RITZ eine Lösung, etwa für die Verschiebungen, mit noch unbestimmten Koeffizienten so an, daß sie die Randbedingungen für beliebige Werte dieser Koeffizienten erfüllt, und bestimmt nun die Koeffizienten so, daß das Minimalprinzip für die Formänderungsarbeit erfüllt wird. Hierbei zeigt es sich, daß man, wenn nur die Randbedingungen für die Verschiebungen in diesen und ihren Ableitungen homogen sind, mit einem sehr einfachen Ansatz oft bemerkenswert gute Resultate erzielen kann: Man setzt die Verschiebungskomponenten u, v, w als Funktionen der Koordinaten proportional zu den Verschiebungen an, die sich beim Hookeschen Gesetz für dasselbe Problem ergeben, und bestimmt den noch unbekannten Proportionalitätsfaktor aus der Minimalforderung für die Formänderungsarbeit. Man kann mit dieser primitiv erscheinenden Methode, zu deren Anwendbarkeit lediglich die Kenntnis der Lösung des entsprechenden Problems für das Hookesche Gesetz nötig ist, besonders bei Biegeproblemen von Stäben und Platten, Ergebnisse erzielen, die beinahe so genau sind wie diejenigen, welche die Störungsrechnung mit einem zweiten Störglied liefert. Wesentlich schlechter sind hingegen die Ergebnisse, die man erhält, wenn man mit einem entsprechenden Proportionalansatz für die Spannungen in den Ausdruck für die konjugierte Formänderungsarbeit eingeht und hiermit das CASTIGLIANOsche Minimalprinzip durch eine Variation der Spannungen zu erfüllen sucht.

6. Anwendungsbeispiele. An einigen typischen Beispielen, die aus einer viel größeren Zahl schon durchgerechneter Probleme herausgegriffen sind, soll nun gezeigt werden, wie sich die Nichtlinearität der Spannungs-Dehnungs-Beziehungen bei kleinen Verzerrungen auswirkt.

Das *erste Beispiel* [9] befaßt sich mit der Torsion zylindrischer Stäbe von beliebigem Querschnitt durch Momente M an den beiden Endflächen. Bekanntlich läßt sich diese Aufgabe beim HOOKEschen Gesetz mit Hilfe einer Spannungsfunktion $\Phi(x, y)$ lösen, aus der sich die Schubspannungskomponenten zu

$$\tau_{zx} = \frac{\partial \Phi}{\partial y}, \quad \tau_{yz} = -\frac{\partial \Phi}{\partial x}$$

berechnen lassen. Diese Spannungsfunktion muß beim HOOKEschen Gesetz mit dem LAPLACEschen Operator Δ die Differentialgleichung

$$\Delta \Phi = -2G\omega$$

erfüllen, falls ω den Verdrillungswinkel des Stabes je Längeneinheit bedeutet; am Rand des Querschnitts muß Φ konstant bleiben. Auch das nichtlineare Elastizitätsgesetz läßt die Einführung dieser Spannungsfunktion zu. Ihre Differentialgleichung lautet jetzt

$$\frac{\partial}{\partial x}\left[g(t_0^2)\frac{\partial \Phi}{\partial x}\right] + \frac{\partial}{\partial y}\left[g(t_0^2)\frac{\partial \Phi}{\partial y}\right] = -2G\omega,$$

wobei

$$t_0^2 = \frac{2}{3G^2}\left[\left(\frac{\partial \Phi}{\partial x}\right)^2 + \left(\frac{\partial \Phi}{\partial y}\right)^2\right]$$

ist.

Hat die Funktion $g(t_0^2)$ speziell die Gestalt $g(t_0^2) = 1 + g_2 t_0^2$, so kann man die Differentialgleichung mittels der Störungsrechnung durch Entwicklung von $\Phi(x, y)$ nach Potenzen von g_2 für einen elliptischen Querschnitt näherungsweise lösen. Mit dem ersten Glied der Entwicklung erhält man z.B. folgende Zahlwerte:

Die Halbachsen der Querschnittsellipse seien 10 cm und 5 cm, ferner sei $G = 0{,}870 \cdot 10^6$ kg/cm², $g_2 = 0{,}085 \cdot 10^6$, $\omega = 2 \cdot 10^{-4}$ cm⁻¹. Dann ist die maximale Torsionsspannung (an den Enden der kleinen Halbachsen)

$$\tau_{max} = 1392 \cdot (1 - 0{,}135)\,\text{kg/cm}^2,$$

und das Torsionsmoment wird $M = 5{,}47 \cdot (1 - 0{,}071)$ m to.

Das HOOKEsche Gesetz hätte bei gleichem Torsionswinkel und gleichem Schubmodul

$$\tau_{max} = 1392\,\text{kg/cm}^2, \quad M = 5{,}47\,\text{m to}$$

geliefert. Die maximale Torsionsspannung verringert sich demnach um 13,5% bei einem um nur 7,1% geringeren Torsionsmoment.

Im *zweiten Beispiel* [10] sei über den ebenen Spannungszustand berichtet. Wie beim HOOKEschen Gesetz läßt sich auch hier die AIRYsche Spannungsfunktion $F(x, y)$ einführen, aus der sich die Spannungen zu

$$\sigma_x = \frac{\partial^2 F}{\partial y^2}, \quad \sigma_y = \frac{\partial^2 F}{\partial x^2}, \quad \tau_{xy} = -\frac{\partial^2 F}{\partial x \partial y}$$

berechnen. Diese Spannungsfunktion, die beim Hookeschen Gesetz die Bipotentialgleichung $\Delta\Delta F = 0$ erfüllen muß, hat hier die Differentialgleichung

$$\Delta\left\{\left[\frac{1}{9\,K}\,k\,(s_0) + \frac{1}{3\,G}\,g\,(t_0^2)\right]\Delta F\right\} -$$
$$-\frac{1}{2\,G}\left(\frac{\partial^2 F}{\partial y^2}\,\frac{\partial^2 g}{\partial x^2} + \frac{\partial^2 F}{\partial x^2}\,\frac{\partial^2 g}{\partial y^2} - 2\,\frac{\partial^2 F}{\partial x\,\partial y}\,\frac{\partial^2 g}{\partial x\,\partial y}\right) = 0$$

zu befriedigen, wobei

$$s_0 = \frac{1}{9\,K}\,\Delta F, \quad t_0^2 = \frac{2}{9\,G^2}\left[(\Delta F)^2 + 3\left(\frac{\partial^2 F}{\partial x\,\partial y}\right)^2 - 3\,\frac{\partial^2 F}{\partial x^2}\,\frac{\partial^2 F}{\partial y^2}\right]$$

zu setzen ist. Diese Differentialgleichung läßt sich für $k(s_0) \equiv 1$ und $g(t_0^2) \equiv 1 + g_2 t_0^2$ mit Hilfe der Störungsrechnung behandeln. So wurde z. B. das ebene Problem des einseitig eingespannten Balkens von der Länge l mit einem schmalen, hohen Rechtecksquerschnitt mit den Seitenlängen h und $2b$, dessen freies Ende eine Last Q trägt, für $l = 50$ cm, $h = 1$ cm, $2b = 10$ cm, $Q = 33,3$ kg und mit den Materialkonstanten für Kupfer durchgerechnet. Man erhält für die Biegespannung an der Einspannstelle die Werte $100,0$ kg/cm² beim Hookeschen Gesetz, $(100,0 - 2,60)$ kg/cm² mit dem ersten Glied der Störungsrechnung und $(100,0 - 2,60 + 0,42) = 97,82$ kg/cm² mit dem zweiten Glied der Störungsrechnung, also eine Verringerung der maximalen Biegespannung um $2,18\%$. Die Durchbiegung in der Mitte des Endquerschnitts wird entsprechend $(13,22 \cdot 10^{-3} + 0,31 \cdot 10^{-3} - 0,47 \cdot 10^{-5})$ cm; sie ist um $2,35\%$ größer als beim Hookeschen Gesetz. Führt man die Rechnung durch, indem man die Minimalforderung für die Formänderungsarbeit für den einfachen, vorher erwähnten Proportionalansatz der Verzerrungen erfüllt, so erhält man mit wesentlich geringerem Rechenaufwand das Ergebnis, daß die Durchbiegung um $2,35\%$ größer und die Maximalspannung um beinahe ebenso viel kleiner ist als beim Hookeschen Gesetz.

Man kann statt der rechtwinkligen auch Polarkoordinaten r, φ einführen. So läßt sich etwa als einfaches Beispiel, bei dem die Spannungen und die Verzerrungen nur noch von einer einzigen Ortskoordinate abhängen, die Kreisringscheibe konstanter Dicke unter gleichmäßigem Außendruck p mit den Halbmessern r_0 (innen) und r_a (außen) berechnen, wobei man jetzt von einem Potenzreihenansatz in $r - r_0$ für dF/dr als Lösungsansatz ausgehen kann. Mit $p = 152,78$ kg/cm² $r_a/r_0 = 1,2$ und den Konstanten für Kupfer werden die Tangentialspannungen $\sigma_{\varphi\,0}$ am Innenrand und $\sigma_{\varphi\,a}$ am Außenrand des Ringes

$$\sigma_{\varphi\,0} = -923,9\,\text{kg/cm}^2, \quad \sigma_{\varphi\,a} = -907,8\,\text{kg/cm}^2.$$

Die entsprechenden Werte, die das Hookesche Gesetz bei der gleichen Belastung liefert, sind

$$\sigma_{\varphi\,0} = -1000,0\,\text{kg/cm}^2, \quad \sigma_{\varphi\,a} = -847,2\,\text{kg/cm}^2.$$

Die Nichtlinearität wirkt sich also hier in dem Sinn aus, daß das Gefälle des Spannungsbetrages vom Innenrand zum Außenrand wesentlich verringert wird.

Im *dritten Beispiel* [11] sei ein kugelsymmetrisches Problem erwähnt, bei dem sich eine ähnliche Erscheinung zeigt wie bei dem soeben erörterten Beispiel des ebenen Rings. Es betrifft die Hohlkugel mit den Halbmessern r_0 und r_a unter gleichmäßigem Innendruck p_0. Für den Fall $k(s_0) \equiv 1$, $g(t_0^2) \equiv 1 + g_2 \cdot t_0^2$ ist hier noch eine geschlossene Integration der Differentialgleichung möglich. Das Ergebnis ist in Abb. 1 für einen typischen Fall wiedergegeben, für den $p_0 = 500$ kg/cm², $r_a/r_0 = 2$ und die hier allein maßgebende Stoffzahl

$$\alpha^2 = \frac{g_2 K}{2 G^2 (3 K + 4 G)} = 10^{-6} \text{ cm}^4/\text{kg}^2$$

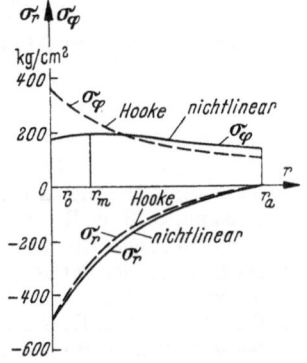

Abb. 1. Spannungsverlauf in der Wand der Hohlkugel bei Innendruck

gesetzt worden ist. Das Diagramm zeigt den Verlauf der Radialspannung σ_r und der Tangentialspannung σ_φ über dem Radius. Zum Vergleich ist der Verlauf der entsprechenden Spannungen beim HOOKE-schen Gesetz gestrichelt eingetragen. Man erkennt wieder die ausgleichende Wirkung der Nichtlinearität auf das Gefälle von σ_φ. Diese Spannung erreicht nun nicht mehr an der Innenfläche, sondern an einer Zwischenstelle $r = r_m$ innerhalb der Kugel ihren Extremwert. Bemerkenswert ist auch die starke Abnahme von σ_φ an der Innenfläche.

Es sei an dieser Stelle erwähnt, daß man ganz ähnliche Verhältnisse antrifft, wenn man den ebenen Verzerrungszustand eines dickwandigen Rohres unter Innen- und Außendruck untersucht [12], was wiederum mit der Methode der Potenzreihenentwicklung möglich ist.

Im *vierten Beispiel* [13] sei noch ein Ergebnis aus der Plattentheorie angeführt. Es bezieht sich auf die am Rand eingespannte Kreisplatte vom Halbmesser R und konstanter Dicke h. Man kann zur Behandlung dieses Problems etwa so vorgehen, daß man eine „Plattengleichung" für die Durchbiegung w aufstellt, die der bekannten Plattengleichung für das HOOKEsche Gesetz entspricht, aber komplizierter aufgebaut ist, und dann diese Gleichung näherungsweise durch ein Verfahren der Reihenentwicklung löst. Wesentlich bequemer führt hier jedoch wieder die Anwendung des Verfahrens von RITZ auf die Minimalforderung für die Formänderungsarbeit zum Ziel. Mit dem Proportionalansatz für die Durchbiegung w erhält man so bei einer gleichmäßigen Belastung $p = 3{,}6$ t/m für $h/R = 2/100$ und den Konstanten für Aluminiumbronze die folgenden Werte für die Durchbiegung w_0 in der Mitte und die Biege-spannungen σ_{ra} am Plattenrand und σ_{r0} in der Mitte

$$w_0 = 0{,}297\,h, \quad \sigma_{ra} = 643{,}7\,\text{kg/cm}^2, \quad \sigma_{r0} = 421{,}8\,\text{kg/cm}^2.$$

Die entsprechenden Werte der linearen Theorie sind zum Vergleich

$$w_0 = 0{,}291\,h, \quad \sigma_{ra} = 675{,}0\,\text{kg/cm}^2, \quad \sigma_{r0} = 452{,}9\,\text{kg/cm}^2.$$

Die Abweichung ist für die zugrunde gelegte Belastung noch ziemlich gering, läßt aber doch schon wieder die für das nichtlineare Elastizitäts-gesetz charakteristische Tendenz erkennen, daß mit der Vergrößerung der Durchbiegung ein Abbau der Spannungsspitzen verbunden ist.

Literatur

[1] KAUDERER, H.: Ing.-Arch. 17, 450 (1949).
[2] BIEZENO, C. B. u. R. GRAMMEL: Technische Dynamik, I. Bd., 2. Aufl., S. 35. Berlin/Göttingen/Heidelberg 1953.
[3] HAMEL, G.: Elementare Mechanik, S. 571. Leipzig und Berlin 1912.
[4] ARGYRIS, J. H.: Aircraft Engineering 26, 347 (1954).
[5] BIEZENO, C. B. u. R. GRAMMEL: Technische Dynamik, I. Bd., 2. Aufl., S. 81 ff. Berlin/Göttingen/Heidelberg 1953.
[6] ROŠ, M. u. A. EICHINGER: Diskussionsbericht der Eidgen. Materialprüfungs-anstalt in Zürich Nr. 34 (1929).
[7] KAUDERER, H.: Ing.-Arch. 17, 464 (1949).
[8] KAUDERER, H.: Ing.-Arch. 17, 468 (1949).
[9] KAUDERER, H.: Ing.-Arch. 17, 476 (1949).
[10] JINDRA, F.: Ing.-Arch. 22, 121 (1954).
[11] JINDRA, F.: Ing.-Arch. 22, 411 (1954).
[12] JINDRA, F.: Ing.-Arch. 23, 122 (1955).
[13] ÖZDEN, K.: Ing.-Arch. 24, 133 (1956).

Second order effects in infinitesimal elasticity

By M. Reiner, Haifa

1. The elastic state of a solid body is described by the tensor of strain e, to which the tensor of stress p is related by the stress-strain equation

$$p = f(e). \tag{1.1}$$

The strain is not a directly observable quantity. When the body is strained, the quantities actually measured are the *displacements* u_i which carry its particles from the *initial* state $_ix$ to the final state x_i, so that

$$u_i = x_i - {}_ix. \tag{1.2}^{[1]}$$

Defining the *displacement-gradient* by either of the two equations

$$_{ij}\gamma = \frac{\partial u_i}{\partial_j x}; \quad \gamma_{ij} = \frac{\partial u_i}{\partial x_j} \tag{1.3}$$

the strain-tensor, as expressed in terms of the displacement-gradient, constitutes what may be named a "measure of strain".

2. The *measure of strain* is not unambiguous. Several different measures have been proposed and used. Two of them have been expressed in tensor form by MURNAGHAN (1937). They were first proposed by GREEN (1841) and ALMANSI (1911) respectively. In the first the extension of a linear element is expressed by relating the elongation Δl to the original length l_0, in the second it is related to the final length

$$l = l_0 + \Delta l. \tag{2.1}$$

For the purpose of our investigation Cartesian coordinates suffice. The two strain-tensors are then

$$\left.\begin{aligned} {}^{G}_{ij}e &= \frac{1}{2}\left({}_{ij}\gamma + {}_{ji}\gamma + {}_{\alpha i}\gamma\, {}_{\alpha j}\gamma\right) \\ {}^{A}e_{ij} &= \frac{1}{2}\left(\gamma_{ij} + \gamma_{ji} - \gamma_{\alpha i}\gamma_{\alpha j}\right). \end{aligned}\right\} \tag{2.2}$$

In the classical theory of infinitesimal elasticity the γ-s are assumed

---

[1] Note that an index to the left refers to the initial, one on the right to the final state.

to be infinitesimal and the products of their components are discarded, a procedure resulting in the linear measure

$$\varepsilon_{ij} = \frac{1}{2} \left(\gamma_{ij} + \gamma_{ji} \right). \tag{2.3}$$

The argument is as follows:

If the γ-s are infinitesimal, $\gamma_{\alpha i} \gamma_{\alpha i}$ is infinitesimal of second order and can be neglected against $\gamma_{ij} + \gamma_{ji}$ which is infinitesimal of the first order. However, this procedure is admissible only if the term $\gamma_{ij} + \gamma_{ji}$ is present; if it is absent, i.e. equal to zero, the $\gamma_{\alpha i} \gamma_{\alpha j}$ evidently cannot be neglected being infinitely great in comparison with 0. For instance, one of the components of Equ. (2.2, 2) is

$$2\,e_{yz} = \left(\frac{\partial u_z}{\partial y} + \frac{\partial u_y}{\partial z} \right) - \left(\frac{\partial u_x}{\partial y}\frac{\partial u_x}{\partial z} + \frac{\partial u_y}{\partial y}\frac{\partial u_y}{\partial z} + \frac{\partial u_z}{\partial y}\frac{\partial u_z}{\partial z} \right). \tag{2.4}$$

Now consider a case where $u_y = u_z = 0$, then the first term in brackets on the right of (2.4) is reduced to zero and the second to $\dfrac{\partial u_x}{\partial y}\dfrac{\partial u_x}{\partial z}$. This cannot be neglected as "infinitely small of the second order". Such expression is in this case meaningless. A quantity can only be neglected in comparison with another one, whose increment it is. If I may be permitted an example from a very different field, one pound sterling may be a negligible quantity for, say, Lord ROTHSCHILD, but it is certainly not one for me.

It will be shown in the following that in this manner *second order effects* may arise in infinitesimal elasticity. In the case of solid metals and similar materials in which the elasticity is of the potential kind, the displacement gradients are always infinitesimal[1], and in such materials these are therefore the only kind of second order effects.

3. We shall now treat a case of homogenous state of stress and strain which will prove our thesis. If the stress is homogeneous, the equilibrium equations are identically complied with in the absence of body forces, and need not be considered. For the purpose of our investigation we may assume that body forces are absent.

The case is one of simple shear.

Simple shear is given for instance by

$$u_x = S\,y_0 = S\,y; \quad u_y = u_z = 0. \tag{3.1}$$

Therefore

$$\gamma_{ij} = \begin{Vmatrix} 0 & S & 0 \\ 0 & 0 & 0 \\ 0 & 0 & 0 \end{Vmatrix} = {}_{ij}\gamma. \tag{3.2}$$

---

[1] Even if the displacements themselves are finite.

The strain tensors are accordingly from Equ. (2.2)

$$\overset{G}{_{ij}e} = \frac{S}{2} \begin{Vmatrix} 0 & 1 & 0 \\ 1 & S & 0 \\ 0 & 0 & 0 \end{Vmatrix} \qquad \overset{A}{e_{ij}} = \frac{S}{2} \begin{Vmatrix} 0 & 1 & 0 \\ 1 & -S & 0 \\ 0 & 0 & 0 \end{Vmatrix}. \tag{3.3}$$

To calculate the stress, we draw upon the classical linear stress-strain relation

$$p_{ij} = \lambda I_e \delta_{ij} + 2\mu e_{ij} \tag{3.4}$$

where I_e is the first invariant of the strain tensor. This relation may be non-linear in displacement-gradient components. Equ. (3.4), while not loosing its *tensorial* linearity, has for such cases been termed "quasi-linear" (compare TRUESDELL [1952]). Furthermore the moduli λ and μ may be functions of the invariants of the strain tensor. Such state has been named *physical* non-linearity. This also does not affect the tensorial linearity of Equ. (3.4).

The stress must be referred to the final state which is the state of equilibrium. The ALMANSI-measure refers to this same state, but when the GREEN-measure, which refers to the initial state, is used, we must carry out a rotation which brings it into the final state.

The rotation matrix is

$$R_{ij} = \begin{Vmatrix} \cos\alpha & \sin\alpha & 0 \\ -\sin\alpha & \cos\alpha & 0 \\ 0 & 0 & 1 \end{Vmatrix} \tag{3.5}$$

where (compare LOVE [1927] Art. 3)

$$\tan\alpha = \frac{S}{2}. \tag{3.6}$$

This makes

$$\overset{G}{e_{ij}} = \frac{S}{2} \begin{Vmatrix} S & 1 & 0 \\ 1 & 0 & 0 \\ 0 & 0 & 0 \end{Vmatrix}. \tag{3.7}$$

The invariants of these two measures are

$$\left.\begin{aligned} &\overset{G}{I} = \frac{S^2}{2}; \quad \overset{A}{I} = -\frac{S^2}{2} \\ &\overset{G}{II} = \overset{A}{II} = -\frac{S^2}{4} \\ &\overset{G}{III} = \overset{A}{III} = 0. \end{aligned}\right\} \tag{3.8}$$

Introducing $\overset{G}{e}_{ij}$ and $\overset{A}{e}_{ij}$ from Equ. (3.7) and (3.3) into Equ. (3.4) yields

$$\overset{G}{p}_{ij} = \frac{\lambda S^2}{2}\delta_{ij} + \mu S \begin{Vmatrix} S & 1 & 0 \\ 1 & 0 & 0 \\ 0 & 0 & 0 \end{Vmatrix}. \qquad (3.9)$$

$$\overset{A}{p}_{ij} = -\frac{\lambda S^2}{2}\delta_{ij} + \mu S \begin{Vmatrix} 0 & 1 & 0 \\ 1 & -S & 0 \\ 0 & 0 & 0 \end{Vmatrix}. \qquad (3.10)$$

In order to maintain simple shear it is therefore necessary to apply an isotropic tension in the first case or an isotropic pressure in the second case, and in addition, a tensional traction parallel to the direction of the displacement in the first case, and a compressional traction normal to the direction of the displacement in the second case. If these second-order stresses and tractions are absent, a simple shearing stress will produce in the first case a cubical shrinkage, and in addition a linear contraction in the direction of the displacement. In the second case it will produce a cubical expansion together with a linear extension in the direction normal to the displacement.

4. Note that in the terms which define the above mentioned phenomena, the first power of S is absent, and the second powers therefore *cannot be neglected* even when S is infinitesimal. There is, of course, nothing to prevent an investigator to *disregard* them — provided he knows of their existence.

If he does so without examining the consequences, such procedure may be dangerous. I would liken him to one driving a car on the street of a Dutch town at the side of a canal. It may be of no importance whether his forward velocity is 50 or 55 miles per hour and he may neglect the additional 5 miles, but if his velocity component in the direction normal to the street is of that "negligible" order, this will land him into the water.

Such elastic effects have actually been observed by POYNTING (1909, 1912) in the torsion of steel and hard copper wires when the maximum shear was of the order of 10^{-3}, and therefore "infinitesimal" in comparison with unity. While only a torsional torque and no pull was applied, the torsional shear was accompanied by extensions, transversal contractions and a cubical expansion, all distinctly observable of the order on 10^{-6}. There may be cases of practical importance where such strains cannot be disregarded in the manner of classical elasticity.

One other remark. Attempts have been made to take into consideration physical non-linearity by assuming that λ changes with I and μ with II. However, from Equ. (3.8) and both (3.9) and (3.10) it seems that this is an effect of third order, incremental to one of the first order. It would therefore appear that the consideration of geometrical non-linearity has priority over that of physical non-linearity.

5. There is, however, something disturbing about the results found. Note that λ and μ, of whatever magnitude and dependence upon invariants, are necessarily positive, and that accordingly the measure of strain chosen prejudices the experimental results. We are therefore forced to conclude that the classical equation (3.4) is not general enough even for infinitesimal strain. It has been generalized by REINER (1948) to

$$p_{ij} = \lambda I_e \delta_{ij} + 2\mu e_{ij} + 4\mu_c e_{i\alpha} e_{\alpha j} \tag{5.1}$$

when the introduction of μ_c, an independent *coefficient of cross-elasticity* solves this difficulty. REINER (1955) has applied Equ. (5.1) to POYNTING's observations.

He has shown that the ratio of μ_c to μ determines whether a cylinder in torsion is lengthened or shortened. While POYNTING found a lengthening in the case of steel and hard copper at room temperature, there are reasons to believe that lead at room temperature and steel and hard copper at elevated temperatures would show shortening. Equ. (5.1) therefore does not prejudice experimental results and may be used with *any measure of strain for e.*

References

[1] ALMANSI, F.: Rend. Lincei (5 A) **201**, 705—714 (1911).
[2] GREEN, G.: Trans. Cambr. Phil. Soc. 7, 121—140 (1841).
[3] LOVE, E. H.: A treatise on the mathematical theory of elasticity, Cambridge 1927 (Dover Publications, New York 1944).
[4] MURNAGHAN, F. D.: Amer. J. Math. 59, 235—260 (1937).
[5] POYNTING, J. H.: Proc. Roy. Soc. A, 82, 546 (1909); Proc. Roy. Soc. A, 86, 534 (1912).
[6] REINER, M.: Amer. J. Math. 70, 433—446 (1948).
[7] REINER, M.: Appl. Scient. Res. A 5, 281—295 (1955).
[8] TRUESDELL, C.: J. Rat. Mech. Anal. 1, 125—300 (1952).

Large Elastic Deformations in Rubberlike Materials

By **L. R. G. Treloar,** Manchester

With 3 figures

1. Introduction. A complete description of the mechanical behaviour of an elastic material should enable the state of strain to be determined for any system of applied stresses. A convenient method of representation, which satisfies this requirement, is in terms of the energy stored elastically in the body as a function of the state of strain. Moreover, if the material is considered to be homogeneous, it is sufficient to determine this function for the case of a pure homogeneous strain. This follows from the considerations 1) that an inhomogeneously strained body may be divided up into a large number of small volume elements such that in any one element the strain is substantially homogeneous, and 2) that a strain which is not pure may be regarded as a pure strain, together with a rotation and translation. Of these, it is only the first which involves the performance of work on the material and which contributes, therefore, to the elastically stored energy.

The essential problem, therefore, is to determine the form of the stored-energy function for the given material in a pure homogeneous strain. In the present paper we shall consider in particular the form of this function for materials which may be classed as rubberlike. These materials have the property that the elastic strains which may be applied are very large — of the order of 1,000 times those normally encountered in elastic solids. These deformations, however, take place substantially at constant volume. This means that rubberlike materials are substantially incompressible, with respect to hydrostatic pressure, but undergo large changes of shape under the action of shear stresses.

By limiting the argument to materials which are incompressible and which in addition are mechanically isotropic, it is possible to express the stored-energy function, and the corresponding stress-strain relations, in terms of two quantities which completely define the strain. In the following pages we shall examine the question of the possible forms which the stored-energy function may assume, subject to these restrictions. We shall also examine the actual behaviour of a vulcanized rubber and see to what extent its properties may be represented in terms of these theoretical concepts.

The above type of reasoning is purely phenomenological, that is to say, it considers the relations between observable properties only, but does not enquire into the physical or molecular mechanisms which may be responsible for those properties. Actually, in the case of rubbers, the molecular type of theory has been the more potent, and has provided the main stimulus to the investigation of their mechanical and other physical properties. It seems desirable, therefore, to give a brief review of the molecular theory of rubber elasticity, and to show how the conclusions derived from it fit in with the more general phenomenological theory.

2. The kinetic theory of rubber elasticity. The unusual physical properties of rubber are associated with equally remarkable thermodynamic effects which have attracted attention for more than a century. Thus JOULE, in 1859 [1] re-examined GOUGH's earlier finding in 1805 [2] that a piece of rubber, stretched by a load, shortens when the temperature is raised, and discovered that the deformation of rubber is accompanied by a reversible evolution of heat, while KELVIN [3] showed that these two effects are thermodynamically related. Little progress was made in their detailed interpretation, however, until the development of the chemistry of high-molecular materials had reached a stage at which it became possible to form a clear picture of the molecular constitution of rubberlike materials. Our present ideas on rubber elasticity are based on the concept originally formulated by MEYER and his associates [4], that this phenomenon arises directly from the statistical properties of the long-chain molecule itself. The molecule of a typical rubber has the form of a very long chain containing some 10,000 basic (C_5H_8) units. Such a chain will not remain in the form of a straight rod, but will assume an irregularly-kinked and continually fluctuating form under the action of the thermal motion of its constituent parts, which are connected together not rigidly, but by single bonds which permit of relatively free rotation. This kinetic interpretation leads to the conclusion that the chain possesses elasticity, since the fully extended state is inherently improbable, while the multiply-kinked or contracted state has the highest probability. The elasticity of the molecule, therefore, is of an entirely different type from the elasticity of a normal (e.g. crystalline) material, in which the applied stress does work against internal attrative forces, thus leading to an increase in the internal energy of the system. In the case of the long-chain molecule, the stretched state has no more internal energy than the unstretched state — it is merely less probable. In thermodynamic terms, this implies that the stretched state has a lower entropy than the normal or contracted state. MEYER drew a parallel between rubber elasticity and the elasticity of a gas. In both cases the internal energy is a function only of temperature and is independent of the "deformation", and in both the work done by an applied force is converted into heat. In both, the magnitude of the applied force should be proportional to the absolute temperature.

In contrast to the complicated and highly artificial hypotheses which had been put forward previously in explanation of rubber elasticity, MEYER's kinetic hypothesis overcame in a simple and natural way the difficulties which had appeared so formidable before his time. Its introduction led very quickly to quantitative mathematical developments designed to account in detail for the mechanical properties of rubber. These will now be considered.

3. The elasticity of a molecular network. Let us consider first a single chain. The simplest molecular structure which can be chosen is the polyethylene or paraffin chain $(CH_2)_n$. This may be represented geometrically as a chain of n links each of length l, such that adjacent links make a constant angle $\pi - \Theta$ to each other. If the distance between the ends of the chain is represented by the vector r the probability $p(x, y, z)$ that this should have components x, y, z, with respect to a fixed coordinate system may be represented in the form

$$p(x,\, y,\, z)\, dx\, dy\, dz = \frac{b^3}{\pi^{3/2}}\, e^{-b^2\,(x^2+y^2+z^2)}\, dx\, dy\, dz \tag{1}$$

$$= \frac{b^3}{\pi^{3/2}}\, e^{-b^2\, r^2}\, dx\, dy\, dz \tag{1a}$$

Equations (1) and (1a) express the fact that the probability is proportional to the size of the volume element $(dx\, dy\, dz)$ within which the end of the vector r lies. The expression (1) has the form of the GAUSSIAN error function. It is characterized by a single parameter b which is related to the geometry of the chain structure. For the type of chain considered

$$b^2 = \frac{3}{2\,n\,l^2}\left(\frac{1-\cos\Theta}{1+\cos\Theta}\right). \tag{1b}$$

The application of BOLTZMANN's relation between entropy S and probability p, i.e.

$$S = k\log p \tag{2}$$

to (1a) leads to the following expression for the entropy of the chain,

$$s = c - k\,b^2\,r^2 \tag{3}$$

where c is an arbitrary constant and k is BOLTZMANN's constant. Since only differences of entropy are significant we may put $s = 0$ when $r = 0$, i.e.

$$s = -k\,b^2\,r^2 \tag{3a}$$

This implies that the entropy of the chain becomes more negative as the distance r between its ends increases.

To deal with the material in bulk it is necessary to take into account the whole assembly of chains, and the forces between them. For the purpose of the theory, it is assumed that the chains are connected together at a few points along their length so as to form a loose three-dimensional network. All other intermolecular forces are neglected. The problem now is to compute the total entropy of the network as a function of its state of strain, i.e. of the position of its boundary surfaces. This calculation may be made on the basis of the expression (3a) for the entropy of the single chain — a chain now being considered as the segment of a molecule between successive points of cross-linkage. Let us consider the case of a pure homogeneous strain, represented by threee principal extension ratios λ_1, λ_2 and λ_3 (the semi-axes of the strain ellipsoid). For this case the entropy of deformation, per unit volume, is found to be [5]

$$\Delta S = -\frac{1}{2} N k (\lambda_1^2 + \lambda_2^2 + \lambda_3^2 - 3) \qquad (4)$$

where N is the number of chains per unit volume. The work of deformation (at constant temperature T) is, by definition, equal to the change in HELMHOLTZ free energy (ΔA) in passing from the unstrained to the strained state. By standard thermodynamics

$$\Delta A = \Delta E - T \Delta S \qquad (5)$$

where ΔE is the change in internal energy. By hypothesis ΔE is zero. The work of deformation per unit volume (W) is therefore

$$W = \Delta A = -T \Delta S = \frac{1}{2} N k T (\lambda_1^2 + \lambda_2^2 + \lambda_3^2 - 3) \qquad (6)$$

This is the desired result. It is interesting to note that it expresses all the deformation characteristics of a rubber in terms of a single material parameter or modulus, which may be written C_1, where

$$C_1 = \frac{1}{2} N k T \qquad (6a)$$

This parameter is independent of the chain geometry and is related solely to the number of chains per unit volume of the network, which is itself dependent on the number of cross-linkages. Thus the form of the stored-energy function, and hence of the stress-strain relations, should be the same for all rubbers, whatever their chemical constitution. This is a conclusion of the utmost significance.

The quantity C_1 may also be expressed in terms of the molecular weight between successive junction points, M_c. The relation is

$$C_1 = \frac{\varrho R T}{2 M_c} \qquad (6b)$$

where ϱ is the density of the material and R is the gas constant per mole.

14*

The consideration of the stress-strain relations corresponding to this stored-energy function (6), and their experimental verification, will be deferred. It will be convenient, however, to refer here to the experimental examination of the relation between the modulus C_1 and the number of cross-linkages ν. For the usual case, in which four network segments or chains terminate on each cross-linkage, $N = 2\nu$. If, therefore, the number of cross-linkages can be estimated chemically, it should be possible, using (6a), to calculate C_1. Actually, the estimation of ν presents considerable difficulty, and is not possible when the ordinary vulcanizing or cross-linking reagents are used, since these produce various uncontrollable side reactions in addition to the main cross-linking reaction. However, by using certain compounds which are believed to give rise only to cross-linkages between chains, FLORY, RABJOHN and SCHAFFER [6] have attempted to estimate quantitatively the number of cross-linkages. Their results show the modulus to be approximately proportional to the number of cross-links, and to agree with the theoretical value to an accuracy of about 25 per cent.

4. Phenomenological theory. We return now to the representation of elastic properties on the basis of the phenomenological type of theory referred to in the introduction. The present discussion is based primarily on the work of RIVLIN [7], who has made extensive contributions to this subject. From considerations of symmetry the stored-energy function for an isotropic material must be symmetrical in the three extension ratios, λ_1, λ_2 and λ_3. It may be shown also, that it must be an even-powered function of these quantities. Any even-powered function which satisfies these conditions may be expressed as a function of the following three quantities, or strain-invariants,

$$\left. \begin{aligned} I_1 &= \lambda_1^2 + \lambda_2^2 + \lambda_3^2, \\ I_2 &= \lambda_1^2 \lambda_2^2 + \lambda_2^2 \lambda_3^2 + \lambda_3^2 \lambda_1^2, \\ I_3 &= \lambda_1^2 \lambda_2^2 \lambda_3^2. \end{aligned} \right\} \tag{7}$$

If the material is incompressible,

$$\lambda_1 \lambda_2 \lambda_3 = 1 \tag{8}$$

and therefore I_3 is constant. The stored energy is then determined by the two quantities I_1 and I_2, in terms of which the strain is completely specified. With the help of (8) these may be written in the form

$$I_1 = \lambda_1^2 + \lambda_2^2 + \lambda_3^2, \quad I_2 = \frac{1}{\lambda_1^2} + \frac{1}{\lambda_2^2} + \frac{1}{\lambda_3^2}. \tag{9}$$

The most general form of stored-energy function for an isotropic, incompressible elastic material may therefore be expressed in the form of the series

$$W = \sum_{i=0,\, j=0}^{\infty} C_{ij} (I_1 - 3)^i (I_2 - 3)^j \tag{10}$$

the quantities $I_1 - 3$ and $I_2 - 3$ being inserted, rather than I_1 and I_2, to ensure that W automatically vanishes at zero strain.

5. Particular forms of stored-energy function. The two simplest forms of stored-energy function, from the purely mathematical standpoint, are therefore

$$W = C_1 (I_1 - 3), \tag{11}$$

$$W = C_2 (I_2 - 3). \tag{12}$$

Of these (11) is the form deduced from the statistical theory [equation (6)]. The form (12) does not appear to have acquired any physical significance. However, the combination of (11) and (12) to give

$$W = C_1 (I_1 - 3) + C_2 (I_2 - 3) \tag{13}$$

yields a formula which was originally discovered by MOONEY [9]. This formula is the most general first-order relation in I_1 and I_2 and it yields, as will be shown, a better approximation to the actual behaviour of a rubber than the single-constant expression (11). It has the property of yielding a linear stress-strain relation in simple shear, and it was on the basis of this property that MOONEY originally derived it.

6. General stress-strain relations. The state of pure homogeneous strain may be maintained by the application of three principal stresses acting normally on surfaces normal to the directions of the principal strains. Expressions for these stresses may be obtained from the stored-energy function by partial differentiation. If t_1, t_2 and t_3 are the principal stresses, the general stress-strain relations for an isotropic incompressible elastic material may be shown to be of the form [8]

$$t_1 - t_2 = 2 (\lambda_1^2 - \lambda_2^2) \left(\frac{\partial W}{\partial I_1} + \lambda_3^2 \frac{\partial W}{\partial I_2} \right) \tag{14}$$

with corresponding expressions for $t_2 - t_3$ and $t_3 - t_1$. It is noteworthy that the absolute magnitude of the stresses is not determined, but only differences between them. This is a direct consequence of the assumption of incompressibility, and implies that state of strain is unaffected by the superposition of a hydrostatic pressure.

The elastic behaviour of a material is completely specified if the dependence of $\partial W/\partial I_1$ and $\partial W/\partial I_2$ on I_1 and I_2 is known for all values of these independent variables. Thus, for example, the statistical theory [equations (6) and (11)] leads to the result

$$\frac{\partial W}{\partial I_1} = C_1, \qquad \frac{\partial W}{\partial I_2} = 0, \tag{15}$$

while the MOONEY equation (13) yields

$$\frac{\partial W}{\partial I_1} = C_1, \qquad \frac{\partial W}{\partial I_2} = C_2. \tag{16}$$

7. Particular stress-strain relations. Before considering the actual behaviour of rubbers, it is desirable to derive the stress-strain relations for certain simple types of strain. The types of strain of particular interest are simple elongation, uni-directional compression and shear.

Simple elongation. In this type of strain we have, for an incompressible material

$$\lambda_1 = \lambda, \quad \lambda_2 = \lambda_3 = \lambda^{-1/2}$$

where λ is the extension ratio. The only stress acting is the tensile stress t. Hence, from (14)

$$t = 2\left(\lambda^2 - \frac{1}{\lambda}\right)\left(\frac{\partial W}{\partial I_1} + \frac{1}{\lambda}\frac{\partial W}{\partial I_2}\right). \tag{17}$$

It is customary to refer the force to the cross-sectional area measured in the unstrained state. Denoting this by f, we have $t = f/(1/\lambda) = \lambda f$, and (17) becomes

$$f = 2\left(\lambda - \frac{1}{\lambda^2}\right)\left(\frac{\partial W}{\partial I_1} + \frac{1}{\lambda}\frac{\partial W}{\partial I_2}\right). \tag{17a}$$

In the case of the statistical theory the introduction of (15) leads to the simple result

$$f = 2C_1\left(\lambda - \frac{1}{\lambda^2}\right). \tag{17b}$$

Uni-directional compression. This case is formally identical with simple elongation, except that λ is less than 1 and f is negative. Equations (17), (17a) and (17b) still apply.

Shear. The principal extension ratios for either pure or simple shear are

$$\lambda_1 = \lambda, \quad \lambda_2 = 1, \quad \lambda_3 = \frac{1}{\lambda}$$

the amount of the shear being

$$\sigma = \lambda - \frac{1}{\lambda}.$$

The strain-invariants then have the form

$$I_1 = I_2 = \left(\lambda^2 + 1 + \frac{1}{\lambda^2}\right).$$

The stored-energy function on the basis of the statistical theory is then, from (11),

$$W = C_1\left(\lambda^2 + \frac{1}{\lambda^2} - 2\right) = C_1\sigma^2 \tag{18}$$

while for the MOONEY form of stored energy function (13)

$$W = (C_1 + C_2)\sigma^2 \tag{19}$$

The shear stresses t_{xy} corresponding to (18) and (19) are respectively

$$t_{xy} = 2C_1\sigma, \tag{18a}$$

$$t_{xy} = 2(C_1 + C_2)\sigma \tag{19a}$$

and are directly proportional to the strain.

It is interesting to note that, according to (18a), (19a), a material may have a linear stress-strain relation in simple shear together with a non-linear relation in elongation or compression (equation 17b).

8. Experimental examination of vulcanized rubber. a) *Elongation and uni-directional compression.* Fig. 1 shows data for a typical vulcanized rubber in extension and compression obtained by the author [10]. It is seen that the points for extension ($\lambda > 1$) and compression ($\lambda < 1$) fall on a single curve. Comparison with the statistical theory shows close agreement in the compression region, but a definite departure at large extensions. The significance of these departures may be better understood by plotting $f/2(\lambda - 1/\lambda^2)$ against $1/\lambda$. This has been done by RIVLIN and SAUNDERS [11] (using an independent set of data) with the result shown in fig. 2. From equation (17a) in conjunction with (11) and (13), it follows that on this form of plot the statistical theory yields a horizontal straight line, while the MOONEY equation yields a straight line of slope C_1. It appears, therefore, that the whole curve cannot be properly represented by either of these simple forms.

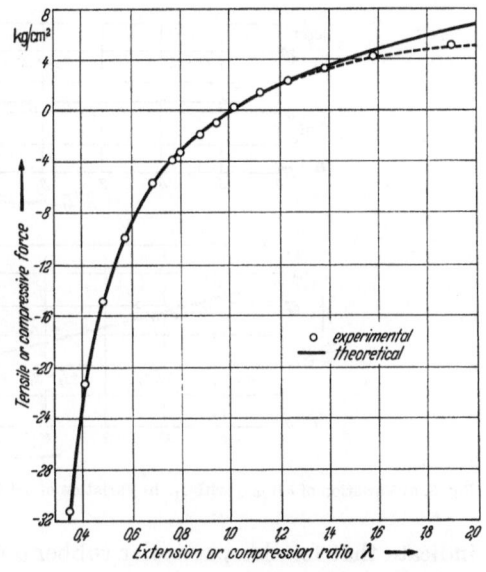

Fig. 1. Complete force-deformation curve for vulcanized rubber in extension and uni-directional compression ○ Experimental ––– Theoretical [equ. (17b)]. (TRELOAR [10])

b) *General strain.* A very complete investigation of the form of the stored-energy function for vulcanized rubber has been carried out by Rivlin and Saunders [*11*]. These workers applied a pure homogeneous strain to a sheet of rubber in such a way that the two principal stresses

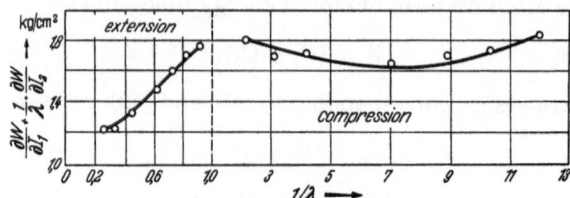

Fig. 2. Alternative representation of data for extension and uni-directional compression (Note change of scale at $1/\lambda = 1$). (Rivlin and Saunders [*11*])

in the plane of the sheet could be measured, the third principal stress (normal to the sheet) being zero. By suitably adjusting the values of λ_1 and λ_2 it was possible to vary either of the two strain invariants I_1 and I_2 independently of the other, and hence to obtain $\partial W/\partial I_1$ and $\partial W/\partial I_2$ directly from the experimental data. Their results, reproduced in fig. 3,

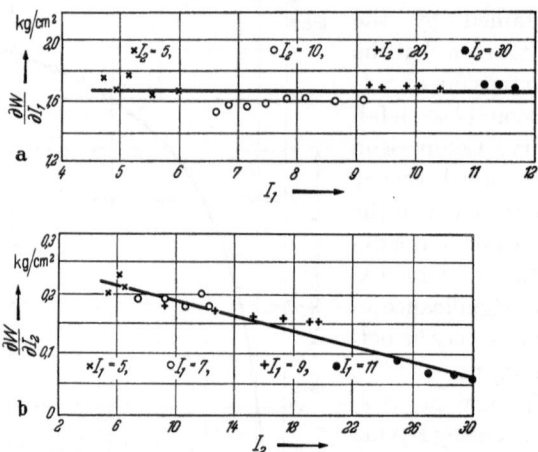

Fig. 3. a) Variation of $\partial W/\partial I_1$ with I_1. b) Variation of $\partial W/\partial I_2$ with I_2. (Rivlin and Saunders [*11*])

indicate that for this particular rubber $\partial W/\partial I_1$ was constant. $\partial W/\partial I_2$, on the other hand, was independent of I_1 but decreased with increasing I_2. This type of variation is consistent with a stored-energy function of the form

$$W = C_1(I_1 - 3) + f(I_2 - 3) \qquad (20)$$

where $f(I_1 - 3)$ is a function involving powers of $(I_1 - 3)$ higher than the first.

Since the effect of higher-order terms becomes smaller and ultimately negligible as the amount of the strain is reduced, it can be understood that the MOONEY equation, involving only first-order terms, should provide a valuable approximation so long as the strains are not too large.

9. General conclusion. It has not been possible in the space available to refer to all the important work on the stress-strain relations and the form of the stored-energy function for vulcanized rubbers. In attempting to evaluate this work it is desirable to bear in mind firstly that rubbers may differ in their detailed properties while still conforming approximately to certain general laws, and secondly, that the finer details of the form of W, as discussed, for example, by RIVLIN and SAUNDERS, can only be observed under carefully controlled experimental conditions specially designed to reveal them. Bearing these considerations in mind, the present position may be summarized as follows.

As a first approximation, and particularly where simplicity of calculation is an important consideration, the behaviour of a rubber may be represented by the 1-constant stored-energy function derived from the statistical theory. If a closer approximation is required, the empirical 2-constant formula of MOONEY is likely to be adequate in most practical cases. Only in special cases is it likely to be necessary to introduce higher approximations, and in such cases a rather elaborate investigation of the properties of the particular material under consideration will be required.

At the present time the molecular mechanisms responsible for the departures from the statistical theory which have been observed are not known, though in view of the complexity of the strucsure and the rather considerable over-simplification introduced into the statistical theory, it is not at all surprising that such departures should exist.

References

[1] JOULE: Phil. Trans. Roy. Soc., Lond. **149**, 91 (1859).
[2] GOUGH: Mem. Lit. Phil. Soc., Manchester **1**, 288 (1805).
[3] KELVIN: Quart. J. Math. 1855.
[4] MEYER, VON SUSICH and VALKO: Kolloid-Z. **59**, 208 (1932).
[5] TRELOAR: Trans. Faraday Soc. **39**, 241 (1943).
[6] FLORY, RABJOHN and SHAFFER: J. Polymer. Sci. **4**, 225 (1949).
[7] RIVLIN: Phil. Trans. Roy. Soc., Lond., Ser. A **240**, 459 and 491 (1948).
[8] RIVLIN: Phil. Trans. Roy. Soc., Lond., Ser. A **241**, 379 (1948).
[9] MOONEY: J. appl. Phys. **11**, 582 (1940).
[10] TRELOAR: Trans. Faraday Soc. **40**, 59 (1944).
[11] RIVLIN and SAUNDERS: Phil. Trans. Roy. Soc., Lond., Ser. A **243**, 251 (1951).

On Physical Effects in Cavitation Damage[1]

By **M. S. Plesset**, Pasadena (Calif.)

With 26 figures

1. Introduction. The damage produced by the collapse of cavitation, or vapor, bubbles near a solid surface is a familiar phenomenon. It is to be expected that the abrupt deceleration of the liquid boundary of such a bubble in the final stage of the collapse would result in the radiation of a sharp pressure pulse. The objectives of this investigation were the study of the physical, or metallurgical, effects of these pulses, and the determination of their duration and magnitude. This paper will summarize the progress made thus far toward these objectives.

2. Measurements of cavitation damage. In a laboratory study of cavitation damage, it is often desirable to accelerate the damage rate. If the physical effects in cavitation damage are of primary concern, short exposure times to cavitation would presumably minimize any chemical, or corrosive, effects. The procedure which has been used in the past to obtain damage in a short time consists in high frequency acceleration of the test specimen relative to the liquid medium. These accelerations are obtained by application of the magnetostrictive effect in a nickel rod to the end of which the test specimen is attached.

A new method for generating cavitation damage has been developed as part of this study which appears to have some significant advantages.

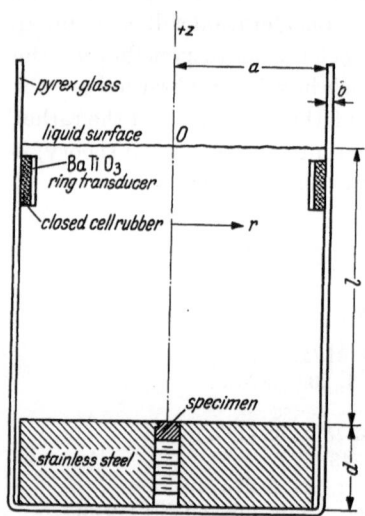

Fig. 1. Diagram of cylindrical beaker and coordinate system

[1] This work was supported by the U. S. Navy, Office of Naval Research. Reproduction in whole or in part is permitted for any purpose of the United States Government.

Since this method has been described in detail elsewhere [1], it will be described only briefly here. The apparatus consists of a cylindrical beaker containing the liquid with a barium titanate ring just below the surface. A diagram of the components is shown in fig. 1. The barium titanate ring has conducting coatings on the inner and outer surfaces and, if an alternating electric field is applied across these surfaces, the volume of the ring oscillates with the applied electric field. At a proper frequency of the field, a standing acoustic wave pattern is produced. Acoustic wave theory describes the pressure pattern accurately, and one may readily show that there is a mode which has the maximum pressure amplitude at the center of the bottom plate in the beaker. Of practical importance is the experimental observation that this mode of acoustic oscillation is the most easily excited. The pressure amplitude is sufficient to produce cavitation in the neighborhood of this point with a sinusoidal voltage of amplitude between 100 and 200 volts. The acoustic resonance frequency for most of these studies was 18,000 cycles; for the remainder of the studies, with a beaker of different dimensions, the oscillation frequency was 22,000 cycles. Fig. 2 shows the well-defined cavitation cloud over the surface of the specimen. With the specimen mounted flush in the base plate the cavitation cloud has a maximum diameter of approximately 0,37 inches. Fig. 3 shows the bubble cloud at two different times in the pressure cycle.

Fig. 2. Beaker in operation showing cavitation cloud

A series of experiments was carried out in which the liquid was distilled water with its surface exposed to air at one atmosphere pressure. The test specimens were annealed and LAUE X-ray patterns were taken before exposure to cavitation and again after exposure for various times. Figs. 4, 5, 6, 7, and 8 show the alterations in the X-ray diffraction patterns which were produced in some of the solids

Fig. 3. The bubble cloud at two different points of the pressure cycle. The upper photograph shows the bubble cloud near the pressure maximum, and the lower photograph shows the bubble cloud near the pressure minimum

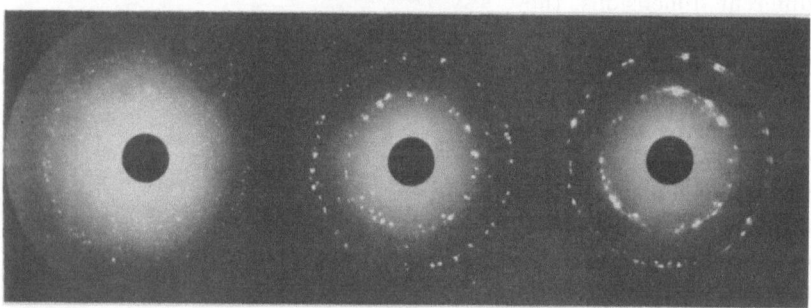

Fig. 4. X-ray diffraction pattern of nickel specimen showing rapid onset of cold work on exposure to cavitation in water

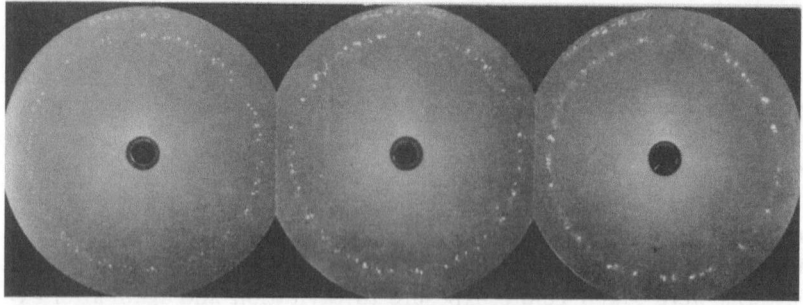

Fig. 5. X-ray diffraction pattern brass specimen showing rapid onset of cold work on exposure to cavitation

studied. The blurring of the LAUE spots following exposure to cavitation shows clearly the plastic deformation produced. In none of these experiments was the exposure time to the cavitation sufficient to produce a loss in specimen weight or a change in the surface which would be evident

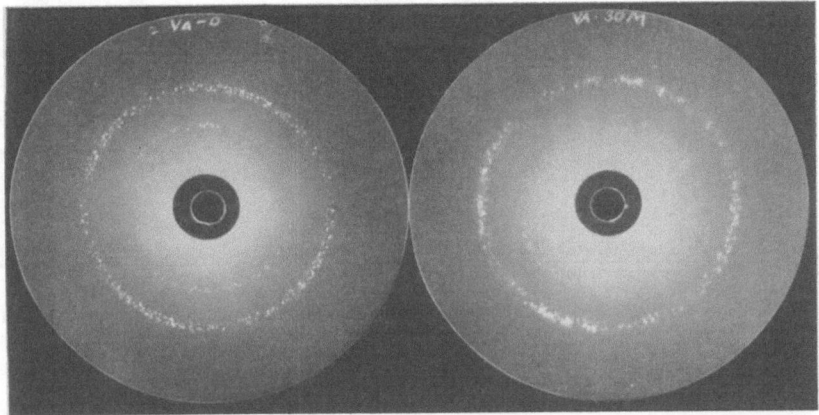

Fig. 6. X-ray diffraction pattern of vanadium specimen showing cold work on exposure to cavitation in water

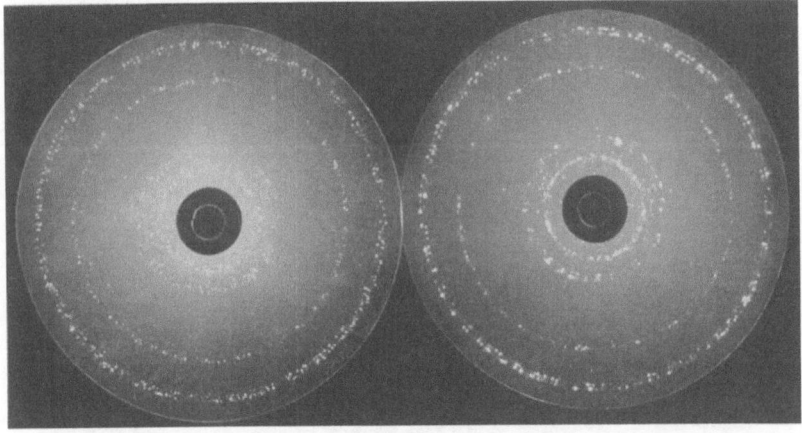

Fig. 7. X-ray diffraction pattern of molybdenum specimen showing cold work on exposure to cavitation in water

from an optical examination. It may be remarked that a significant amount of plastic deformation takes place in nickel or brass in only a few seconds while a roughly similar amount of cold work takes several minutes for vanadium, approximately an hour for molybdenum and several hours for tungsten. This resistance of tungsten to cavitation damage might

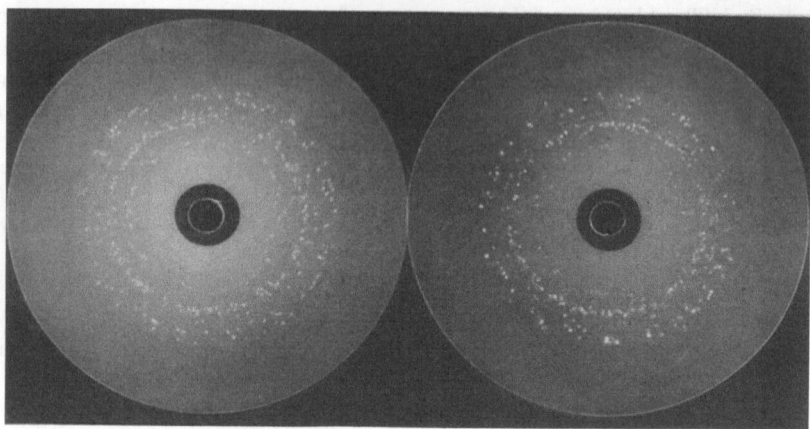

Fig. 8. X-ray diffraction pattern of tungsten specimen showing cold work on exposure to cavitation in water

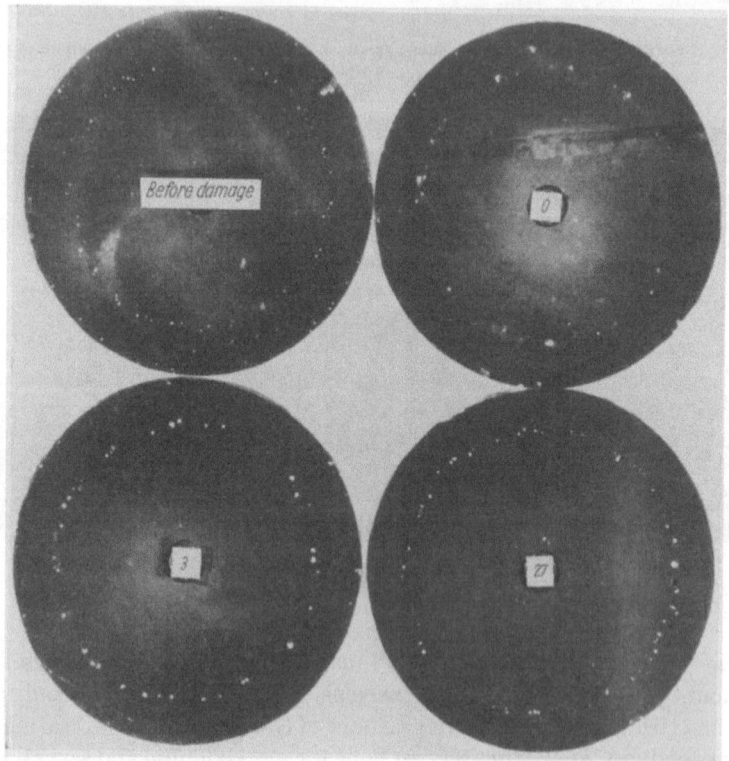

Fig. 9. Annealed nickel. Damaged 30 secs. Depth of layer removed after damage is shown in microns

have been expected in view of its great hardness and high ultimate tensile strength (cf. Table 1). The experiments with molybdenum, on the other hand, show that these properties are not necessary for high resistance to cavitation damage. The molybdenum specimens have a

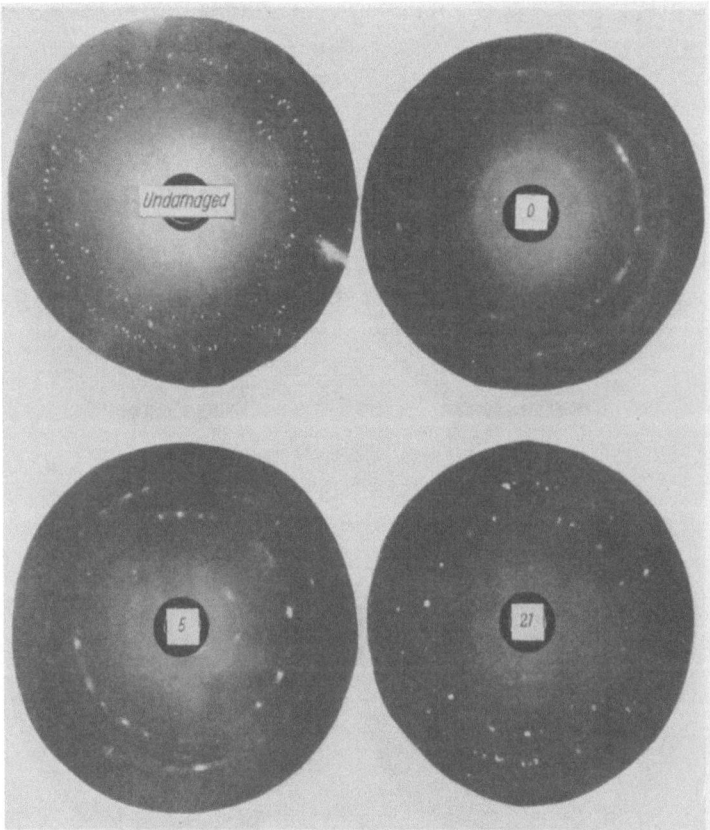

Fig. 10. Annealed brass. Damaged 30 min. Depth of layer removed after damage is shown in microns

hardness and ultimate tensile strength similar to those of brass, and yet they were much more resistant to cavitation damage. The appearance of plastic deformation upon exposure to cavitation indicates that an important property for resistance to cavitation damage is the resistance of the solid to fatigue. Further, the fatigue behaviour of a solid is related in a complex way to its usually measured properties. It must also be kept in mind that the stresses applied to the solid are of very short duration so that the fatigue behaviour for low frequency stress applications is not necessarily a measure of resistance to cavitation damage.

M. S. Plesset

The resistance of molybdenum to cavitation damage, which is exceptional in view of its hardness and tensile strength, might be explained by the lag in this solid to yielding under an applied stress. It would follow that the plastic deformation in molybdenum would be reduced for the short duration stress pulses in cavitation.

Fig. 11. Pure annealed titanium. Damaged 30 minutes. Depth of layer removed after damage is shown in microns

In order to obtain some information on the depth below the surface of the plastically deformed region, thin layers were removed successively from the surface until the X-ray pattern returned to its original configuration. The layers were removed uniformly by the process of electrolytic polishing. The results are illustrated in fig. 9 for nickel and fig. 10 for brass. A similar study was made with a pure titanium specimen and is illustrated in fig. 11. Here the X-ray diffraction spots are replaced by

Table 1

Material (arranged in order of increasing hardness)	Composition percent	Hardness (Brinell)	Ultimate tensile strength psi 10³	Modulus of elasticity psi 10⁶	Depth of cavitation damage hole in microns (10^{-4} cm)						
					10 sec	1 min	15 min	30 min	1 hr	2 hrs	3 hrs
Aluminium (soft)		16	16	10	10	80					
Titanium (annealed)		58	79	16			43	78			
Nickel		90	50	30			80	115			
Molybdenum		120	57	50				10	25	60	100
Brass	Cu 70, Zn 30	123	56	13		1	85	128			
Stainless Steel	Cr 18, Ni 8	163	102	29			15	28			
Titanium 75-A		203	80	16					30	66	
Steel (4130)		258	130	30					32	55	
Tungsten		350	597	51				0	0	3	12
Titanium 130-A	Ti 92, Mn 7.9	351	130	16				0	3	16	34
Colmonoy		400	61					0	3	18	26
Titanium 150-A	Ti 96, Cr 2.7, Fe 1.3	437	150	16				0	0	3	
Stellite	Co 55, Cr 33, W 6	495	100	36				0	3	14	29
Pyrex		Moh 5		10			120				
Fused Quartz		Moh 7		9			100				

circular lines since the specimen was rotated while being exposed to the X-ray beam. The broadening of the circular lines after exposure to cavitation shows again the plastic deformation which takes place. The depth of penetration was determined by etching off layers of the metal until the lines returned to their original sharpness.

The view has often been expressed that cavitation damage is primarily the result of chemical, or corrosive, action. This view is contradicted by the evidence presented here that the primary mechanism for the damage is cold work. In order to get further information on this point, experiments were performed in which the cavitation cloud was generated in liquid toluene from which any dissolved air was removed. Helium was kept over the liquid at one atmosphere pressure. Toluene is known to be a very inert liquid chemically, and helium is, of course, an inert gas.

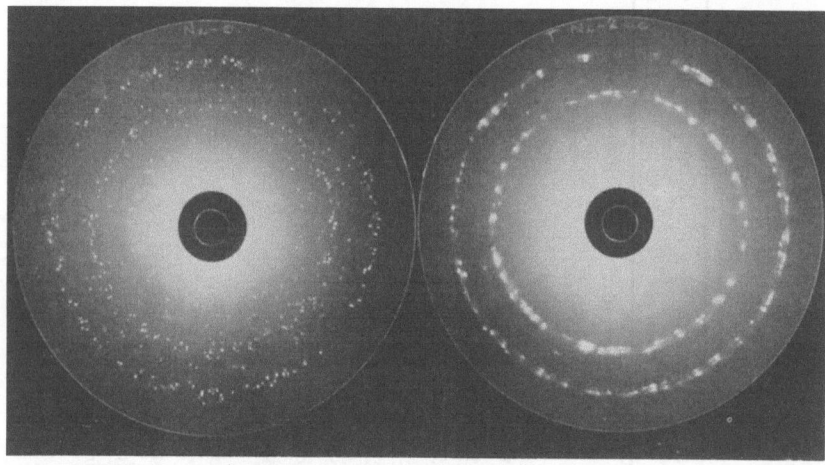

Fig. 12. X-ray diffraction pattern of nickel specimen showing rapid onset of cold work on exposure to cavitation in toluene in a helium atmosphere

Fig. 12 shows an X-ray diffraction pattern from a nickel specimen before exposure to cavitation and after 2 seconds exposure in liquid toluene with a helium atmosphere. A comparison of this pattern with that obtained in water with an air atmosphere (fig. 4) shows the similarity in the plastic deformation obtained. Similar results have been found with other kinds of specimens. It is indicated that, so far as the basic process of cavitation damage is concerned, chemical effects are not of primary significance. It should not be concluded that there are no possible chemical effects; in fact, it is well known that a chemically active environment can affect the fatigue properties of a solid.

3. Rate of material loss in cavitation damage. Experiments were carried out to determine the rate of material loss in a number of solids so as to secure a rough measure of their resistance to cavitation damage. Such measurements are of some engineering interest. To obtain significant removal of material in a short time, the specimens were made with a small cylindrical tip extending above the level of the bottom flat plate. This tip had a diameter of 0.07 inches and the cavitation cloud is concentrated in a roughly hemispherical region of this same diameter over the end of this small cylinder. The damage rate is considerably increased and the cavitation cloud drills nearly cylindrical holes in the specimen.

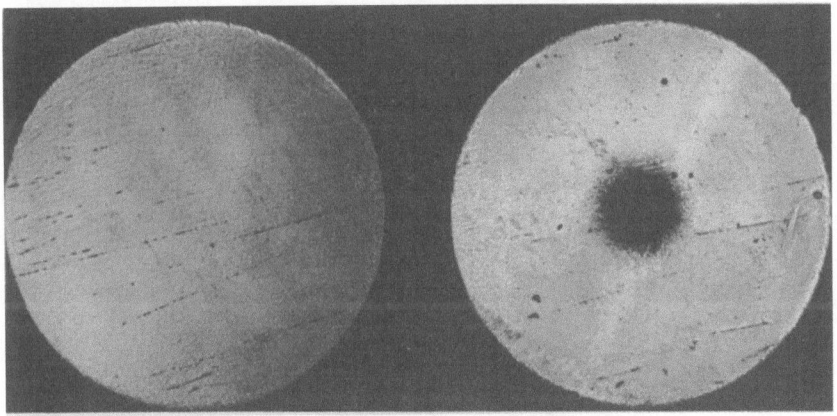

Fig. 13. Stellite specimen – magnification 40 X

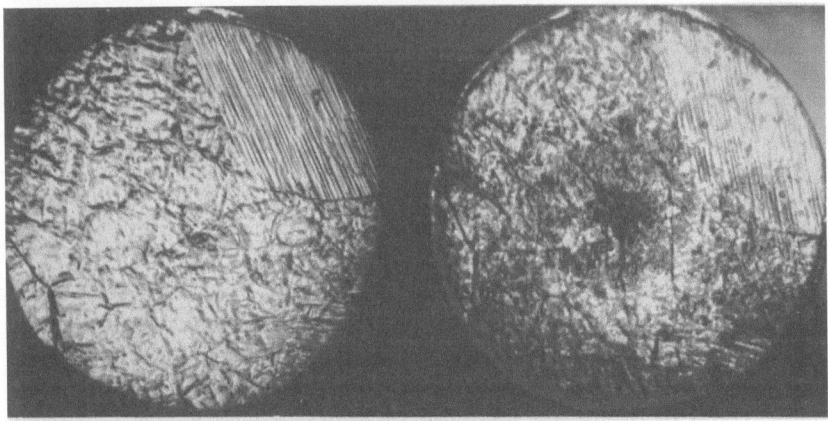

Fig. 14. Titanium 150-A specimen – magnification 40 X

A photograph of a stellite specimen exposed to this concentrated bubble cloud is shown in fig. 13, and a corresponding photograph of a titanium 150-A specimen in fig. 14. The depth of these cylindrical holes for a fixed

exposure time may be taken as a measure of the resistance of the material
to damage, and values obtained in this way are summarized in Table 1.

4. Photoelastic study of strain waves produced by cavitation. Photoelastic techniques have been applied to the problem of direct observation
of the duration and magnitude of the transient stresses or strains produced in a solid when a cavitation bubble collapses on its surface. The
material selected for these experiments is a common photoelastic plastic,

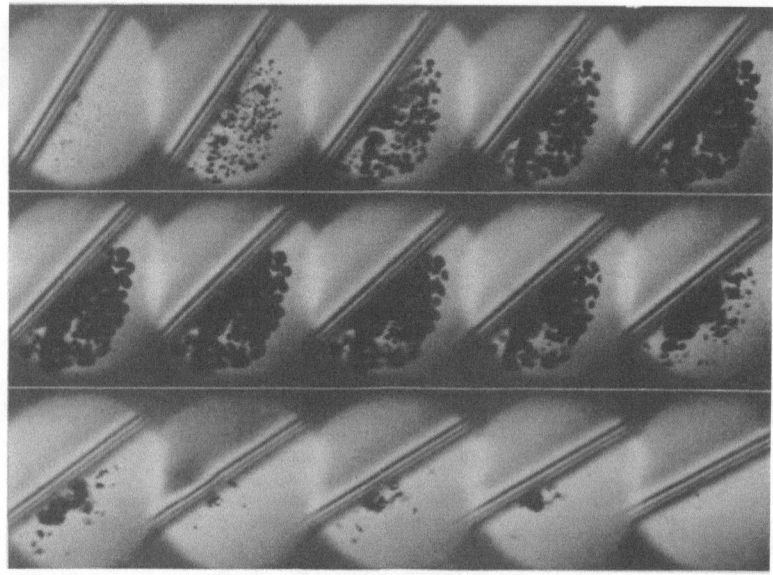

Fig. 15a. 10,000 cycle acoustic cavitation bubbles in water collapsing on CR-39 photoelastic plastic.
Shift of residual fringe due to bubble collapse is shown in frame No. 12. Picture rate 150,000 per
second. Magnification 10x. Time progresses from left to right beginning with the picture in the upper
left corner

CR-39, which is manufactured by the Cast Optic Corporation of River-
side, Conn. This plastic is a thermosetting polymer of allyl diglycol car-
bonate, and it is colorless, clear, and moderately hard. Since it is not
obvious that a photoelastic material has sufficient sensitivity to detect
strains arising from cavitation, some qualitative observations were under-
taken. A bar of CR-39, with a square cross section 1/4 inch on a side,
was exposed to cavitation produced by a 10,000 cycle acoustic field in
water. The bubble cloud formed over the end of the bar, and photoelastic
pictures were taken with a high-speed camera which has been described
elsewhere [2]. The field of view was illuminated so that the isochromatics,
as well as the cavitation bubbles in the liquid, would be visible. The
photographs shown in fig. 15a were obtained in this way; these pictures

were taken at a rate of 150,000 frames per second. A different series of photographs at 200,000 frames per second is shown in fig. 15 b. It may be noted that residual fringe lines show just below the solid surface; these fringes were the result of a long immersion of the plastic bar in water. Only one frame shows the transient cavitation strain as would be expected from the fact that the wave propagates in the solid at a speed of 0.06 inch per microsecond and the diameter of the field of view is 0.14 inch. Also, since the time interval between pictures is 5 or 6 microseconds, on ecannot expect to find a cavitation transient fringe shift in every bubble collapse cycle. Actually, six cases of fringe shifts have been observed in fifty sequences of bubble collapse. For all of these cases of observed fringe shift, the bubble which produced the shift collapsed immediately adjacent to the solid surface. While the collapsing bubble did not have spherical symmetry, the fringe pattern was propagated in the solid symmetrically from the point of collapse.

In order that quantitative results on cavitation transients may be obtained from photoelastic observations, more complete information is required regarding dynamic photoelastic behavior. Such a study has recently been initiated with CR-39 [3], and is continuing. While the dynamic properties are of primary concern, the static properties have also been determined for comparison. The results of the measurements of the static properties are illustrated in figs. 16 through 21. Fig. 16 shows the indicated strain for a typical creep test, and it is evident that CR-39 is viscoelastic. The ratio of the stress, based on the original area, to strain is given in fig. 17 as a function of time, and it is evident that

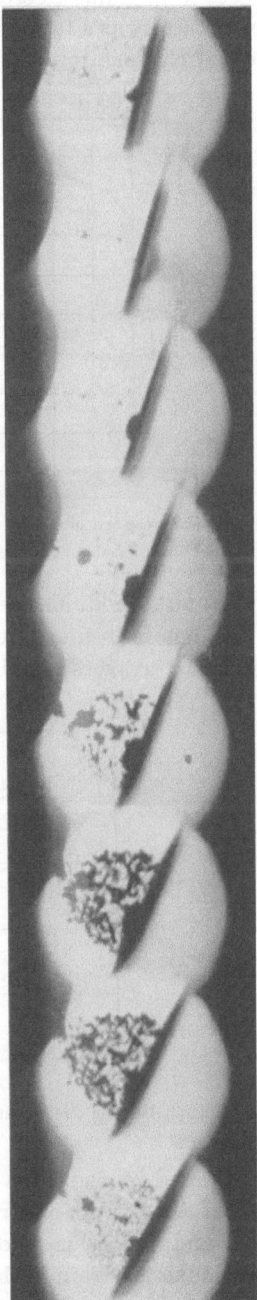

Fig. 15b. 20,000 cycle acoustic cavitation bubbles in water collapsing on CR-39 photoelastic plastic. Picture rate is 200,000 per second in this sequence

the secant modulus decreases both with time and with stress. Fig. 18
gives the stress as a function of the strain at 10 minutes after the application of the load, at which time the YOUNG's modulus is $3.36 \cdot 10^5$ psi.

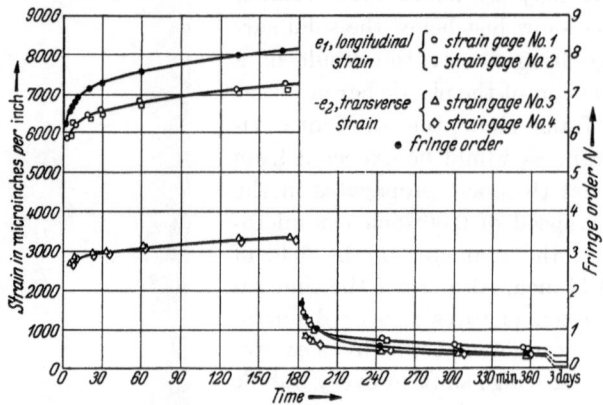

Fig. 16. Typical creep test of CR-39, showing strain and birefringence for 2100 psi tension, specimen
0.252 inches thick

The YOUNG's modulus measured immediately after the application of
the load has been found to be $3.76 \cdot 10^5$ psi. The relation between axial
strain and transverse strain is shown in fig. 19; POISSON's ratio, which

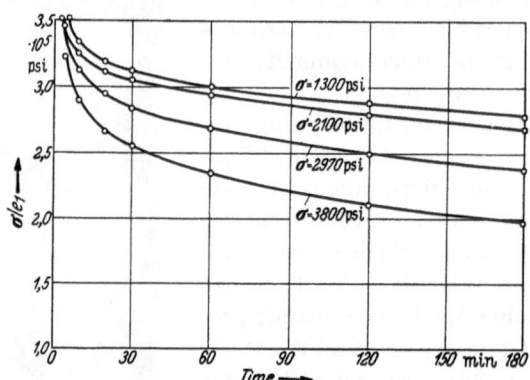

Fig. 17. Ratio of stress to strain during creep tests of CR-39 tension specimen

is determined by the slope of this line, is found to be 0.443. The measurements of the static birefringent properties of CR-39 are presented in
figs. 20 and 21. It is evident that the birefringence of this plastic is
proportional to strain rather than to stress. The strain fringe constant,
G, as defined by

$$G = (\varepsilon_1 - \varepsilon_2)\, W/N,$$

where ε_1 and ε_2 are the principal strains normal to the light path, W is the specimen width, and N is the fringe order, has the value $3.48 \cdot 10^{-4}$ inch/fringe. The so-called stress optic constant, F, is determined from the slope of the curve shown in fig. 21. By extrapolation of the fringe order to the instant of loading, the value of F so obtained is 90.8 psi-inch/fringe. F decreases from this value with time as the plastic creeps.

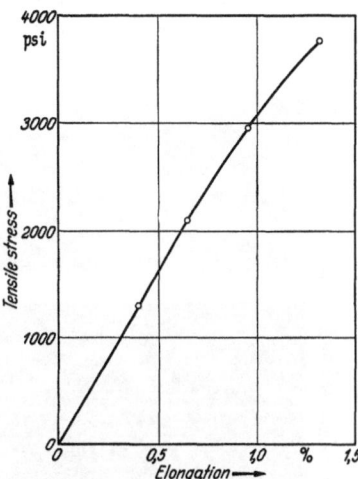

To determine the dynamic strain fringe constant, a transient strain was produced in a CR-39 specimen by the impact of a small hammer.

Fig. 18. Tensile stress vs. longitudinal strain after 10 minutes of creep for CR-39 tensile specimen

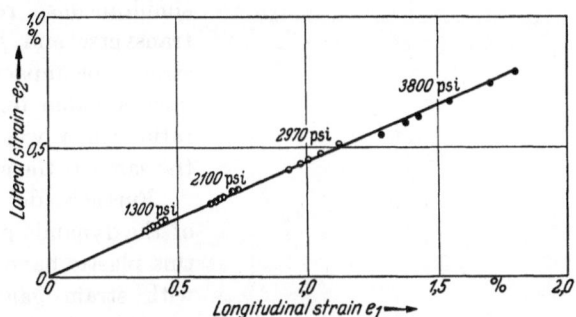

Fig. 19. Lateral vs. longitudinal strain during creep tests of CR-39

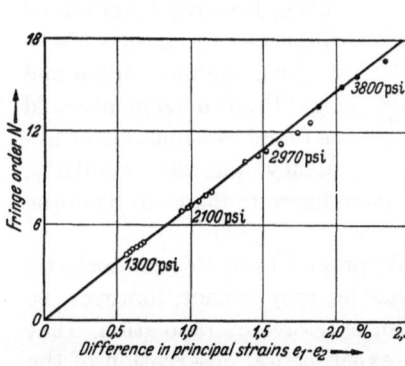

Fig. 20. Birefringence vs. strain during creep for CR-39

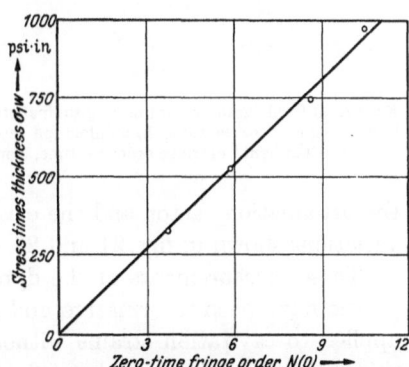

Fig. 21. Zero-time fringe order vs. stress for CR-39

Strain gages were cemented to the specimen and their outputs could be displayed on an oscilloscope. At the same time, monochromatic circularly polarized light passing through the specimen was detected with a multiplier phototube. The output of this phototube, after amplification, was displayed on a second oscilloscope. Fig. 22 shows examples of the oscillograms obtained in this way; the fringe order, N, is also shown as a function of time. The rise time of this strain wave is approximately 25 microseconds. The dynamic strain fringe constant was found by these experiments to be the same as the static value within experimental error. It may also be remarked that simultaneous readings of transverse and longitudinal strains for impact transients gave a value for POISSON's ratio which was essentially the same as the static value.

Further determinations of the dynamic properties of this plastic have been made with strain gage measurements on bars of this material of various lengths executing free-free longitudinal vibrations. From these observations, the wave speed and logarithmic decrement could be found as a function of frequency (fig. 23); similarly,

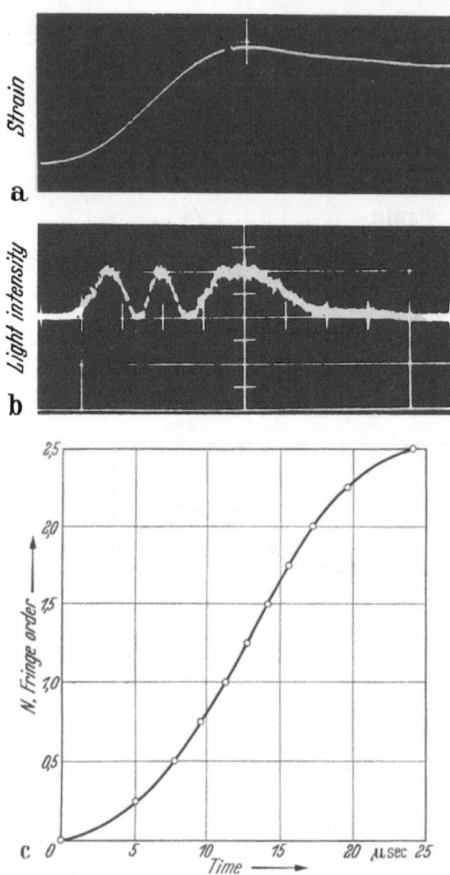

Fig. 22. a) Oscillograph record of longitudinal strain on leading edge of stress wave, b) oscillograph record of transmitted light, c) fringe order vs. time, from b)

the attenuation factor and the elastic modulus were found to have the variations shown in figs. 24 and 25.

These measurements of the dynamic properties of this photoelastic plastic have been informative and valuable; they cannot, however, be applied to cavitation strains without an excessive extrapolation. That this is the case has been shown by an experimental observation of the time duration of the strains produced by cavitation. A small bar of CR-39 was attached to the center of the base plate of the acoustic cavitation generator described above. The beaker was placed in a polariscope and only the light from an area 0.004 inches on a side of the specimen passed through the slit system. The light slit was centered 0.006 inches below the surface of the specimen exposed to the cavitation. The light transmitted by this region of the specimen was detected with an IP 21 multiplier phototube, which was immersed in liquid nitrogen to reduce the noise level. The phototube output was amplified and recorded with an oscilloscope. The combined frequency response of the phototube, amplifier, and oscilloscope was above 10 megacycles. For low voltage amplitudes on the barium titanate ring a sinusoidal wave

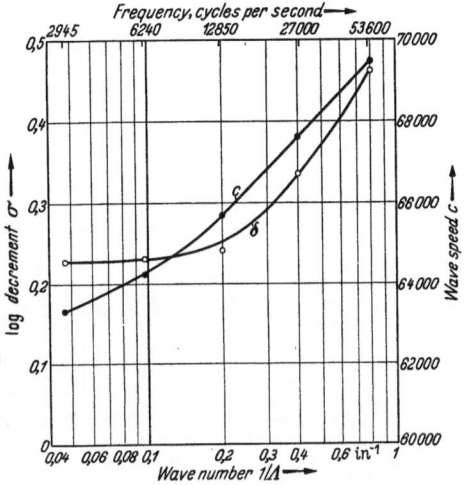

Fig. 23. Wave speed and logarithmic decrement vs. frequency and wave number for CR-39

was observed corresponding to the sinusoidal pressure variations in the water over the specimen. When this voltage was sufficiently high to produce cavitation, the sinusoidal wave was distorted by the presence of the bubble cloud into a sawtooth form. There were also observed spikes of large amplitude and short duration in the light pattern transmitted through the specimen. These spikes appeared only when the cavitation cloud was present. If the light slit was moved down from the exposed upper end of the specimen or toward the side of the

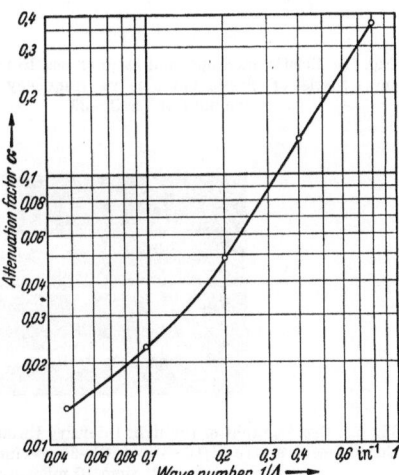

Fig. 24. Attenuation factor α vs. wave number

specimen, these spikes gradually decreased in amplitude without great change in their time duration. The amplitude of the spikes corresponds to approximately 1/4 of a fringe line. These spikes clearly arise from the cavitation strains produced in the specimen by the collapse of the vapor bubbles on its exposed surface. The shift of about 1/4 of a fringe is consistent with fringe shift

estimates made from the high-speed camera pictures such as are shown
in fig. 15. It may be seen in fig. 26, where typical oscillograms are shown,
that the time duration of the
spikes is of the order of one
microsecond.

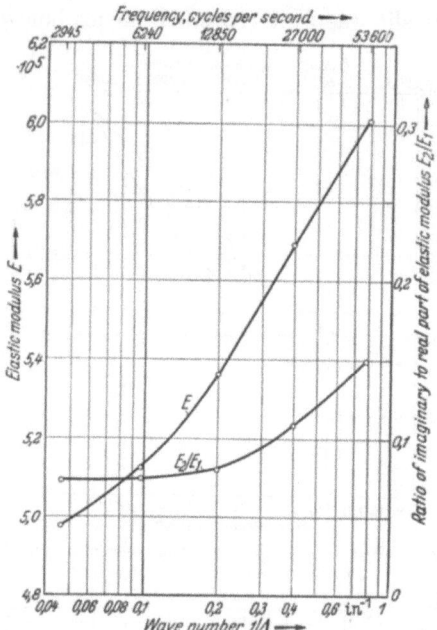

Fig. 25. Elastic modulus and ratio of real to imaginary parts of elastic modulus vs. frequency and wave number for CR-39

While the present techniques
appear adequate for measurement of the time duration of
cavitation transients, an accurate determination of their absolute magnitudes still presents
a difficult experimental problem. A very rough estimate
may be made in the following
way. If it is assumed that the
lateral elongations of the specimen during the initial pulse are
zero, then the fringe order is

$$N = \frac{1}{G} \int \varepsilon_1 \, d\,x \,,$$

where ε_1 is the strain normal to
the surface. For a plane strain,
the normal stress is given by

$$\sigma_1 = \frac{(1 - \nu)\, E}{(1 + \nu)\,(1 - 2\,\nu)} \, \varepsilon_1 \,,$$

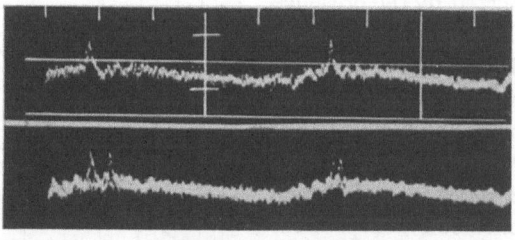

Fig. 26. Oscillographs of the light intensity through a 0.004 · 0.004 in. slit centered 0.006 in. below the surface of a 1/16 · 1/16 · 1/8 in. CR-39 specimen when cavitation is occurring. Horizontal sweep-time 10 microseconds per division

where ν is POISSON's ratio. If d is the diameter of the region over which
the stress acts initially, one gets for an average value of the stress

$$\bar{\sigma}_1 = \frac{1}{d} \int\limits_{-d/2}^{d/2} \sigma_1 \, d\,x = \frac{E\,G\,N\,(1 - \nu)}{d\,(1 + \nu)\,(1 - 2\,\nu)} \,.$$

The elastic modulus, E, is not known in the high frequency range required. For an order of magnitude estimate, an extrapolated value may be taken to be 10^6 psi; further, G is taken to be $3.42 \cdot 10^{-4}$ inch/fringe, and N is taken to be $1/4$. It has not been possible to get a value of d from photographs like that shown in fig. 15 because of the blurring of the edge of the specimen. It has been observed that cavitation pits in the specimen with relatively light damage have a diameter of about 0.001 inches. If d is assumed to be of this size, then one gets an average stress, $\bar{\sigma}_1$, of $3 \cdot 10^5$ psi. Clearly, this value is only approximate in view of the assumptions which have been required.

The author wishes to thank his colleague, Dr. ALBERT ELLIS, for his valuable contributions to all phases of this study, and Dr. GEORGE SUTTON for his contributions to the photoelastic observations. He is also greatly indebted to Professor POL DUWEZ for his helpful interest and encouragement throughout the program.

References

[1] ELLIS, A. T.: J. acoust Soc. Amer. **27**, 913 (1955). — M. S. PLESSET and A. T. ELLIS: Trans. Amer. Soc. mech. Engrs. **77**, 1055 (1955).
[2] ELLIS, A. T.: California Institute of Technology, Hydrodynamics Laboratory Report No. 21—20 (1955).
[3] SUTTON, GEORGE W.: A Study of the Application of Photoelasticity to the Investigation of Stress Waves, California Institute of Technology, Thesis 1955.

Some Applications of the Method of "Internal Constrains" to Dynamic Problems

By **Enrico Volterra**, Troy (N.Y.)

With 5 figures

1. Introduction. In the present paper a method of discussing the problem of vibrations of straight or curved bars in which the effects of shear and of rotatory inertia are taken into account will be presented.

The elastic isotropic homogeneous elongated solid is supposed to have a curvilinear baricentric axis. It will be referred to a fixed system of orthogonal cartesian coordinates Ω, ξ, η, ζ and to a system of curvilinear right-handed coordinate system $(OXYZ)$ in which the origin O coincides with a generic point on the central axis of the bar, X with the central axis (and has as positive direction the positive direction of the central axis), Y coincides with the principal normal $(\bar{n})$ and is directed positively towards the center of curvature, and Z coincides with the binormal $(\bar{b})$ at O (see fig. 1).

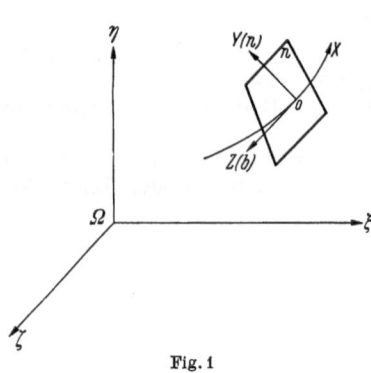

Fig. 1

The fundamental hypothesis on which the method is based is that during motion the sections originally normal to the axis of the bar remain plane.

These constraints are expressed by the following vector equation:

$$\bar{V}(x, y, z; t) = \bar{v}(x, t) + y\bar{\lambda}(x, t) + z\bar{\mu}(x, t) \tag{1}$$

when $\bar{V}$ is the displacement vector at a generic point $\bar{v}$, $\bar{\lambda}$ and $\bar{\mu}$ are three vectors, a priori unknown, functions of the variables x and t.

This approximate theory which will be referred to as "Method of Internal Constraints" is based upon equation (1) and the application of HAMILTON's Principle.

The geometric significance of the equation of constraints (1) has been discussed in detail in a previous paper [1]. It has been shown there that equation (1) implies two geometric successive linear transformations of the cross-section of the bar which transformations can be expressed in terms of the components v_x, v_y, v_z; λ_x, λ_y, λ_z; μ_x, μ_y, μ_z of the three vectors $\bar{v}$, $\bar{\lambda}$ and $\bar{\mu}$ in the coordinate system of reference $(OXYZ)$. The first transformation is a transformation which operates in the (y, z) plane normal to the axis of the bar and is composed of:

a) a pure deformation:

$$f = \frac{1}{2} \left[\mu_z z^2 + (\lambda_z + \mu_y) z y + \lambda_y y^2 \right]$$

accompanied by a rigid motion composed of:

b) a translation of components on the y and z axis:

$$a_y = v_y, \quad a_z = v_z$$

c) and of a rotation:

$$\omega = \frac{1}{2} (\mu_y - \lambda_z)$$

around an axis perpendicular to the yz plane.

The second transformation produces a rotation

$$x' - x_1 = v_x + \mu_x z + \lambda_x y$$

of the transformed section around an axis r in the yz plane the axis r being defined by the equation

$$v_x + \mu_x z + \lambda_x y = 0$$

Although the method of "Internal Constraints" can be used to discuss vibrations of an elongated elastic solid in the more general case in which its central axis possesses both first and second variable curvatures, the particular case, in which the central axis is a plane curve will only be considered here.

Under the above assumption the square of the line element ds^2 in the orthogonal curvilinear system xyz is expressed by [1][1]

$$ds^2 = a_{11} dx^2 + a_{22} dy^2 + a_{33} dz^2 = (1 - 2yc) dx^2 + dy^2 + dz^2 \quad (2)$$

The equations of motion and the corresponding boundary conditions will be derived in the particular case in which the cross-section area of the bar is constant. The equations of motion and corresponding boundary conditions for a straight bar reduce from these equations when $c = 0$. From the equations of motion of straight bars so obtained, it is possible to discuss their eigenvibrations and to derive an approximate one-dimen-

---

[1] In deriving equation (2) it was supposed that, in a first approximation, the second powers of the quantities (pure numbers) yc and zc could be neglected.

sional theory of wave propagation in elastic rods of infinite lengths which theory takes into account the effects of shear and of rotatory inertia and unifies a number of separate engineering treatments of the problem. This theory shows a good agreement with the exact theory derived from the use of the equations of the Mathematical Theory of Elasticity [2].

2. The equations of motion. The components of strain are calculated from the formulae [3]

$$
\left.
\begin{aligned}
\varepsilon_i &= \frac{\partial}{\partial x_i}\left(\frac{u_i}{\sqrt{a_{ii}}}\right) + \frac{1}{2 a_{ii}} \sum_{k=1}^{3} \frac{\partial a_{ii}}{\partial x_k} \frac{u_k}{\sqrt{a_{kk}}}, \\
\gamma_{ij} &= \frac{1}{\sqrt{a_{ii}\,a_{jj}}}\left[a_{ii}\frac{\partial}{\partial x_j}\left(\frac{u_i}{\sqrt{a_{ii}}}\right) + a_{jj}\frac{\partial}{\partial x_i}\left(\frac{u_j}{\sqrt{a_{jj}}}\right)\right], \\
(u_1 &= u,\ u_2 = v,\ u_3 = w;\ x_1 = x,\ x_2 = y,\ x_3 = z)
\end{aligned}
\right\}
\tag{3}
$$

and the Total Potential Energy of the bar from the formula

$$
U = -\int_V \sqrt{a}\left\{\frac{1}{2}\lambda\,[e^2] + G[\varepsilon_x^2 + \varepsilon_y^2 + \varepsilon_z^2] + \frac{G}{2}[\gamma_{xy}^2 + \gamma_{yz}^2 + \gamma_{zx}^2]\right\}dx\,dy\,dz,
\tag{4}
$$

where V is the volume of the bar, λ is Lame's constant and G the modulus of rigidity of the material of the bar.

From formulae (3) and (4) it follows

$$
\begin{aligned}
U = -\int_0^l \Bigg[&\left(\frac{\lambda}{2}+G\right)I_z\left(\frac{\partial\lambda_z}{\partial x}\right)^2 + \left(\frac{\lambda}{2}+G\right)I_y\left(\frac{\partial\mu_x}{\partial x}\right)^2 + \\
&+ \left(\frac{\lambda}{2}+G\right)A\left(\frac{\partial v_x}{\partial x}\right)^2 + \left(\frac{\lambda}{2}+G\right)A\,(\lambda_y)^2 + \left(\frac{\lambda}{2}+G\right)A\,(\mu_z)^2 + \\
&+ \frac{G}{2}I_z\left(\frac{\partial\lambda_y}{\partial x}\right)^2 + \frac{G}{2}I_y\left(\frac{\partial\mu_y}{\partial x}\right)^2 + \frac{G}{2}A\left(\frac{\partial v_y}{\partial x}\right)^2 + \frac{G}{2}A\,(\lambda_x)^2 + \\
&+ \frac{G}{2}I_z\left(\frac{\partial\lambda_z}{\partial x}\right)^2 + \frac{G}{2}I_y\left(\frac{\partial\mu_z}{\partial x}\right)^2 + \frac{G}{2}A\left(\frac{\partial v_z}{\partial x}\right)^2 + \frac{G}{2}A(\mu_x)^2 + \\
&+ \frac{G}{2}A\,(\mu_y)^2 + \frac{G}{2}A\,(\lambda_z)^2 + \lambda A\,\lambda_y\frac{\partial v_z}{\partial x} + \lambda A\,\mu_z\frac{\partial v_x}{\partial x} + \\
&+ \lambda A\,\lambda_y\mu_z + GA\,\mu_x\frac{\partial v_z}{\partial x} + GA\,\mu_y\lambda_z + \\
&+ (\lambda+2G)I_z c\frac{\partial\lambda_x}{\partial x}\frac{\partial v_x}{\partial x} - (\lambda+2G)c I_z\lambda_y\frac{\partial\lambda_z}{\partial x} - \\
&- (\lambda+2G)I_y c\mu_y\frac{\partial\mu_x}{\partial x} - \lambda A\,c\lambda_y v_y - \lambda A\,c\mu_z v_y - \\
&- (\lambda+2G)c A\,v_y\frac{\partial v_x}{\partial x} + GI_z c\lambda_x\frac{\partial\lambda_y}{\partial x} + GI_y c\mu_x\frac{\partial\mu_y}{\partial x} + \\
&+ GA\,c v_x\frac{\partial v_y}{\partial x} + GA\,\lambda_x\frac{\partial v_y}{\partial x} + GI_z c\frac{\partial v_z}{\partial x}\frac{\partial\lambda_z}{\partial x} + \\
&+ GI_z c\frac{\partial v_y}{\partial x}\frac{\partial\lambda_y}{\partial x} + GA\,c\lambda_x v_x\Bigg]dx.
\end{aligned}
\tag{5}
$$

The Total Kinetic Energy of the bar is expressed by

$$T = \frac{\varrho}{2} \int\limits_{V} \sqrt{a} \left[\left(\frac{\partial u}{\partial t} \right)^2 + \left(\frac{\partial v}{\partial t} \right)^2 + \left(\frac{\partial w}{\partial t} \right)^2 \right] dx\, dy\, dz$$

where ϱ is the density of the material of the bar or by

$$
\left.
\begin{aligned}
T = \frac{\varrho}{2} \int\limits_{0}^{l} &\left[I_z \left(\frac{\partial \lambda_x}{\partial t} \right)^2 + I_y \left(\frac{\partial \mu_x}{\partial t} \right)^2 + A \left(\frac{\partial v_x}{\partial t} \right)^2 - \right. \\
&- 2 I_z c \frac{\partial v_x}{\partial t} \frac{\partial \lambda_x}{\partial t} + I_z \left(\frac{\partial \lambda_y}{\partial t} \right)^2 + I_y \left(\frac{\partial \mu_y}{\partial t} \right)^2 + \\
&+ A \left(\frac{\partial v_y}{\partial t} \right)^2 - 2 I_z c \frac{\partial v_y}{\partial t} \frac{\partial \lambda_y}{\partial t} + I_z \left(\frac{\partial \lambda_z}{\partial t} \right)^2 + \\
&\left. + I_y \left(\frac{\partial \mu_z}{\partial t} \right)^2 + A \left(\frac{\partial v_z}{\partial t} \right)^2 - 2 I_z c \frac{\partial v_z}{\partial t} \frac{\partial \lambda_z}{\partial t} \right] dx.
\end{aligned}
\right\}
\tag{6}
$$

In equations (5) and (6) l represents the length of the bar, I_y and I_z the principal moments of inertia of the cross-section and A the area of the cross-section of the bar.

On the assumption that no external forces (body and surface forces) are applied to the bar, and supposing also, for simplicity sake, that the inertia characteristics of the cross-section are constant, by applying HAMILTON's Principle under the form

$$\delta \int\limits_{t_1}^{t_2} (U + T)\, dt = 0$$

to the whole bar, supposing the displacements to be zero at the instants t_1 and t_2, one obtains the following nine equations of motion in the nine unknown functions $v_x, v_y, v_z; \lambda_x, \lambda_y, \lambda_z; \mu_x, \mu_y, \mu_z$ of the variables x and t:

$$
\left.
\begin{aligned}
&(\lambda + 2G) r_z^2 \frac{\partial^2 \lambda_x}{\partial x^2} - G\lambda_x + (\lambda + 2G) r_z^2 c \frac{\partial^2 v_x}{\partial x^2} - (\lambda + 3G) c r_z^2 \frac{\partial \lambda_x}{\partial x} - \\
&\quad - G \frac{\partial v_y}{\partial x} - G c v_x = \varrho\, r_z^2 \frac{\partial^2 \lambda_x}{\partial t^2} - \varrho\, r_z^2 c \frac{\partial^2 v_x}{\partial t^2}, \\
&G \frac{\partial^2 v_y}{\partial x^2} + G \frac{\partial \lambda_x}{\partial x} + (\lambda + 3G) c \frac{\partial v_x}{\partial x} + \lambda c \mu_z + G c \lambda_y + \\
&\quad + G r_z^2 c \frac{\partial^2 \lambda_y}{\partial x^2} = \varrho \frac{\partial^2 v_y}{\partial t^2} - \varrho\, r_z^2 c \frac{\partial^2 \lambda_y}{\partial t^2}, \\
&G r_z^2 \frac{\partial^2 \lambda_y}{\partial x^2} - (\lambda + 2G) \lambda_y - \lambda \frac{\partial v_x}{\partial x} - \lambda \mu_z + (\lambda + 3G) c r_z^2 \frac{\partial \lambda_x}{\partial x} + \\
&\quad + \lambda c v_y + G r_z^2 c \frac{\partial^2 v_y}{\partial x_z} = \varrho\, r_z^2 \frac{\partial^2 \lambda_y}{\partial t^2} - \varrho\, r_z^2 c \frac{\partial^2 v_y}{\partial t^2}, \\
&G r_y^2 \frac{\partial^2 \mu_z}{\partial x^2} - \lambda \lambda_y - (\lambda + 2G) \mu_z - \lambda \frac{\partial v_x}{\partial x} + \lambda c v_y = r_y^2 \varrho \frac{\partial^2 \mu_z}{\partial t^2}, \\
&(\lambda + 2G) \frac{\partial^2 v_x}{\partial x^2} + \lambda \frac{\partial \lambda_y}{\partial x} + \lambda \frac{\partial \mu_z}{\partial x} + (\lambda + 2G) r_z^2 c \frac{\partial^2 \lambda_x}{\partial x^2} - G c \frac{\partial v_y}{\partial x} - \\
&\quad - (\lambda + 2G) c \frac{\partial v_y}{\partial x} - G c \lambda_x = \varrho \frac{\partial^2 v_x}{\partial t^2} - \varrho\, r_z^2 c \frac{\partial^2 \lambda_x}{\partial t^2}
\end{aligned}
\right\}
\tag{7}
$$

and

$$\left.\begin{array}{l}
G\,r_z^2\,\dfrac{\partial^2\lambda_z}{\partial x^2} + G\,r_z^2\,c\,\dfrac{\partial^2 v_z}{\partial x^2} - G\,\lambda_z - G\,\mu_y = \varrho\,r_z^2\,\dfrac{\partial^2\lambda_z}{\partial t^2} - r_z^2\varrho\,c\,\dfrac{\partial^2 v_z}{\partial t^2}\,, \\[2ex]
G\,r_y^2\,\dfrac{\partial^2\mu_y}{\partial x^2} - G\,\mu_y - G\,\lambda_z + (\lambda+3G)\,c\,r_y^2\,\dfrac{\partial\mu_x}{\partial x} = \varrho\,r_y^2\,\dfrac{\partial^2\mu_y}{\partial t^2}\,, \\[2ex]
(\lambda+2G)\,r_y^2\,\dfrac{\partial^2\mu_x}{\partial x^2} - G\,\mu_x - G\,\dfrac{\partial v_z}{\partial x} - (\lambda+3G)\,c\,r_y^2\,\dfrac{\partial\mu_y}{\partial x} = \varrho\,r_y^2\,\dfrac{\partial^2\mu_x}{\partial t^2}\,, \\[2ex]
G\,\dfrac{\partial^2 v_z}{\partial x^2} + G\,\dfrac{\partial\mu_x}{\partial x} + G\,r_z^2\,c\,\dfrac{\partial^2\lambda_z}{\partial x^2} = \varrho\,\dfrac{\partial^2 v_z}{\partial t^2} - r_z^2\varrho\,c\,\dfrac{\partial^2\lambda_z}{\partial t^2}
\end{array}\right\} \qquad (8)$$

with the following boundary conditions at $x=0$, $x=l$:

$$\left.\begin{array}{l}
\left[\dfrac{\partial\lambda_z}{\partial x} + c\,\dfrac{\partial v_x}{\partial x} - c\,\lambda_x\right]\delta\lambda_x = 0\,, \\[2ex]
\left[\dfrac{\partial v_y}{\partial x} + r_z^2\,c\,\dfrac{\partial\lambda_y}{\partial x} + c\,v_x + \lambda_x\right]\delta v_y = 0\,, \\[2ex]
\left[\dfrac{\partial\lambda_y}{\partial x} + c\,\dfrac{\partial v_y}{\partial x} + c\,\lambda_x\right]\delta\lambda_y = 0\,, \\[2ex]
\left[\dfrac{\partial\mu_z}{\partial x}\right]\delta\mu_z = 0\,, \\[2ex]
\Big[(\lambda+2G)\dfrac{\partial v_x}{\partial x} + (\lambda+2G)\,r_z^2\,c\,\dfrac{\partial\lambda_x}{\partial x} - \\[2ex]
\quad - (\lambda+2G)\,c\,v_y + \lambda\,\lambda_y + \lambda\,\mu_z\Big]\delta v_x = 0
\end{array}\right\} \qquad (9)$$

and

$$\left.\begin{array}{l}
\left[\dfrac{\partial\lambda_z}{\partial x} + c\,\dfrac{\partial v_z}{\partial x}\right]\delta\lambda_z = 0\,, \\[2ex]
\left[\dfrac{\partial\mu_y}{\partial x} + c\,\mu_x\right]\delta\mu_y = 0\,, \\[2ex]
\left[\dfrac{\partial\mu_x}{\partial x} - c\,\mu_x\right]\delta\mu_x = 0\,, \\[2ex]
\left[\dfrac{\partial v_z}{\partial x} + \mu_x + c\,r_z^2\,\dfrac{\partial\lambda_z}{\partial x}\right]\delta v_z = 0\,.
\end{array}\right\} \qquad (10)$$

In these equations appear together with the nine unknown functions: $\lambda_x,\ \lambda_y,\ \lambda_z;\ \mu_x,\ \mu_y,\ \mu_z;\ v_x,\ v_y,\ v_z$ the radii of inertia r_y and r_z of the cross-section of the bar, and the curvature c of the axis of the bar. The more general case in which the inertia characteristics of the cross-section are functions of the variable x can be dealt with by the use of the same variational principle. As equations (7), (8), (9), and (10) show, during free vibrations of a curved bar the functions $\lambda_x,\ \lambda_y,\ \mu_z,\ v_x$ and v_y, and $\lambda_z,\ \mu_x,\ \mu_y$ and v_z are coupled.

3. Eigenvibrations of straight bars of finite lengths. In the case of a straight rod $c = 0$, and the above equations become

$$
\left.
\begin{aligned}
G r_z^2 \frac{\partial^2 \lambda_y}{\partial x^2} - (\lambda + 2G)\lambda_y - \lambda \frac{\partial v_x}{\partial x} - \lambda \mu_z &= \varrho\, r_z^2 \frac{\partial^2 \lambda_y}{\partial t^2}, \\
G r_y^2 \frac{\partial^2 \mu_z}{\partial x^2} - \lambda \lambda_y - (\lambda + 2G)\mu_z - \lambda \frac{\partial v_x}{\partial x} &= \varrho\, r_y^2 \frac{\partial^2 \mu_z}{\partial t^2}, \\
(\lambda + 2G)\frac{\partial^2 v_x}{\partial x^2} + \lambda \frac{\partial \lambda_y}{\partial x} + \lambda \frac{\partial \mu_z}{\partial x} &= \varrho \frac{\partial^2 v_x}{\partial t^2},
\end{aligned}
\right\}
\tag{11}
$$

$$
\left.
\begin{aligned}
(\lambda + 2G) r_z^2 \frac{\partial^2 \lambda_x}{\partial x^2} - G\lambda_x - G\frac{\partial v_y}{\partial x} &= \varrho\, r_z^2 \frac{\partial^2 \lambda_x}{\partial t^2}, \\
G \frac{\partial^2 v_y}{\partial x^2} + G \frac{\partial \lambda_x}{\partial x} &= \varrho \frac{\partial^2 v_y}{\partial t^2},
\end{aligned}
\right\}
\tag{12}
$$

$$
\left.
\begin{aligned}
(\lambda + 2G) r_y^2 \frac{\partial^2 \mu_x}{\partial x^2} - G\mu_x - G\frac{\partial v_z}{\partial x} &= \varrho\, r_y^2 \frac{\partial^2 \mu_x}{\partial t^2}, \\
G \frac{\partial^2 v_z}{\partial x^2} + G \frac{\partial \mu_x}{\partial x} &= \varrho \frac{\partial^2 v_z}{\partial t^2},
\end{aligned}
\right\}
\tag{13}
$$

$$
\left.
\begin{aligned}
G r_z^2 \frac{\partial^2 \lambda_z}{\partial x^2} - G\lambda_z - G\mu_y &= \varrho\, r_z^2 \frac{\partial^2 \lambda_z}{\partial t^2}, \\
G r_y^2 \frac{\partial^2 \mu_y}{\partial x^2} - G\mu_y - G\lambda_z &= \varrho\, r_y^2 \frac{\partial^2 \mu_y}{\partial t^2}
\end{aligned}
\right\}
\tag{14}
$$

with the following boundary conditions at $x = 0$, $x = l$:

$$
\left.
\begin{aligned}
\left[\frac{\partial \lambda_y}{\partial x}\right]\delta\lambda_y = 0, \qquad \left[\frac{\partial \mu_z}{\partial x}\right]\delta\mu_z = 0, \\
\left[(\lambda + 2G)\frac{\partial v_x}{\partial x} + \lambda\lambda_y + \lambda\mu_z\right]\delta v_x = 0,
\end{aligned}
\right\}
\tag{15}
$$

$$
\left[\frac{\partial \lambda_x}{\partial x}\right]\delta\lambda_x = 0, \qquad \left[\frac{\partial v_y}{\partial x} + \lambda_x\right]\delta v_y = 0,
\tag{16}
$$

$$
\left[\frac{\partial \mu_x}{\partial x}\right]\delta\mu_x = 0, \qquad \left[\frac{\partial v_z}{\partial x} + \mu_x\right]\delta v_z = 0,
\tag{17}
$$

$$
\left[\frac{\partial \mu_y}{\partial x}\right]\delta\mu_y = 0, \qquad \left[\frac{\partial \lambda_z}{\partial x}\right]\delta\lambda_z = 0.
\tag{18}
$$

In order to study the eigenvibrations of the bar the method of separation of variables will be used.

In the case of longitudinal vibrations expressed by equations (11) and (15) let

$$
v_x(x, t) = X(x)f(t), \quad \lambda_y(x, t) = Y(x)f(t), \quad \mu_z(x, t) = Z(x)f(t). \tag{19}
$$

16 Grammel, Madrid-Kolloquium

Substituting equations (19) into equations (11) one obtains

$$
\left.
\begin{aligned}
\frac{G r_z^2 \dfrac{d^2 Y}{d x^2} - (\lambda + 2 G) Y - \lambda \dfrac{d X}{d x} - \lambda Z}{\varrho\, r_z^2 Y} &= \frac{1}{f(t)} \frac{d^2 f(t)}{d t^2} = -\omega^2, \\[2ex]
\frac{G r_y^2 \dfrac{d^2 Z}{d x^2} - \lambda Y - (\lambda + 2 G) Z - \lambda \dfrac{d X}{d x}}{\varrho\, r_y^2 Z} &= \frac{1}{f(t)} \frac{d^2 f(t)}{d t^2} = -\omega^2, \\[2ex]
\frac{(\lambda + 2 G) \dfrac{d^2 X}{d x^2} + \lambda \dfrac{d Y}{d x} + \lambda \dfrac{d Z}{d x}}{\varrho X} &= \frac{1}{f(t)} \frac{d^2 f(t)}{d t^2} = -\omega^2.
\end{aligned}
\right\} \quad (20)
$$

Since the constant ω is a real quantity[1], the longitudinal eigenvibrations are defined by the following equations which descend from equations (20) and the boundary conditions (15):

$$
\left.
\begin{aligned}
\frac{d^2 f(t)}{d t^2} + \omega^2 f(t) &= 0, \\[1ex]
G r_z^2 \frac{d^2 Y}{d x^2} - (\lambda + 2 G) Y - \lambda \frac{d X}{d x} - \lambda Z + \omega^2 \varrho\, r_z^2 Y &= 0, \\[1ex]
G r_y^2 \frac{d^2 Z}{d x^2} - \lambda Y - (\lambda + 2 G) Z - \lambda \frac{d X}{d x} + \omega^2 \varrho\, r_y^2 Z &= 0, \\[1ex]
(\lambda + 2 G) \frac{d^2 X}{d x^2} + \lambda \frac{d Y}{d x} + \lambda \frac{d Z}{d x} + \omega^2 \varrho X &= 0,
\end{aligned}
\right\} \quad (21)
$$

$$
\left.
\begin{aligned}
\left[\frac{d Y}{d x}\right] Y &= 0, \\[1ex]
\left[\frac{d Z}{d x}\right] Z &= 0, \\[1ex]
\left[(\lambda + 2 G) \frac{d X}{d x} + \lambda Y + \lambda Z\right] X &= 0
\end{aligned}
\right\} \ \text{at } x = 0,\ x = l \quad (22)
$$

The equations of motion (21) together with the boundary conditions (22) are satisfied for a triple infinite set of discrete eigenfrequencies ω_n to each of which corresponds a fundamental mode given by the eigenfunctions X_n, Y_n and Z_n. From equations (21) and (22) the following orthogonality property for the eigenfunctious X, Y, Z can be derived:

$$
\int_0^l [X_n X_m + r_z^2 Y_n Y_m + r_y^2 Z_n Z_m]\, d x = 0 \quad \text{for} \quad \omega_n \neq \omega_m \quad (23)
$$

---

[1] In fact from equations (20) and (15) the following relationship is derived:

$$
\varrho\, \omega^2 \left[r_z^2 \int_0^l [Y]^2 d x + r_y^2 \int_0^l [Z]^2 d x + \int_0^l [X]^2 d x \right] =
$$

$$
G r_z^2 \int_0^l \left[\frac{d Y}{d x}\right]^2 d x + G r_y^2 \int_0^l \left[\frac{d Z}{d x}\right]^2 d x + 2 G \int_0^l \left[Y^2 + Z^2 + \left(\frac{d X}{d x}\right)^2\right] d x +
$$

$$
+ \lambda \int_0^l \left[Y + Z + \frac{d X}{d x}\right]^2 d x.
$$

where X_n, Y_n, Z_n are the eigenfunctions corresponding to the eigenvalue ω_n.

In the case of flexural vibrations expressed by equations (12), (13), (15) and (17) let

$$\left.\begin{aligned} \lambda_x &= (x, t) = X_1(x) f_1(t), & \mu_x(x, t) &= X_1(x) f_1(t), \\ \nu_y &= (x, t) = Y_1(x) f_1(t), & \nu_z(x, t) &= Y_1(x) f_1(t). \end{aligned}\right\} \tag{24}$$

or

Substituting equations (24) into equations (12) or (13) after changing r_z or r_y with r one obtains

$$\left.\begin{aligned} \frac{(\lambda + 2G) r^2 \dfrac{d^2 X_1}{d x^2} - G X_1 - G \dfrac{d Y_1}{d x}}{\varrho\, r^2 X_1} &= \frac{1}{f_1(t)} \frac{d^2 f_1(t)}{d t^2} = - \omega_1^2, \\[2ex] \frac{G \dfrac{d^2 Y_1}{d x^2} + G \dfrac{d X_1}{d x}}{\varrho\, Y_1} &= \frac{1}{f_1(t)} \frac{d^2 f_1(t)}{d t^2} = - \omega_1^2, \end{aligned}\right\} \tag{25}$$

and since the constant ω_1 is a real quantity the flexural eigenvibrations are defined by the following equations which are obtained from equations (25) and the corresponding boundary conditions

$$\left.\begin{aligned} \frac{d^2 f_1(t)}{d t^2} + \omega_1^2 f_1(t) &= 0, \\[1ex] (\lambda + 2\,G) r^2 \frac{d^2 X_1}{d x^2} - G X_1 - G \frac{d Y_1}{d x} + \omega_1^2 r^2 \varrho X_1 &= 0, \\[1ex] G \frac{d^2 Y_1}{d x^2} + G \frac{d X_1}{d x} + \omega_1^2 \varrho\, Y_1 &= 0, \end{aligned}\right\} \tag{26}$$

$$\left.\begin{aligned} \left[\frac{d X_1}{d x_1}\right] X_1 &= 0, \\[1ex] \left[\frac{d Y_1}{d x} + X_1\right] Y_1 &= 0 \end{aligned}\right\} \qquad \text{at} \quad x = 0, \quad x = l. \tag{27}$$

The eigenfunctions X_1, Y_1 satisfy the following orthogonality property:

$$\int_0^l [Y_n Y_m + r^2 X_n X_m]\, d x = 0, \quad \text{for} \quad \omega_{1\,n} \neq \omega_{1\,m}. \tag{28}$$

In the case of torsional vibrations expressed by equations (14) and (18) let

$$\left.\begin{aligned} \mu_y(x, t) &= T(x) f_2(t), \\ \lambda_z(x, t) &= t(x) f_2(t). \end{aligned}\right\} \tag{29}$$

Substituting equations (29) into equations (14), one obtains

$$\left.\begin{aligned} \frac{G r_z^2 \dfrac{d^2 t(x)}{d x^2} - G t(x) - G T(x)}{\varrho\, r_z^2 t(x)} &= \frac{1}{f_2(t)} \frac{d^2 f_2(t)}{d t^2} = - \omega_2^2, \\[2ex] \frac{G r_y^2 \dfrac{d^2 T(x)}{d x^2} - G T(x) - G t(x)}{\varrho\, r_y^2 T(x)} &= \frac{1}{f_2(t)} \frac{d^2 f_2(t)}{d t^2} = - \omega_2^2, \end{aligned}\right\} \tag{30}$$

and since the constant ω_2 is a real quantity the torsional eigenvibrations are defined by the following equations which are deduced from equations (30) and the corresponding boundary conditions (18):

$$\left.\begin{array}{l} \dfrac{d^2 f_2(t)}{dt^2} + \omega_2^2 f_2(t) = 0, \\[2ex] G r_z^2 \dfrac{d^2 t(x)}{dx^2} - G t(x) - G T(x) + \varrho\, \omega_2^2 r_z^2\, t(x) = 0, \\[2ex] G r_y^2 \dfrac{d^2 T(x)}{dx^2} - G T(x) - G t(x) + \varrho\, \omega_2^2 r_y^2\, T(x) = 0, \end{array}\right\} \tag{31}$$

$$\left.\begin{array}{l} \left[\dfrac{dT}{dx}\right] T = 0, \\[2ex] \left[\dfrac{dt}{dx}\right] t = 0 \end{array}\right\} \qquad \text{at} \quad x = 0, \quad x = l. \tag{32}$$

The eigenfunctions $T(x)$, $t(x)$ satisfy the following orthogonality property:

$$\int_0^l [r_z^2 t_n t_m + r_y^2 T_n T_m]\, dx = 0 \quad \text{for} \quad \omega_{2n} \neq \omega_{2m}, \tag{33}$$

4. Wave propagation in straight bars of infinite lengths. In order to study the propagation of longitudinal waves in straight bars of infinite lengths, let

$$\left.\begin{array}{l} \nu_x(x, t) = N\, e^{i\gamma(x - c_p t)}, \\[1ex] \lambda_y(x, t) = L\, e^{i\gamma(x - c_p t)}, \\[1ex] \mu_z(x, t) = M\, e^{i\gamma(x - c_p t)}, \end{array}\right\} \tag{34}$$

where N, L, M are three constants, $\gamma = 2\pi/L$, L the wave length, c_p the phase velocity and $i = \sqrt{-1}$. Substituting equations (34) into equations (11), one obtains three homogeneous algebraic equations in N, L and M whose determinant set equal to zero, yields the following frequency equation:

$$\left.\begin{array}{l} - x^6 + x^4\left[4 + \dfrac{a+2}{r_x^2\gamma^2} + \dfrac{a+2}{r_y^2\gamma^2} + a\right] - \\[2ex] - x^2\left[\dfrac{5a+6}{r_x^2\gamma^2} + \dfrac{5a+6}{r_y^2\gamma^2} + \dfrac{4a+4}{r_x^2 r_y^2\gamma^4} + 2a + 5\right] + \\[2ex] + 4\left[\dfrac{1+a}{r_x^2\gamma^2} + \dfrac{1+a}{r_y^2\gamma^2} + \dfrac{3a+2}{r_x^2 r_y^2\gamma^4}\right] + a + 2 = 0. \end{array}\right\} \tag{35}$$

In equation (35)

$$x = \frac{c_p}{c_e}, \qquad c_e = \sqrt{\frac{G}{\varrho}}, \qquad a = \frac{\lambda}{G}.$$

If the bar has a square cross-section $r_x = r_y = r$, and equation (35) takes the following form:

$$
\left.
\begin{aligned}
x^6 - x^4 \frac{1}{2\,(1+\nu)\,(1-2\,\nu)} \left[\frac{1-\nu}{\pi^2\,\delta^2} + 4 - 6\,\nu \right] + \\
- x^2 \frac{1}{4\,(1+\nu)^2\,(1-2\,\nu)} \cdot \left[\frac{1}{4\,\pi^4\,\delta^4} + \frac{3-\nu}{\pi^2\,\delta^2} + 5 - 6\,\nu \right] - \\
- \frac{1}{8\,(1+\nu)^3\,(1-2\,\nu)} \left[\frac{1-\nu}{2\,\pi^4\,\delta^4} + \frac{2}{\pi^2\,\delta^2} + 2 - 2\,\nu \right] = 0.
\end{aligned}
\right\}
\qquad (36)
$$

In equation (36) $x = c_p/c_0$, $c_0 = \sqrt{E/\varrho}$, $\delta = r/L$ and ν is POISSON's ratio of the material of the bar.

In fig. 2 the ratio c_p/c_0 between phase velocities of longitudinal waves and velocities of longitudinal waves of infinite wave length are given

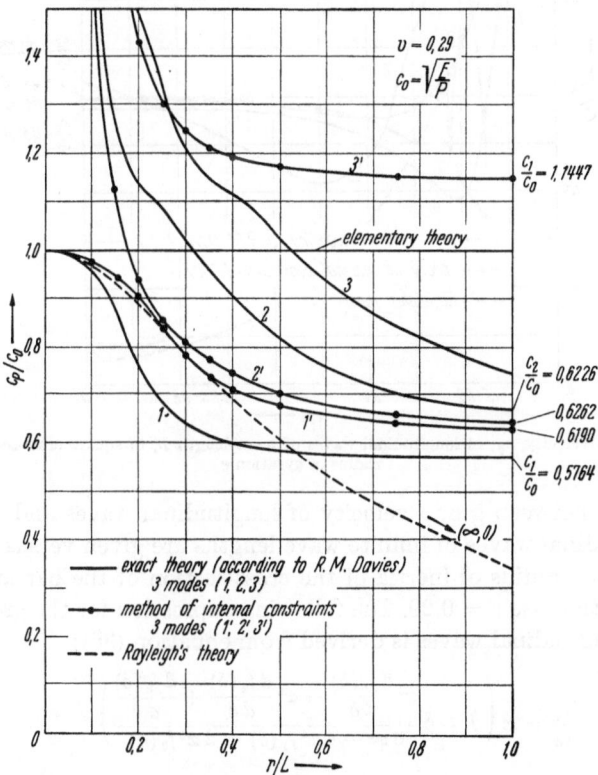

Fig. 2. Phase velocity, c_p, of longitudinal waves of wave length L in square cross-section bars of radius of gyration r

versus the ratio r/L between radius of inertia of the cross-section of the bar and wave-length in the case $\nu = 0.29$. The curves obtained from equation (36) in the case of a bar of square cross-section are compared with the curves calculated by R. M. DAVIES [4] using the POCHHAMMER's solution and with the RAYLEIGH's solution [5] for a cylindrical bar.

According to recent experiments by R. W. MORSE [6, 7, 8] a bar of square cross-section should give dispersion curves very close to those obtained with a bar of circular cross-section if the ratio of the diameter of the cylinder to the side of the square cross-section is 1.13. In fig. 3 the

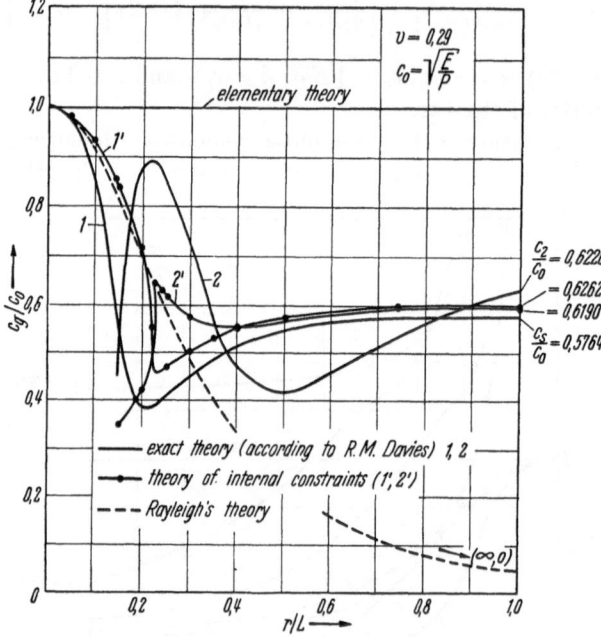

Fig. 3. Group velocity, c_g, of longitudinal waves of wave length L, in square cross-section bars of radius of gyration r

ratio c_g/c_0 between group velocity of longitudinal waves and velocities of longitudinal waves of infinite wave lengths are given versus the ratio r/L between radius of inertia of the cross-section of the bar and wavelength in the case $\nu = 0.29$. The following expression for the group velocity of longitudinal waves is derived from equation (36):

$$\frac{c_g}{c_0} = x \left(1 + \delta \frac{x^4 \dfrac{d f_1(\delta)}{d\delta} - x^2 \dfrac{d f_2(\delta)}{d\delta} + \dfrac{d f_3(\delta)}{d\delta}}{6 x^6 - 4 x^4 f_1(\delta) + 2 x^2 f_2(\delta)} \right)$$

with

$$f_1(\delta) = \frac{1}{2(1+\nu)(1-2\nu)} \left(\frac{1-\nu}{\pi^2 \delta^2} + 4 - 6\nu \right),$$

$$f_2(\delta) = \frac{1}{4(1+\nu)^2(1-2\nu)} \left(\frac{1}{4\pi^4 \delta^4} + \frac{3-\nu}{\pi^2 \delta^2} + 5 - 6\nu \right),$$

$$f_3(\delta) = \frac{1}{4(1+\nu)^3(1-2\nu)} \left(\frac{1+\nu}{4\pi^4 \delta^4} + \frac{1}{\pi^2 \delta^2} + 1 - \nu \right).$$

(37)

If the bar has a circular cross-section of radius a, $\lambda_y = \mu_z$ [2] and equations (11) reduce to Professors R. D. MINDLIN and G. HERRMANN's wave equation [9, 10]:

$$\left.\begin{aligned} -\frac{\partial^2 W}{\partial t^2} + \frac{E}{\varrho}\frac{\partial^2 W}{\partial x^2} + (3 - 4\nu)\frac{a^2}{8}\frac{\partial^4 W}{\partial x^2 \partial t^2} - \\ -\frac{a^2}{8\varrho}\frac{E(1-\nu)}{(1+\nu)}\frac{\partial^4 W}{\partial x^4} - \frac{(1+\nu)(1-2\nu)a^2}{4E}\frac{\partial^4 W}{\partial t^4} = 0. \end{aligned}\right\} \tag{38}$$

The frequency equation corresponding to the waves defined by equation (38) is

$$x^4 - \frac{2 + (3-4\nu)\pi^2\delta_1^2}{2(1+\nu)(1-2\nu)\pi^2\delta_1^2}x^2 + \frac{2 + \pi^2(1-\nu)\delta_1^2}{2(1+\nu)(1-2\nu)\pi^2\delta_1^2} = 0 \tag{39}$$

where

$$x = \frac{c_p}{c_0}, \quad c_0 = \sqrt{\frac{E}{\varrho}}, \quad \delta_1 = \frac{a}{L}.$$

The corresponding group velocity is given by

$$\frac{c_g}{c_0} = x\left\{\frac{2(1-x^2)}{4(1+\nu)(1-2\nu)\pi^2\delta_1^2 - x^2[2 + (3-4\nu)\pi^2\delta_1^2]}\right\}. \tag{40}$$

In the case of plane strain the longitudinal wave equation derived by applying the "Method of Constrained Elasticity" is [2]

$$\left.\begin{aligned} -\frac{\partial^2 W}{\partial t^2} + \frac{E}{\varrho}\frac{\partial^2 W}{\partial x^2} - \varrho r^2\frac{(1-\nu^2)}{E}\frac{\partial^4 W}{\partial t^4} - \\ -\frac{E}{2(1+\nu)}r^2\frac{\partial^4 W}{\partial x^4} + \frac{(3-\nu)}{2}r^2\frac{\partial^4 W}{\partial x^2 \partial t^2} = 0. \end{aligned}\right\} \tag{41}$$

The frequency equation corresponding to waves defined by equation (41) is [2]

$$x^4 - \frac{1 + 2\pi^2\delta^2(3-\nu)}{4\pi^2\delta^2(1-\nu^2)}x^2 + \frac{1 + \dfrac{2\pi^2\delta^2}{(1+\nu)}}{4\pi^2\delta^2(1+\nu^2)} = 0 \tag{42}$$

where

$$x = \frac{c_p}{c_0}, \quad c_0 = \sqrt{\frac{E}{\varrho}}, \quad \delta = \frac{r}{L}.$$

The corresponding group velocity is given by

$$\frac{c_g}{c_0} = x\left\{1 + \frac{1 - x^2}{8\pi^2(1-\nu^2)\delta^2 x^4 - x^2[1 + 2\pi^2\delta^2(3-\nu)]}\right\}. \tag{43}$$

From equations (12) or (13) the following wave equation for flexural waves is derived [2]:

$$\left.\begin{aligned} \frac{E}{\varrho}\frac{(1-\nu)}{(1+\nu)(1-2\nu)}r^2\frac{\partial^4 W}{\partial x^4} - r^2\left(\frac{3-4\nu}{1-2\nu}\right)\frac{\partial^4 W}{\partial x^2 \partial t^2} + \\ + \frac{2(1+\nu)}{E}r^2\varrho\frac{\partial^4 W}{\partial t^4} + \frac{\partial^2 W}{\partial t^2} = 0. \end{aligned}\right\} \tag{44}$$

The frequency equation corresponding to waves defined by equation (44) is [2]

$$x^4 - \frac{(1 - 2\nu) + 4\pi^2 \delta^2 (3 - 4\nu)}{8(1 - 2\nu)(1 + \nu)\pi^2 \delta^2} x^2 + \frac{(1 - \nu)}{2(1 + \nu)^2 (1 - 2\nu)} = 0. \tag{45}$$

The corresponding group velocity is given by [2]

$$\frac{c_g}{c_0} = x \left[1 + \frac{1}{1 + 4\pi^2 \delta^2 \left(\dfrac{3 - 4\nu}{1 - 2\nu}\right) - 4(1 + \nu)x^2} \right]. \tag{46}$$

In fig. 4 and 5 the curves obtained from equations (45) and (46) are compared with the corresponding curves obtained from the Elementary Theory, from the Rayleigh's Theory, from the Timoshenko's Theory [11], and with the results obtained by R. M. Davies [4] interpolating the

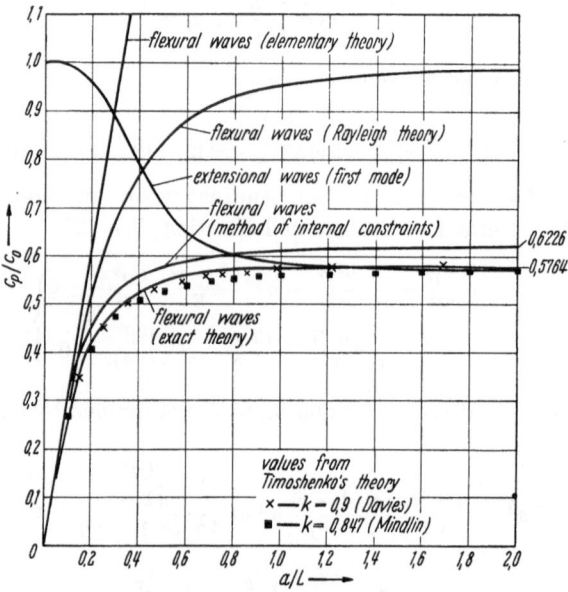

Fig. 4. Phase velocity of flexural waves in cylindrical bars ($\nu = 0.29$)

values given by Hudson [12] from the Exact Theory of Flexural Wave Propagation. In the case of plane strain the flexural wave equation derived by applying the "Method of Internal Constraints" is [2]

$$\frac{E}{\varrho}\frac{r^2}{(1 - \nu^2)}\frac{\partial^4 W}{\partial x^4} - r^2 \left(\frac{3 - \nu}{1 - \nu}\right)\frac{\partial^4 W}{\partial x^2 \partial t^2} + \frac{2(1 + \nu)}{E}\varrho r^2 \frac{\partial^4 W}{\partial t^4} + \frac{\partial^2 \overline{W}}{\partial t^2} = 0. \tag{47}$$

The frequency equation corresponding to waves defined by equation (47) is [2]

$$x^4 - \frac{1 + 4\pi^2 \delta^2 \left(\dfrac{3 - \nu}{1 - \nu}\right)}{8(1 - \nu)\pi^2 \delta^2} x^2 + \frac{1}{2(1 - \nu)^3} = 0. \tag{48}$$

The corresponding group velocity is given by [2]:

$$\frac{c_g}{c_0} - \left[1 + \frac{1}{1 + 4\pi^2 \delta^2 \dfrac{3 - \nu}{1 - \nu} - 4(1 - \nu)x^2} \right].$$ (49)

Finally, from equations (14) the following wave equation for torsional waves in the case of a bar of circular cross-section is derived.

Since in this case [2]

$$- \mu_y = \lambda_z = \omega; \quad r_y = r_z$$

equations (14) become

$$G \frac{\partial^2 \omega}{\partial x^2} = \frac{\partial^2 \omega}{\partial t^2}.$$ (50)

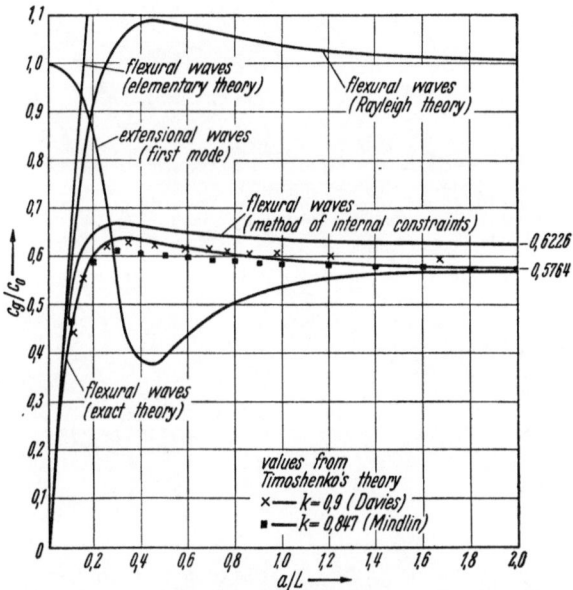

Fig. 5. Group velocity of flexural waves in cylindrical bars ($\nu = 0.29$)

From equation (50) it follows for the velocity c_e of propagation of torsional waves:

$$c_e = \sqrt{\frac{G}{\varrho}}.$$

References

[1] VOLTERRA, E.: Ing.-Arch. **23**, 402 (1955).
[2] VOLTERRA, E.: Ing.-Arch. **23**, 410 (1955).
[3] LOVE, A. E. H.: The Mathematical Theory of Elasticity, 4. Ed., pp. 51—58. Cambridge 1934.

[4] DAVIES, R. M.: Trans. Roy. Soc., Lond., Ser. A **240**, 375 (1948).
[5] LORD RAYLEIGH: The Theory of Sound.
[6] MORSE, R. W.: J. Acoust. Soc. Amer. **20**, 833 (1948).
[7] MORSE, R. W.: J. Acoust. Soc. Amer. **22**, 219 (1950).
[8] KOLSKY, H.: Stress Waves in Solids. Oxford 1953.
[9] MINDLIN, R. D. and G. HERRMANN: Proc. of the First Nat. Congr. Appl. Mech. Chicago 1951, p. 187.
[10] MINDLIN, R. D. and G. HERRMANN: Proc. of the Second Nat. Congr. Appl. Mech. Ann Arbor 1954, p. 233.
[11] TIMOSHENKO, S. P.: Phil. Mag. **41**, 744 (1921).
[12] HUDSON, G. E.: Phys. Rev. **63**, 46 (1943).

III

Viskoelastizität und Relaxation
Viscoelasticity and Relaxation

Variational and Lagrangian Methods in Viscoelasticity

By **M. A. Biot**, New York (N. Y.)

1. Introduction. The time history of a thermodynamic system perturbed from equilibrium under the assumption of linearity obeys certain differential equations. Starting from ONSAGER's reciprocity relations we have shown [1] how they may be derived from generalized concepts of free energy and dissipation function. This provides a most fruitful link between physical chemistry, thermodynamics, and mechanics, and leads to a very general formulation of relations between stress and strain in linear viscoelasticity in operational form. The outline of this development is given in section 2. The matrix relating stress and strain is formally identical with the matrix of twenty-one distinct coefficients in the theory of Elasticity. In linear viscoelasticity the elements of this matrix are functions of the differential time operator. The form of this operator is also derived from the theory. Variational formulation of deformation and stress field problems are outlined in section 3 and lead to generalizations of LAGRANGE's equations with operational coefficients. We also introduce a general principle expressing the formal correspondence between problems in viscoelasticity and Elasticity. By the latter it is possible to carry over almost all solutions of the theory of Elasticity into that of a corresponding problem of viscoelasticity. Thus we uncover in one stroke a vast area of solved problems for viscoelastic media. This approach provides a compact and synthetic formulation of linear viscoelasticity. As an example we derive in section 4 some general properties of a medium with a uniform relaxation spectrum. An outline is also given of a new approach to the dynamics of plates or shells for isotropic or anisotropic media. It includes the classical theories for elastic materials as first order approximation. The method is also, of course, applicable to improving the theory of plates and shells in the purely elastic case when the effect of increasing thickness is taken into account.

The section 5 deals with large deformations and it is shown that the same methods are also applicable to this case. The approach to the non-linear problem is different from the traditional one followed by the mathematician in the elastic case. Thus it is possible to separate the nonlinear effects of purely geometric origin from those arising from the physical relations between stress and strain. This leads to a treatment of plates and shells with large deformations which parallels the one outlined above for the linear case.

2. Viscoelastic stress-strain relations derived from thermodynamics. A thermodynamic derivation of the stress-strain relations in linear visco-elasticity for the most general case of anisotropy has been established by the writer. It is based on ONSAGER's reciprocity relations. We have shown [1, 2] that a thermodynamic system in the vicinity of equilibrium is in its linear ranges of behaviour entirely defined by two quadratic invariants. A generalized free energy

$$V = \frac{1}{2} \sum^{ij} a_{ij} q_i q_j \tag{2.1}$$

and a generalized dissipation function

$$D = \frac{1}{2} \sum^{ij} b_{ij} \dot{q}_i \dot{q}_j. \tag{2.2}$$

Both are positive definite forms and the q's are incremental state variables defining the deviations of the thermodynamic system from equilibrium. The function is a generalization of HELMHOLTZ's free energy concept to include the case of nonuniform temperature. It is defined in references [1] and [2] as

$$V = TS, \tag{2.3}$$

where S is the entropy of a total isolated system, by the adjunction of a large heat reservoir at the equilibrium temperature T. The dissipation function is defined as

$$D = \frac{1}{2} T \dot{S}, \tag{2.4}$$

where $\dot{S}$ is the rate of entropy production, in the total isolated system — expressed in terms of rate variables $\dot{q}$.

When a system is under the action of perturbing generalized forces Q similar concepts may be introduced by adding large energy reservoirs to the isolated systems, the total energy of the perturbing reservoirs being $\sum Q q$.

The total entropy S' of this new system is then given by

$$TS' = V - \sum Q q. \tag{2.5}$$

This constitutes a generalization of the GIBBS free energy concept to nonuniform temperature. Application of the ONSAGER reciprocity relation to the total system with the "forces" $\partial S'/\partial q$ and conjugate "fluxes" $\dot{q}$ leads to the differential equation

$$\frac{\partial V}{\partial q_i} + \frac{\partial D}{\partial \dot{q}_i} = Q_i. \tag{2.6}$$

These are the equations of a spring dashpot system or in the electric analogy a resistance capacity network. An interesting case is that of a system with a great number of unobserved coordinates with conjugate forces equal to zero. This is the analogue of a large R. C. network with a small number of outlets or pair terminals. The voltages of the terminals and the total quantities of electricity flowing at these terminals are related by an impedance matrix. Consider now an elementary cube of viscoelastic material oriented along the coordinate axes. If we attribute its viscoelastic properties to a large number of unobserved internal state variables associated with chemical, electrical, thermal effects, etc., we may assimilate this element to an impedance for which the observed input forces are the nine stress components σ_{ij} and the associated coordinates the nine strain components, e_{ij}. They are, therefore, related by

$$\sigma_{\mu\nu} = \sum^{ij} P^{ij}_{\mu\nu} e_{ij}, \tag{2.7}$$

where $P^{ij}_{\mu\nu}$ is an operator analogous to an impedance matrix. The strain sensor is defined in terms of the displacement vector u_i as

$$e_{ij} = \frac{1}{2}\left(\frac{\partial u_i}{\partial x_j} + \frac{\partial u_j}{\partial x_i}\right). \tag{2.8}$$

We have shown that if we consider the system represented by equations (2.6) with a generalized free energy and dissipation function and assume a large number of coordinates the operator is

$$P^{ij}_{\mu\nu} = \int_0^\infty \frac{p}{p+r}\, D^{ij}_{\mu\nu}(r)\, \gamma(r)\, dr + D^{ij}_{\mu\nu} + p\, D'^{ij}_{\mu\nu} \tag{2.9}$$

with

$$p = \frac{d}{dt}.$$

The operator has the following symmetry properties:

$$P^{\mu\nu}_{ij} = P^{\mu\nu}_{ji} = P^{\nu\mu}_{ij} = P^{ij}_{\mu\nu}. \tag{2.10}$$

These are the same as in the case of the elastic moduli of the theory of elasticity. There are twenty-one distinct operators which constitute the formal analogue of these moduli. We have also demonstrated [1] that expression (2.9) is a general formula valid whether the coefficient matrices of V or D are singular or not. The variable r represents a distribution of internal relaxation constants with a density or relaxation spectrum $\gamma(r)$. The integral expression may be considered as containing formally the case of a discreet spectrum when it is replaced by a finite or infinite summation. The reader will recall the significance of these operators. Take for instance a relation such as

$$\sigma = \left(\frac{p}{p+r} + \alpha + \beta p\right)\varepsilon,\tag{2.11}$$

where ε is a strain and σ a stress. If the deformation ε jumps from zero to unity at $t=0$ and remains constant thereafter, it is represented by

$$\varepsilon = 1(t).\tag{2.12}$$

The first term in (2.11) is

$$\sigma = \frac{p}{p+r}\, 1(t) = e^{-rt}\, 1(t).\tag{2.13}$$

This represents an exponential relaxation of stress. The term α represents an elasticity while the last one represents Newtonian viscosity

$$\sigma = \beta p \varepsilon = \beta \dot{\varepsilon}.\tag{2.14}$$

If ε varies in an arbitrary way the first term is the integral transform

$$\sigma = \frac{p}{p+r}\,\varepsilon(t) = e^{-rt}\int\limits_0^t e^{r\tau}\,d\varepsilon(\tau).\tag{2.15}$$

Attention is also called to the general significance of the term viscoelasticity in the present theory. It is to be taken in a very general sense and includes for instance the thermoelastic effect in which the energy dissipation of a perfectly elastic body arises through temperature variations associated with the volume changes and the resulting exchange of heat through conduction. The theory of thermoelastic dissipation has already been developed by Zener [3] but from a less general viewpoint.

A thermodynamic approach to relaxation phenomena was also given by Staverman [4] and Meixner [5].

3. Variational principles and Lagrangian methods. It is possible to formulate the fundamental laws of dissipative phenomena by means of variational principles. One such principle refers to minimum rate of entropy production for the stationary or nonstationary state. We have established [2] that the rate of dissipation is a minimum for a given power input of the disequilibrium forces. Although this statement was proved only for linear thermodynamics, there are indications that it is a particular case of a much more general principle. In the case of viscoelasticity where the particular stress-strain relations has been expressed operationally we have also shown [2] that it is possible to formulate a different variational principle which establishes a powerful tool for the calculations of stress fields or deformations. We define an operational invariant

$$I = \frac{1}{2} \sum^{\mu\nu} \sum^{ij} P_{\mu\nu}^{ij} e_{\mu\nu} e_{ij}, \tag{3.1}$$

which is the formal equivalent of the elastic potential energy, per unit volume and we introduce the volume integral

$$J = \iiint_\tau I \, d\tau. \tag{3,2}$$

The variational principle is then stated as

$$\delta J = \iiint_\tau \bar{G} \, \delta \bar{u} \, d\tau + \iint_S \bar{F} \, \delta \bar{u} \, dS. \tag{3.3}$$

This is an identity valid for all variations of the displacement field $\bar{u}$. The integrals on the right-hand side are respectively the virtual work of the body force $\bar{G}$ and the surface boundary force $\bar{F}$. Proof of the variational equations may be established by evaluating the variation of J integrating by parts and showing that this leads to the equations of equilibrium for the stress field. It may also be derived as in reference [2] as a particular case of a variational principle of interconnected thermodynamic systems. The systems in this case are the elements of the continuum considered as infinitesimal cells.

The variational equation (3.3) may be extended readily to include dynamics by using D'ALEMBERT's principle. The inertia force is then included in the body force. The amounts to replacing $\bar{G}$ by $\bar{G} - \varrho \frac{\partial^2 \bar{u}}{\partial t^2}$, ϱ being the mass density. The variational equation becomes

$$\delta J + \iiint_\tau \varrho \frac{\partial^2 u}{\partial t^2} \, \delta \bar{u} \, d\tau = \iint_\tau \bar{G} \, \delta \bar{u} \, d\tau + \iint_S \bar{F} \, \delta \bar{u} \, dS. \tag{3.4}$$

With the operator $p = \frac{d}{dt}$ we introduce a kinetic energy invariant

$$T = \frac{1}{2} p^2 \iiint_\tau \varrho \, u^2 \, d\tau. \tag{3.5}$$

The variational principle is then written

$$\delta J + \delta T = \iiint_\tau \bar{G} \, \delta \bar{u} \, d\tau + \iint_S F \, \delta \bar{u} \, dS. \tag{3.6}$$

An interesting application is obtained by the use of generalized coordinates. If the field is expressed in terms of n discreet coordinates q_i we write

$$\bar{u} = \sum^i \bar{u}_i q_i, \tag{3.7}$$

where $\bar{u}_i$ are field distributions represented by fixed function of x, y, z
then

$$\delta J = \sum^i \frac{\partial J}{\partial q_i} \delta q_i, \left.\begin{array}{l}\\[2ex]\end{array}\right\}$$
$$\delta T = \sum^i \frac{\partial T}{\partial q_i} \delta q_i.$$

(3.8)

Defining a generalized force by

$$Q_i = \iiint_\tau \bar{G}\,\bar{u}_i\,d\tau + \iint_S \bar{F}\,\bar{u}_i\,dS$$

(3.9)

the variational equation (3.6) leads to the n equations for q_i

$$\frac{\partial J}{\partial q_i} + \frac{\partial T}{\partial q_i} = Q_i.$$

(3.10)

Note that this is an operational expression since J and T contain the operator p. Hence these equations are integro-differential equations in LAGRANGEian form.

If we define the kinetic energy in the usual way, i.e.

$$T = \frac{1}{2} \iiint_\tau \varrho\,\dot{u}^2\,d\tau$$

(3.11)

we may replace $\dfrac{\partial T}{\partial q_i}$ above by $\dfrac{d}{dt}\left(\dfrac{\partial T}{\partial \dot{q}_i}\right)$ and the equations assume the more familiar form

$$\frac{\partial J}{\partial q_i} + \frac{d}{dt}\left(\frac{\partial T}{\partial \dot{q}_i}\right) = Q_i.$$

(3.12)

If the operators in J correspond to pure elasticity and NEWTONian viscosity then

$$P_{\mu\nu}^{ij} = D_{\mu\nu}^{ij} + p\,D_{\mu\nu}'^{\,ij}$$

(3.13)

hence

$$J = \frac{1}{2} \sum^{ij} a_{ij}q_i q_j + \frac{1}{2}\,p \sum^{ij} b_{ij}q_i q_j$$

(3.14)

putting

$$V = \frac{1}{2} \sum^{ij} a_{ij}q_i q_j, \left.\begin{array}{l}\\[2ex]\end{array}\right\}$$
$$D = \frac{1}{2} \sum^{ij} b_{ij}\dot{q}_i \dot{q}_j$$

(3.15)

equations (3.12) become

$$\frac{\partial V}{\partial q_i} + \frac{\partial D}{\partial \dot{q}_i} + \frac{d}{dt}\left(\frac{\partial T}{\partial \dot{q}_i}\right) = Q_i.$$

(3.16)

This is the usual form of LAGRANGE's equations for an elastic system with NEWTONian damping. An important principle may also be formulated relating to the formal correspondance between a large class of equations of the theory of elasticity and viscoelasticity. Because of the identical properties of the operators $P_{\mu\nu}^{ij}$ and the elastic moduli we may generally extend the formulas of the classical theory of elasticity to visco-elasticity provided the elastic moduli are replaced by the corresponding operators. This leads immediately to a large class of solutions of problems.

We shall refer to this as the *correspondence rule*.

In particular the various cases of geometric symmetry are the same. For instance a system with cubic symmetry is characterized by three operators and an isotropic medium by two.

The variational formulation above indicates that the correspondence rule also applies to the approximate solutions of Elasticity derived by energy methods.

It should be noticed that the above principles do not depend on the particular form (2.9) of the operators but only on their symmetry property (2.10). Hence they are valid also in the case when there are internal dynamic degrees of freedom, i.e., microscopic kinetic energy. The latter appears in the particular operational form of J.

4. Some specific applications. We shall now consider some specific applications of the principles formulated above. We shall first derive a general theorem for a viscoelastic medium with a homogeneous relaxation spectrum. This is a case for which the operators are of the form

$$P_{\mu\nu}^{ij} = C_{\mu\nu}^{ij} P, \tag{4.1}$$

where $C_{\mu\nu}^{ij}$ are constants and P is an invariant operator

$$P = \int_0^\infty \frac{p}{p+r} \gamma(r)\, dr + \alpha + p\beta. \tag{4.2}$$

In this case

$$J = P J' \tag{4.3}$$

with

$$J' = \frac{1}{2} \int\!\!\int\!\!\int_\tau \sum^{\mu\nu} \sum^{ij} C_{\mu\nu}^{ij} e_{\mu\nu}\, e_{ij}\, d\tau. \tag{4.4}$$

This invariant is the potential energy of an elastic medium of moduli $C_{\mu\nu}^{ij}$. We consider the deformation for the case of negligible inertia effect. The variational equation (3.3) may be written after multyplying by the inverse operator P^{-1}

$$\delta J = \int\!\!\int\!\!\int_\tau \bar{G}'\, \delta\bar{u}\, d\tau + \int\!\!\int_S \bar{F}'\, \delta\bar{u}\, dS. \tag{4.5}$$

The forces $\bar{G}'$ and $\bar{F}'$ are body forces and boundary forces obtained by applying to their actual values the operator P^{-1}

$$\left.\begin{array}{l} \bar{G}' = P^{-1}\bar{G}, \\[4pt] \bar{F}' = P^{-1}\bar{F}. \end{array}\right\} \tag{4.6}$$

We conclude from (4.5) that the *deformations are the same as for an elastic medium under the action of the transformed forces* $\bar{G}'$, $\bar{F}'$. The stresses are given by

$$\sigma_{\mu\nu} = P \sum^{ij} C_{\mu\nu}^{ij} e_{ij}, \tag{4.7}$$

where e_{ij} are the elastic strains due to the transformed forces $\bar{G}'$, $\bar{F}'$. Because of linearity they may also be written

$$e_{ij} = P^{-1} e_{ij}' \tag{4.8}$$

where e_{ij}' is the elastic strain due to the original forces $\bar{G}$, $\bar{F}$. Hence

$$\sigma_{\mu\nu} = \sum^{ij} C_{\mu\nu}^{ij} e_{ij}'. \tag{4.9}$$

The stress is therefore the same as for the elastic case under the original forces.

This applies in particular to an incompressible isotropic medium by considering the invariant made up of the product of the stress deviator by the strain. A single operator is then factorized and the above reasoning may be repeated leading to a known theorem by Alfrey [6].

The theory of deformation of viscoelastic media furnish a fertile field for the application of Lagrangeian and variational methods. The use of generalized coordinates constitutes a powerful method of approach to the dynamics of plates and shells. It is also of great flexibility and permits the gradual introduction of thickness corrections with a degree of accuracy adjusted to the practical requirements.

This general method was introduced by the writer in references [2, 7]. It applies also of course to purely elastic plates and shells since this a particular case of viscoelasticity.

The method is best illustrated by an example. It is valid for the most general case of anisotropy of the material and applied without difficulty. However, for simplicity we shall assume an isotropic material. In this case we have seen [1] that the stress-strain law is

with
$$\left. \begin{aligned} \sigma_{\mu\nu} &= 2Q\,e_{\mu\nu} + \delta_{\mu\nu}\,Re, \\ e &= e_{xx} + e_{yy} + e_{zz}, \\ \delta_{\mu\nu} &= \begin{cases} 1 & \mu = \nu, \\ 0 & \mu \neq \nu. \end{cases} \end{aligned} \right\} \tag{4.10}$$

The operators are the two invariations

$$\left. \begin{aligned} Q &= \int_0^\infty \frac{p}{p+r}\, Q(r)\,\gamma(r)\,dr + Q + p\,Q', \\ R &= \int_0^\infty \frac{p}{p+r}\, R(r)\,\gamma(r)\,dr + R + p\,R'. \end{aligned} \right\} \tag{4.11}$$

We consider a plate of uniform thickness h, the x, y plane coinciding with the plane of symmetry, and the z axis being directed across the thickness. The boundaries of the plate are at $z = \pm h/2$. We propose to find the deformation of the plate by expaunding the displacement field u, v, w, into a TAYLOR series in z. We put

$$\begin{aligned} u &= \sum^n u_n z^n, \\ v &= \sum^n v_n z^n, \\ w &= \sum^n w_n z^n. \end{aligned} \right\} \tag{4,12}$$

The coefficients $u_n(x, y)$, $v_n(x, y)$, $w_n(x, y)$ are unknown functions of x, y and implicitly also of the time operator p. In order to find the equations for these unknown functions we apply the variational equation (3.6). In this case

$$\left. \begin{aligned} I &= \frac{1}{2}\,(2Q + R)\,(e_{xx}^2 + e_{zz}^2) + R e_{zz}\, e_{xx} + 2Q\, e_{zx}^2. \\ J &= \iiint_\tau I\, d\tau, \\ T &= \frac{1}{2}\, p^2 \varrho \iiint_\tau (u^2 + v^2 + w^2)\, d\tau. \end{aligned} \right\} \tag{4.13}$$

Since we aim principally to illustrate the method let us further simplify the problem by assuming a cylindrical deformation parallel with the x, z plane. Hence $v = 0$ and $u_n v_n$ are functions only of x. Finally we expand u, w, to the third order in the thickness and put

$$\left. \begin{aligned} u &= u_1 z + u_3 z^3, \\ w &= w_0 + w_2 z^2. \end{aligned} \right\} \tag{4.14}$$

If we introduce the condition that the shear stress is zero at $z = \pm h/2$ we find

$$u_3 = -\frac{1}{3}\frac{d w_2}{d x} - \frac{4}{3 h^2}\left(\frac{d w_0}{d x} + u_1\right). \tag{4.15}$$

This leaves only three unknown functions u, w_0, w_2 of x. The variational principle is applied by first integrating along z

$$\left. \begin{aligned} \delta \int_0^l d x \int_{-h/2}^{+h/2} I\, d z + \frac{1}{2}\, p^2 \varrho\, \delta \int_0^l d x \int_{-h/2}^{+h/2} (u^2 + w^2)\, d z \\ = f \int_0^l \delta w_0\, d x + \frac{f h^2}{12} \int_0^l \delta w_2\, d x. \end{aligned} \right\} \tag{4.16}$$

In these expressions quantities of order higher than z^3 are neglected. The right-hand side represents the virtual work of applied load f uniformly distributed along the thickness. The integrals along z are readily evaluated We are then left with single integrals with respect to x and a variational problem yielding three EULER differential equations obtained by cancelling the variation due to δu_1, δw_0, δw_2. These three equations are

$$\left.\begin{array}{l} -(2Q+R)\dfrac{h^3}{12}\dfrac{d^2 u_1}{dx^2}-\dfrac{Rh^3}{6}\dfrac{dw_2}{dx}+\dfrac{Qh}{3}\left(\dfrac{dw_0}{dx}+u_1\right)+p^2\dfrac{h^3}{12}\varrho\,u_1=0, \\[2ex] 4(2Q+R)w_2+2R\dfrac{du_1}{dx}+p^2\varrho\,w_2=\dfrac{f}{h}, \\[2ex] -\dfrac{Qh}{3}\dfrac{d}{dx}\left(\dfrac{dw_0}{dx}+u_1\right)+p^2\varrho\,h\,w_0+p^2\varrho\dfrac{h^3}{12}w_0=f. \end{array}\right\} \quad (4.17)$$

Elimating $u_1\,w_2$ we find

$$\left.\begin{array}{l} \dfrac{d^4 w_0}{dx^4}-\dfrac{p^2\varrho}{Q}\left[3\beta+\dfrac{4Q+R}{4(Q+R)}\right]\dfrac{d^2 w_0}{dx^2}=\dfrac{\beta\gamma}{B_1}(f-p^2\varrho\,h\,w_0)-\dfrac{3\beta}{Qh}\dfrac{d^2 f}{dx^2} \\[1ex] \text{with} \\[1ex] \beta=1-\dfrac{p^2\varrho\,h^2}{48(2Q+R)}, \quad \gamma=1+\dfrac{p^2\varrho\,h^2}{4Q}, \quad B_1=\dfrac{4Q(Q+R)}{2Q+R}\dfrac{h^3}{12}. \end{array}\right\} \quad (4.18)$$

In conformity with the correspondence rule, for the purely elastic case, Q and R become LAMÉ constants. Putting $p=0$ i.e., for the static case at zero frequency we obtain the classical equation of the elastic beam with the addition of a shear deflection term.

A more direct method of deriving some simplified plate equations is to apply the correspondence rule to the classical equation of flexural deformation of plates which may be written

$$\nabla^4 w_0=\frac{12}{h^3}\frac{2\mu+\lambda}{4\mu(\lambda+\mu)}f, \quad (4.19)$$

where λ and μ are LAMÉ constants. Replacing these constants by the corresponding operators R and Q we obtain the equations for the visco-elastic plate

$$\nabla^4 w_0=\frac{12}{h^3}\frac{2Q+R}{4Q(R+Q)}f. \quad (4.20)$$

We see that the deflection is proportional to that of an elastic plate under a load derived from a transformation of the actual load f to which the time operator on the right-hand side has been applied. For instance, if the operator is expanded in partial fractions

$$\frac{2Q+R}{4Q(R+Q)}=\sum^{n}\frac{A_n}{p+\alpha_n}+A_0, \quad (4.21)$$

the transformed load is

$$\frac{2Q+R}{4Q(R+Q)}f=\sum^{n}A_n e^{-\alpha_n t}\int_{0}^{t}e^{\alpha_n\tau}f(\tau)\,d\tau+A_0 f(t). \quad (4.22)$$

Because of some general theorems derived in reference [1] the roots α are real. Equation (4.20) is for the nondynamical case. An inertia term $p^2 \varrho h w_0$ could be added on the left-hand side.

5. Nonlinear problems associated with large deflections. In dealing with nonlinear problems of deformation of solids it is important to distinguish between the nonlinearity arising from geometric properties of the deformation field and that due to the nonlinearity of the stress-strain relations. The former is essentially a mathematical problem while the latter is closely related to the physical nature of the material. An approach to a nonlinear theory of Elasticity which emphasizes this separation was developed by the writer in a series of publications some years ago [8, 9, 10, 11]. This constitutes a departure from the traditional approach to finite strain by the mathematician.

It was found that the equilibrium equations for the stress field are

$$\frac{\partial}{\partial x^\nu}\left[(1+e)\,\sigma_{\nu i}\right] + \frac{\partial}{\partial x^\nu}\left[\sigma_{i\nu}\,\omega_{i\mu} - \sigma_{i\mu}\,e_{\mu\nu}\right] + X_i\varrho = 0. \qquad (5.1)$$

In these equations the stress components σ_{ij} are referred to axes which rotate locally with the material. The rotation tensor is

$$\omega_{i\mu} = \frac{1}{2}\left(\frac{\partial u_i}{\partial x^\mu} - \frac{\partial u_\mu}{\partial x^i}\right). \qquad (5.2)$$

The stress is a function of the strain components referred to the same rotated axes namely

$$\varepsilon_{\mu\nu} = e_{\mu\nu} + \frac{1}{2}\left(\omega_{i\mu}e_{i\nu} + \omega_{i\nu}\,e_{i\mu}\right) + \frac{1}{2}\,\omega_{i\mu}\omega_{i\nu}. \qquad (5.3)$$

The body force is X_i per unit mass and summation signs are omitted. The gradients $\dfrac{\partial\,\sigma_{\nu i}}{\partial\,x^i}$ in equations (5.1) contain the second order terms of physical origin due to the nonlinearity of the stress-strain relations. The others contain the second order terms of geometric origin which arise from the product of the stress by the strain and rotation components. When introducing the stress-strain relations in the latter terms, it is only necessary to use the first order effects.

The above equations were derived for an elastic body but are applicable to solids in general. For instance, we may consider a viscoelastic solid and assume that the linear stress-strain relations in operational form

$$\sigma_{\mu\nu} = P^{ij}_{\mu\nu}\,e_{ij} \qquad (5.4)$$

are valid throughout the range of strain involved. In that case the three equations for the displacement field are obtained by replacing in the three equilibrium equations (5.1) the value of $\sigma_{\mu\nu}$ by its operational expression (5.4). Since we are now dealing with nonlinear equations care must be exercised to locate the operators in the proper place so that they operate only on the quantities to which they were originally attached. Since the equations contain time operators we thus obtain three nonlinear integro-differential equations for the displacement field $\overline{u}$ of the solid.

The present approach also leads immediately to the theory of incremental stress and deformation for a body under initial stress. By linearizing the equilibrium equations (5.1) in the vicinity of an initial state of stress $S_{\mu\nu}$ we obtain linear equilibrium equations for the incremental stresses. These equations are

$$\frac{\partial}{\partial x^\nu}\,\sigma_{\nu i} + \frac{\partial}{\partial x^\nu}\,(S_{\nu i}e) + \frac{\partial}{\partial x^\nu}\,(S_{i\nu}\omega_{i\mu} - S_{i\mu}e_{\mu\nu}) + \varrho\,\varDelta\,X_i = 0\,. \quad (5.5)$$

The incremental body force is $\varDelta X_i$. These equations, if we formulate a relation between incremental stress and deformation, lead to the solution of stability problems of elastic or plastic prestressed fields. In particular, it leads to solutions of incremental stability problems of viscoelastic prestressed fields if we assume incremental stress-strain relations of the type (5.4). The nature of the incremental stress-strain relation appropriate for various materials is essentially a physical problem which still remains to be investigated.

Of special interest here is the possible introduction of variational and Lagrangeian methods in the formulation of problems of large deformation of elastic and anelastic solids. To this effect we follow a procedure which we have introduced in the elastic theory. We define a variational invariant

$$\delta J = \iiint_\tau \tau_{\mu\nu}\,\delta\,\varepsilon_{\mu\nu}\,d\tau\,, \quad (5.6)$$

where

$$\tau_{\mu\nu} = (1 + \varepsilon)\,\sigma_{\mu\nu} - \frac{1}{2}\,(\sigma_{\alpha\mu}\varepsilon_{\alpha\nu} + \sigma_{\alpha\nu}\varepsilon_{\alpha u}) \quad (5.7)$$

and $\varepsilon_{\mu\nu}$ is given by (5.3). In these expressions the $\sigma_{\mu\nu}$ components may be expressed in terms of $e_{\mu\nu}$ by means of operators. In the case of a viscoelastic material it is, therefore, an operational invariant of the same type as (3.2). The variational identity in the present case is

$$\delta J = \iiint_\tau \overline{G}\,\delta\,\overline{u}\,d\tau + \iint_S \overline{F}\,\delta\,\overline{u}\,dS\,, \quad (5.8)$$

which must be valid for all variations of the displacement field.

This variational principle is derived from [9] and [10] where we have shown that it is equivalent to the equilibrium equations (5.1) of the stress field. Making use are before of d'Alembert's principle we may introduce the inertia effect and write

$$\delta J + \delta T = \iiint_\tau \overline{G}\,\delta\,\overline{u}\,d\tau + \iint_S \overline{F}\,\delta\,\overline{u}\,dS\,. \quad (5.9)$$

This equation opens the way to a systematic treatment of nonlinear problems by methods of generalized coordinates entirely analogous to the linear case. Since the equations are now nonlinear we must take care that the operators remain attached to those quantities upon which they initially operate. We could treat problems of vibrations of plates and shells by expanding the displacement field in a power series of the transverse thickness coordinate, and obtain simplified equations with any order of approximation desired as exemplified above. We could for instance generalize the KÁRMÁN-FOEPPL equations for the finite deflection of elastic plates to include higher order effects of the thickness and auy viscoelastic stress-strain law whether isotropic or anisotropic. Finally, it should be pointed out that a modified correspondence rule may be applied in the nonlinear case. Consider for instance the KÁRMÁN-FOEPPL equations of finite deformation of an elastic plate. If we follow the derivation of these equations and replace the LAMÉ constants λ and μ by the operators R and Q we obtain

$$\left.\begin{array}{c} \dfrac{4Q(R+Q)}{2Q+R}\ \dfrac{h^3}{12}\ \nabla^4 w_0 = f + \dfrac{\partial^2 F}{\partial y^2}\dfrac{\partial^2 w_0}{\partial x^2} - 2\dfrac{\partial^2 F}{\partial x\partial q}\dfrac{\partial^2 w_0}{\partial x\partial q} + \dfrac{\partial^2 F}{\partial x^2}\dfrac{\partial^2 w_0}{\partial y^2}, \\[4mm] \dfrac{Q+R}{Q(2Q+3R)h}\ \nabla^4 F = \left(\dfrac{\partial^2 w_0}{\partial x\partial y}\right)^2 - \dfrac{\partial^2 w_0}{\partial x^2}\dfrac{\partial^2 w_0}{\partial y^2}. \end{array}\right\} \quad (5.10)$$

These are two nonlinear integro-differential equations for the finite deflection of an isotropic viscoelastic plate.

References

[1] BIOT, M. A.: J. appl. Phys. 25, 1385 (1954).
[2] BIOT, M. A.: Phys. Rev. 97, 1463 (1955).
[3] ZENER, C.: Phys. Rev. 53, 90 (1938).
[4] STAVERMAN, A. J.: Proceedings of the Second International Congress on Rheology 1954, 134.
[5] MEIXNER, J.: Z. Naturforsch. 9a, 654 (1954).
[6] ALFREY, T.: Mechanical Behaviour of High Polymers. New York 1948.
[7] BIOT, M. A.: Phys. Rev. 98, 1869 (1955).
[8] BIOT, M. A.: Proceedings Fifth International Congress of Applied Mechanics. Cambridge, U.S.A. 1938, 117.
[9] BIOT, M. A.: Phil. Mag. 27, 468 (1939).
[10] BIOT, M. A.: Z. angew. Math. Mech. 20, 89 (1940).
[11] BIOT, M. A.: J. appl. Phys. 11, 522 (1940).

Viscosité et déformations irréversibles

Par **H. Le Boiteux,** Châtillon-sous-Bagneux

Avec 13 figures

1. Introduction. Les déformations irréversibles dans les matériaux ductiles ont fait l'objet d'un grand nombre de travaux, dont certains très anciens (MAXWELL-TRESCA).

Le problème prend d'ailleurs une importance technique croissante en raison de l'utilisation toujours plus poussée de ces matériaux, dans des conditions de température et de charge qui ne permettent plus d'admettre que l'on reste dans la zone élastique (purement réversible).

Diverses théories de la Plasticité ont ainsi été élaborées et des critères d'apparition des phénomènes irréversibles ont été dégagés.

Par contre, il a été le plus souvent admis que pour les matériaux fragiles aucune transition n'existe entre la zone élastique et la rupture, autrement dit qu'il ne peut apparaître de déformation irréversible notable dans ce cas.

Cette absence, jointe à l'aspect particulier de la rupture, a même servi pendant longtemps à définir le caractère de fragilité.

Les travaux de M. CAQUOT sur la courbe intrinsèque ont permis de préciser cette notion.

Ayant étudié le comportement du Plexiglass, substance présentant les caractères de fragilité, en vue de son application aux méthodes optiques de détermination des contraintes (photoélasticité) j'ai été amené à constater l'apparition de phénomènes irréversibles parfaitement mesurables pour une sollicitation nettement inférieure à la rupture.

Ce travail a fait l'objet d'une communication au Colloque de Photoélasticité et Photoplasticité organisé par l'IUTAM à Bruxelles en Juillet 1954 [1].

Le présent mémoire est relatif à de nouvelles expériences prolongeant et complétant celles qui servaient de base à cette communication et à une tentative d'explication de ces phénomènes utilisant la notion de courbe intrinsèque.

Les premières constatations se sont trouvées entièrement confirmées apportant la preuve que des déformations irréversibles d'un type particulier peuvent apparaître dans un matériau possédant les caractères généraux de la fragilité.

2. Experiences anterieures. La matière utilisée est du Plexiglass M 222 optiquement actif, c'est-à-dire présentant de la biréfringence accidentelle par déformation.

L'éprouvette est soumise à un effort de traction pure maintenu rigoureusement constant et la biréfringence est mesurée en fonction du temps.

La matière présente des phénomènes de déformation différée très importants (réactivité) et l'expérience doit être prolongée très longtemps.

Lors de la décharge, la réactivité se manifeste de la même façon et le retour à l'équilibre est très lent.

Ce comportement est du à l'existence d'un coefficient de viscosité important et j'ai montré, dans une communication au 8ème Congrès International de Mécanique à Istambul en 1952 [2] qu'il est possible d'en faire la théorie en complétant les équations de LAME par des termes de contraintes visqueuses. La concordance avec les résultats expérimentaux est tout à fait satisfaisante.

En opérant ainsi pour des charges croissantes on constate que

a) En dessous d'une certaine valeur, de l'ordre de 1,5 kg/mm², toute biréfringence disparaît après une décharge prolongée, la matière revient exactement à son état initial.

On ne peut toutefois parler de phénomènes rigoureusement réversibles, les courbes de charge et de décharge n'étant pas superposables en raison de la viscosité, le type de déformation correspondant peut être appelé «élastique différé» ou «pseudo-réversible».

b) Au-dessus de la valeur précédente, il apparaît une biréfringence résiduelle qui subsiste indéfiniment. La matière est devenue anisotrope. La biréfringence irréversible (ou figée) ainsi constatée croît avec la charge. Dans la première série d'expériences celle-ci a été poussée jusqu'à 3 kg/mm² alors que la rupture doit être attendue vers 7 kg/mm².

c) A partir de 2 ou 2,5 kg/mm² la biréfringence au cours de la charge croît constamment avec le temps et ne semble pas se stabiliser, il y a fluage vrai de la matière.

d) Pour les fortes valeurs, on voit apparaître sur l'échantillon des stries rappelant les lignes de LÜDERS des corps ductiles mais orientées perpendiculairement à la direction de tension maximum et qui traduisent une irréversibilité mécanique liée à celle de la biréfringence.

3. Nouvelles experiences. Mon élève, Mme PAUTHIER, et moi-même avons repris sur la même substance une nouvelle série d'expériences en poussant la sollicitation de traction pure jusqu'à 3,6 kg/mm².

Le processus d'application de la charge est rendu parfaitement constant par un dispositif identique à celui que l'on trouvera décrit dans la référence [2] et la température maintenue constante à 1/10 degré près pendant toute la durée des mesures[1].

Pour chaque valeur de la charge on procède à une série d'expériences correspondant à des durées totales croissantes sur des éprouvettes aussi identiques que possible et pour chacune on mesure en outre la biréfringence résiduelle.

Bien que, pour de tels matériaux, la biréfringence soit liée directement à la déformation, celle-ci a été mesurée directement au moins pour la partie irréversible, en traçant sur l'éprouvette deux traits de repère très fins dont la distance est lue au micromètre avant et après l'expérience.

Cette façon de procéder nous a permis de mettre en évidence une déformation résiduelle qui se conserve indéfiniment.

Le phénomène déjà constaté des stries normales à la tension maximum a été retrouvé. Il prend, aux fortes charges, une importance telle que l'éprouvette devient presque opalescente, en même temps que les stries présentent un aspect plus largement ouvert et se propagent en profondeur tandis qu'aux faibles charges elles sont seulement superficielles.

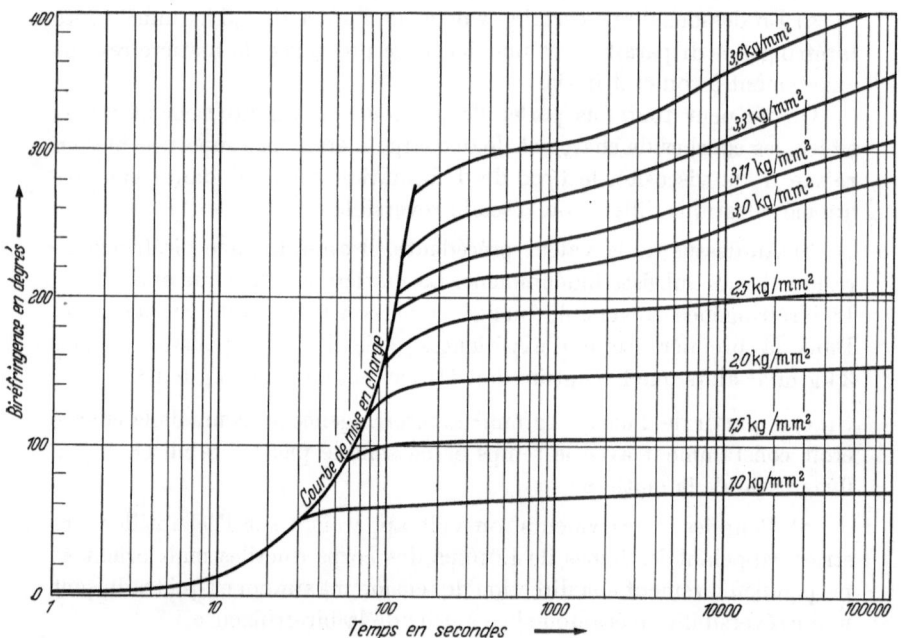

Fig. 1. Biréfringence en fonction du temps pour diverses charges

La fig. 1 donne l'allure générale du phénomène. On a tracè en coordonnées semi-logarithmiques les courbes donnant la variation de la biré-

[1] L'influence de la superposition de cycles de température n'a pas encore été étudiée.

fringence en fonction du temps pendant les 48 premières heures. (Les expériences étaient en réalité poussées jusqu'à plusieurs centaines d'heures.)

Le processus de mise en sollicitation adopté (effort linéairement croissant jusqu'à la valeur choisie) correspond à la courbe qui limite le tracé à gauche.

A partir de 3 kg/mm² la biréfringence croît constamment sans qu'il soit possible de parler d'une valeur asymptotique fixe.

Les mesures au micromètre révèlent l'apparition d'un allongement irréversible qui, par exemple, atteint 2,5% pour une sollicitation de 3,3 kg/mm² maintenue pendant 450 heures.

La fig. 2 montre simultanément l'allure des courbes de charge et de décharge pour les valeurs 3 3,3 et 3,6 kg/mm².

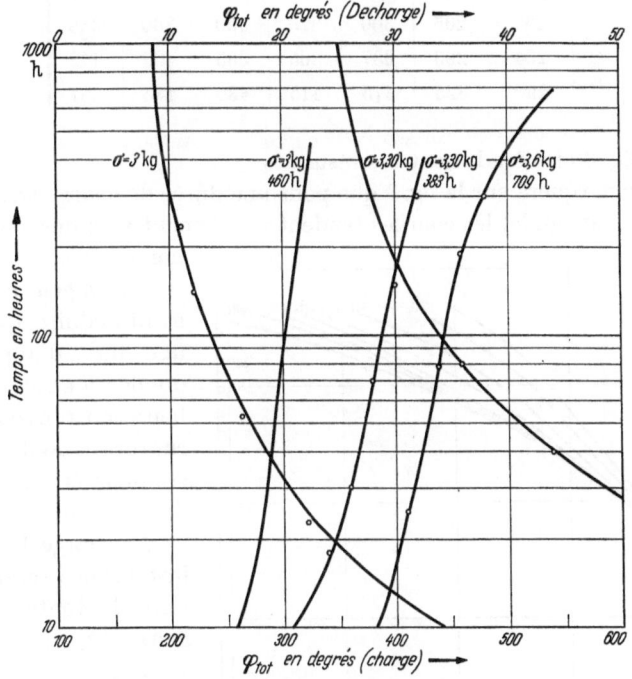

Fig. 2. Courbes de charge et décharge

$$\varphi_{tot} = f(t)$$

On voit nettement apparaître la biréfringence résiduelle irréversible vers laquelle tendent les courbes de décharge.

Des très nombreux essais effectués dans ces conditions nous avons pu déduire le table 1 qui donne l'évolution de la biréfringence en fonction des deux facteurs: valeur de la sollicitation et durée de l'opération. La

dernière colonne indique en outre la biréfringence résiduelle après décharge complète et pour la durée de charge maximum.

Table 1. *Biréfringence en degrés en fonction de la sollicitation et de la durée*

Durée ＼ Sollicitation kg/mm²	6 minutes	1^h	4^h	24^h	48^h	100^h	200^h	Biréfringence résiduelle (durée de charge 700^h)
1	64	68	69^5	70	70	70	70	0
1,5	104	109	109^5	110	110	110	110	0
2	142	150	156	160	162	163	163	2°20
2,5	182	192	203	215	225	230	233	7°
3	217	235	260	285	310	320	325	18°
3,2	237	265	300	325	350	360	368	24°
3,4	264	295	337	365	385	400	412	36°
3,6	290	325	375	415	430	445	461	50°

Les fig. 3 et 4 traduisent ces résultats.

On remarquera sur la fig. 3 que pour une durée de charge très grande
(700 h par exemple) les courbes tendent rapidement vers une asymptote
horizontale.

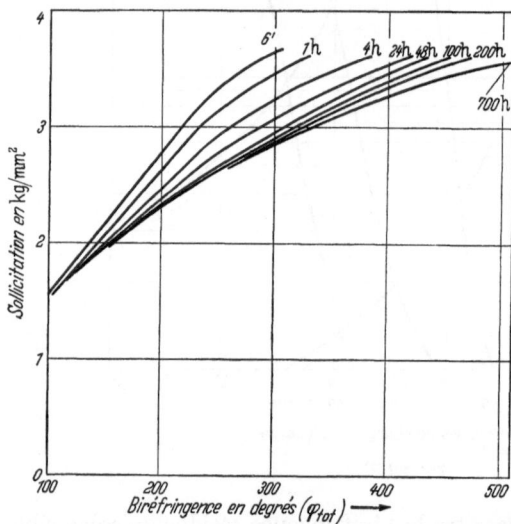

Fig. 3. Biréfringence totale en fonction de la charge pour diverses durées

On en peut probable-
ment déduire que, pour
une durée d'application
donnée il existe une sol-
licitation provoquant la
rupture et dont la valeur
est précisément fonction
de cette durée.

La charge de rupture
instantanée serait la li-
mite de toutes ces va-
leurs.

**4. Criteres de defor-
mation irreversible.** Un
grand nombre d'études
ont été consacrées à
l'étude des déformations
plastiques dans les mé-
taux [3—9] et des critères d'apparition des phénomènes irréversibles en
ont été déduits, dont l'utilisation est maintenant courante.

Rappelons seulement que les deux courbes de la fig. 5 sont le plus souvent utilisées. En 5 a on a représenté la courbe tension-allongement du solide plastique idéal pour une sollicitation de traction pure et en 5 b celle relative aux corps présentant des phénomènes d'écrouissage non négligeables.

Dans les deux cas σ_0 est le «seuil de plasticité». Les phénomènes irréversibles apparaissent donc en traction pure dès que la tension maximum σ dépasse ce seuil et sont caractérisés par des glissements.

Dans le cas plus général 5 b la relation contrainte-déformation dans la zone plastique devient

$$\sigma = \sigma_0 + B\left(\varepsilon - \frac{\sigma_0}{E}\right).$$

Le cisaillement maximum ayant pour valeur $\sigma/2 = \tau_{max}$ le critère de plasticité s'écrit encore

$$\tau_{max} \geqq \bar{\tau}$$

$\bar{\tau}$ étant un seuil traduit en terme de cisaillement.

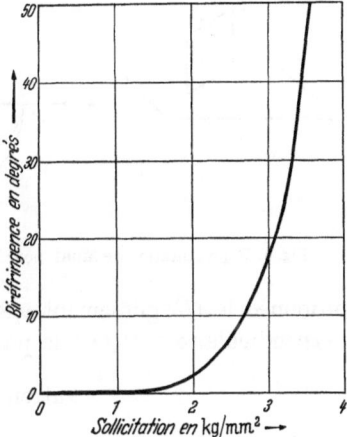

Fig. 4. Biréfringence irréversible en fonction de la sollicitation Durée de charge 700 h

Nous avons vu que l'expérience révèle l'existence de déformations irréversibles dans des corps fragiles mais qu'elles sont alors caractérisées par des extensions et non par des glissements, ce qui est d'ailleurs en accord avec les conclusions antérieures de M. Caquot.

Nous nous proposons de discuter de l'existence d'un critère relatif à ce cas.

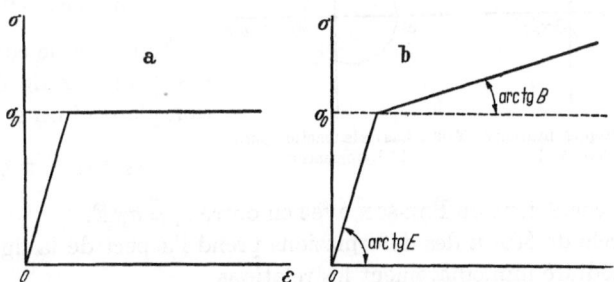

Fig. 5a et b. Schémas types des relations contrainte-déformation pour des solides plastiques
a) Solide plastique idéal   b) Solide plastique avec écrouissage

Pour y parvenir il nous a paru plus logique de raisonner, non pas sur les contraintes comme on le fait habituellement mais sur les déformations qui constituent le phénomène physique, les contraintes n'étant introduites que pour la commodité des raisonnements mathématiques.

Il suffit, pour y parvenir, de substituer la représentation de MOHR des déformations à celle des contraintes.

On sait que celle-ci a lieu dans un plan en prenant en ordonnée la moitié du glissement $\gamma/2$ et en abscisse l'extension ε et qu'elle comporte l'ensemble de 3 cercles coupant l'axe en $\varepsilon_1 \varepsilon_2 \varepsilon_3$ valeurs des déformations principales (fig. 6).

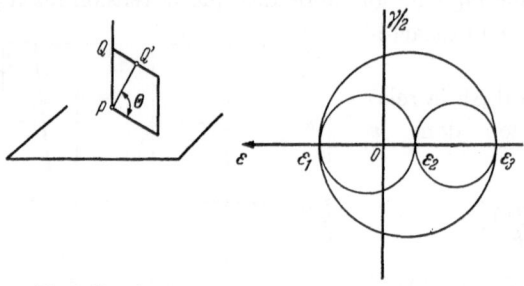

Dans ces conditions, le point représentatif relatif à une direction quelconque PQ du solide est obtenu en portant en abscisse l'extension ε dans la direction PQ en

Fig. 6. Représentation de MOHR pour les déformations

ordonnée le 1/2 glissement $\gamma/2$ entre PQ et la projection sur le plan π perpendiculaire à PQ de la position finale PQ' après déformation

$$\text{autrement dit } \gamma = \cos\Theta.$$

Toutefois, le passage de l'une à l'autre des représentations de MOHR entraîne un changement d'aspect, parfois important.

Limitons-nous au cas de la traction pure, qui seule nous intéresse ici.

Il s'agit d'un cas de contrainte plane, c'est-à-dire que le diagramme de MOHR pour les contraintes se réduit à un seul cercle (fig. 7 a) autrement dit

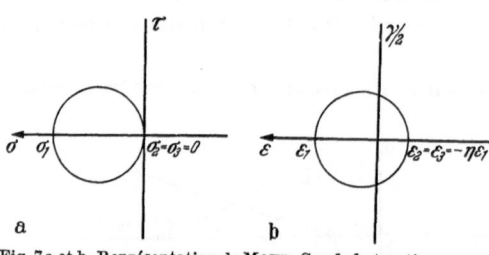

$$\sigma_2 = \sigma_3 = 0.$$

Mais ceci ne correspond pas à un cas de déformation plane et on a

Fig. 7 a et b. Représentation de MOHR. Cas de la traction pure
a) Contraintes   b) Déformations

$$\varepsilon_2 = \varepsilon_3 = -\eta\,\varepsilon_1$$

si η est le coefficient de POISSON avec en outre $\varepsilon_1 = \sigma_1/E$.

Le cercle de MOHR des déformations prend l'aspect de la fig. 7 b. On peut en déduire immédiatement les relations

$$\gamma_{\max} = \frac{1+\eta}{E}\,\sigma_1,$$

$$\varepsilon_{\max} = \frac{\sigma_1}{E}.$$

Transposons maintenant dans cette représentation la courbe intrinsèque de M. CAQUOT qui définit la relation $\gamma = f(\varepsilon)$ caractéristique de la rupture.

Le sommet s de cette courbe pour lequel $\varepsilon_1 = \varepsilon_2 = \varepsilon_3$ c'est-à-dire aussi $\sigma_1 = \sigma_2 = \sigma_3$ correspond encore à la décohésion.

Pour les corps ductiles la courbe est très fermée tandis que pour les corps fragiles elle est très ouverte.

Admettons maintenant que des phénomènes irréversibles peuvent apparaître de deux façons différentes

a) par glissement lorsque $\gamma_{\max}$ dépasse une valeur donnée $\bar\gamma$ caractéristique de la matière,

b) par extension lorsque $\varepsilon_{\max} \geqq \bar\varepsilon$ et considérons d'abord le cas d'un corps ductile (fig. 8).

En accroissant progressivement l'effort de traction pure il arrive un moment où

$$\gamma_{\max} = \frac{1+\eta}{E}\,\sigma_1$$

devient égal à $\bar\gamma$ alors que $\varepsilon_{\max}$ reste inférieur à $\bar\varepsilon$.

Il apparaît alors une déformation irréversible de glissement. Si celle-ci s'accompagne d'un écrouissage on sait que celui-ci accroît la fragilité, c'est-à-dire que la courbe intrinsèque s'ouvre (tracé en pointillé). En même temps, le seuil de plasticité $\bar\gamma$ s'accroît et la déformation irréversible cesse. C'est la zone des déformations permanentes avec équilibre.

Si on accroît suffisamment la charge le processus peut à nouveau se produire.

A un moment cependant $\bar\varepsilon$ est devenu suffisamment petit pour que $\varepsilon_1 = \bar\varepsilon$. On a

$$\sigma_1 = \bar\varepsilon\,E\,,$$

et

$$\sigma_1 \frac{1+\eta}{E} < \dot{\bar\gamma}\,.$$

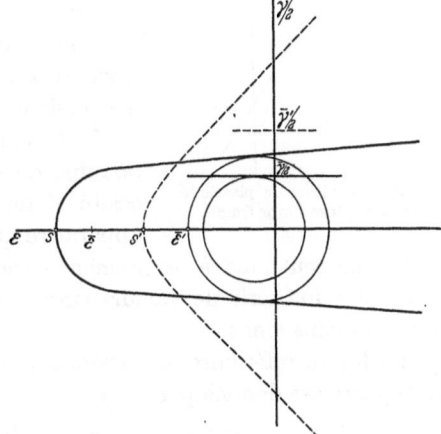

Fig. 8. Mécanisme de plasticité et de rupture Corps Ductile

Il doit alors apparaître une déformation irréversible d'extension qui entraîne une rupture presque immédiate avec arrachement.

Un corps ductile se romprait donc au moment où il est devenu fragile ce qui, traduit sous cette forme, paraît presque une évidence. Ceci explique la brutalité, toujours constatée, de la fin du processus.

La condition pour que la fragilité apparaisse en traction est donc

$$\bar{\varepsilon} < \frac{\bar{\gamma}}{1 + \eta}.$$

Considérons maintenant le cas d'un corps fragile à l'état naturel, c'est-à-dire pour lequel la condition ci-dessus est remplie.

L'application d'un effort de traction pure suffisant provoque une extension $\varepsilon > \bar{\varepsilon}$ et il doit apparaître une déformation irréversible d'extension (fig. 9). Si celle-ci s'établit, la courbe intrinsèque est déformée comme ci-dessus mais cette fois le mécanisme s'accélère et la rupture intervient presque instantanément.

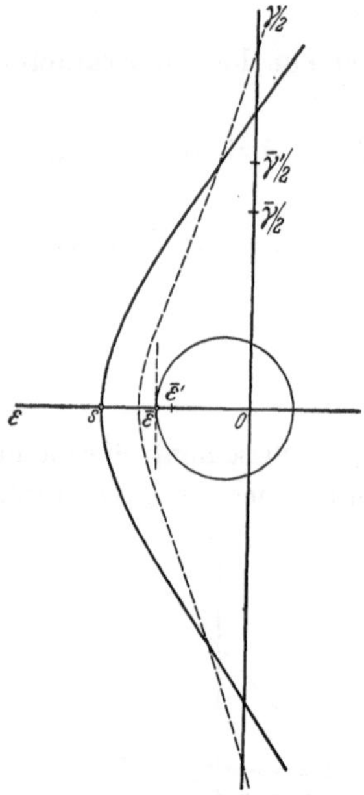

Autrement dit, et si le solide n'a pas de viscosité appréciable, dès que

$$\sigma_1 = E\,\bar{\varepsilon}$$

le corps se rompt sans qu'il soit possible d'apercevoir de déformation permanente mesurable.

C'est le cas de substances comme le verre par exemple.

Si par contre la matière présente une viscosité importante (cas du plexiglass) la rupture ne se produira qu'au bout d'un temps plus ou moins long et en interrompant l'expérience on pourra mesurer un allongement irréversible.

Il devient donc, dans ce cas, presque impossible de séparer les phénomènes de viscosité et de plasticité, le second ne pouvant apparaître que grâce au premier.

Fig. 9. Mécanisme de plasticité et de rupture Corps fragile

On ne peut plus, à proprement parler, considérer une charge de rupture mais une série de valeurs entrainant la rupture au bout de temps plus ou moins longs.

La limite inférieure, en dessous de laquelle il ne se produira jamais de rupture est donnée par

$$\sigma_{\min} = E\,\bar{\varepsilon}$$

et coïncide avec celle qui amorce la déformation irréversible.

Pour le corps ductile au contraire la déformation plastique apparaît dès que

$$\sigma = \frac{E}{1 + \eta}\,\bar{\gamma}$$

mais la charge de rupture lui est supérieure.

Remarquons que, dans ce qui précède, nous avons implicitement supposé que E et η ne varient pas pendant l'écrouissage ce qui semble bien vérifié par la pratique, au moins pour les métaux.

Remarque. Les raisonnements ci-dessus permettent peut-être aussi de comprendre pourquoi les stries apparaissent toujours sur la surface de l'éprouvette.

Il est très probable, en effet, qu'en raison d'un certain nombre de causes (polymérisation plus poussée — évaporation de surface, etc. ...) la couche superficielle présente une viscosité beaucoup plus réduite que le centre. Les phénomènes s'y produisent par suite beaucoup plus rapidement et des ruptures apparaissent alors que l'intérieur n'a pas encore atteint des déformations suffisantes.

Ces criques se propagent à l'intérieur lorsque la charge est suffisamment grande, c'est bien ce que révèle l'expérience.

Même en surface, les manques d'homogénéité jouent un rôle important. Nous avons constaté par exemple que les stries sont très abondantes sur les rayures de surface. Elles s'amorcent sur ces rayures mais gardent toujours une orientation perpendiculaire à la tension maximum.

5. Ductilité et fragilité. Un certain nombre de conséquences peuvent se déduire de ce qui précède:

a) Les notions de ductilité et de fragilité se trouvent précisées. Une relation déterminée entre les valeurs limites $\bar{\varepsilon}$ et $\bar{\gamma}$ constitue, pour un type de sollicitation donnée, le critère de fragilité. Pour un autre mode de sollicitation le critère peut être différent.

b) En l'absence de viscosité, la déformation irréversible par extension ne peut pas être mise en évidence, la rupture brutale survient toujours. Mais pour des substances présentant une réactivité importante il est possible d'arrêter le phénomène à un stade intermédiaire et de mesurer une déformation irréversible.

c) Un corps, normalement ductile, soumis à une charge supérieure à sa limite élastique évolue vers un état fragile qui est en définitive responsable de la rupture.

Il n'y aurait donc aucune différence sur le mécanisme de la rupture finale.

A ce sujet, il est intéressant de signaler [10] que des idées très voisines ont été émises il y a fort longtemps par TRESCA, RESAL et HARTMANN en se basant sur les expériences de VICAT et THURSTON.

Ces derniers expérimentateurs ont montré la possibilité d'obtenir la rupture d'un fil de fer sous une charge de traction pure seulement égale à 0,65 fois la résistance à la traction. Cette rupture se produit, à charge rigoureusement constante au bout d'une année.

HARTMANN de son coté est conduit à préciser que, pour de tels matériaux la déformation élastique est quasi immédiate mais que la déformation plastique est très lente (influence de la viscosité). C'est bien ce que révèlent les expériences de fluage à froid et le développement de la striction.

Il constate, en outre, que la rupture finale est toujours normale aux lignes de tension maximum alors que les lignes de glissement, à la limite de la phase élastique, suivent sensiblement le réseau de cisaillement maximum.

Si l'on se rappelle que le plexiglass a un module d'YOUNG très faible (de l'ordre de 360 kg/mm²) on comprend que les déformations réversibles, contrairement au cas des métaux, seront considérables par rapport aux déformations irréversibles. Elles sont, en outre, elles-mêmes affectées par la viscosité.

Mais qualitativement les phénomènes qui provoquent la rupture sont exactement les mêmes.

Il convient d'ajouter que, compte tenu de la viscosité, les notions de fragilité et de ductilité sont très relatives, une même substance se comportant de façon très différent suivant la rapidité avec laquelle on lui impose une déformation donnée [2].

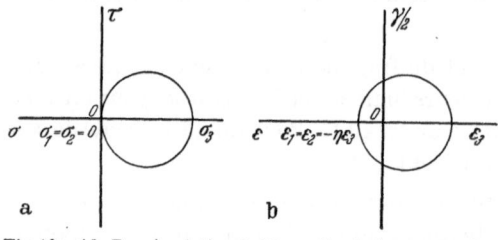

Fig. 10 a et b. Représentation de MOHR. Cas de la compression pure
a) Contraintes   b) Déformations

Nos expériences montrent enfin que pour le plexiglass les phénomènes irréversibles commencent pour des charges relativement faibles, 2 kg/mm² environ, c'est-à-dire le tiers de la valeur généralement admise pour la résistance à la rupture.

6. Cas de la compression pure. Le cercle de MOHR des déformations est alors celui de la fig. 10 b correspondant à la fig. 10 a pour les contraintes. On a encore les relations:

$$\sigma_3 = E\,\varepsilon_3\,, \quad \tau_{max} = \frac{\sigma_3}{2}\,,$$

$$\varepsilon_2 = \varepsilon_1 = -\eta\,\varepsilon_3\,, \quad \gamma_{max} = (1+\eta)\,\varepsilon_3 = -\frac{1+\eta}{\eta}\,\varepsilon_1\,.$$

Dans ce cas c'est la déformation principale ε_1 qui intervient pour le critère de fragilité.

Un matériau est donc fragile en compression pure si

$$\bar{\varepsilon} \leqq \frac{\eta}{1+\eta}\,\bar{\gamma}\,.$$

Comme le coefficient de Poisson η est toujours inférieur à l'unité on voit que cette limite $\bar{\varepsilon}$ est inférieure à celle trouvée pour la tension pure.

Une substance donnée peut donc être fragile en traction sans l'être en compression.

Par exemple pour $\eta = 0{,}25$ les 2 limites sont

$$\bar{\varepsilon} < 0{,}8\,\bar{\gamma} \ \text{en traction,}$$
$$\bar{\varepsilon} < 0{,}2\,\bar{\gamma} \ \text{en compression.}$$

Pour des corps ductiles les phénomènes en compression pure seront exactement les mêmes qu'en traction pure, on constatera l'existence d'une déformation irréversible de glissement avec équilibre possible précédant une zone de fragilité entrainant la rupture.

Pour le plexiglas, que nous avons reconnu fragile en traction, il apparaît que $\bar{\varepsilon}$ satisfait à la première condition mais non à la seconde.

Il ne doit donc pas apparaître de stries (déformation irréversible) en compression pure.

L'expérience confirme ce point de vue. Si l'on soumet une éprouvette de plexiglas à de la flexion circulaire on voit apparaître des fissures sur les fibres tendues, mais même en poussant l'expérience jusqu'à la rupture, aucune fissure ne se manifeste du côté comprimé.

7. Influence de la viscosité. Elle entraîne, pour une charge constante, une variation au cours du temps de la déformation réversible (réactivité) et aussi de la déformation irréversible.

J'ai cherché s'il était possible d'écrire la loi de cette variation.

La fig. 11 montre l'allure des phénomènes pour une éprouvette soumise à une sollicitation de traction pure de 3 kg/mm² (variation de la biréfringence).

La partie réversible est déduite d'une série de mesures s'étendant sur des durées croissantes, en re-

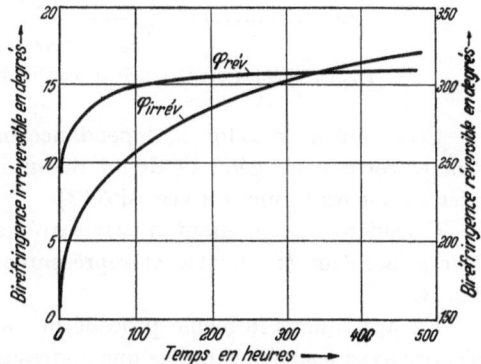

Fig. 11. Biréfringence réversible et irréversible en fonction du temps. Traction pure — Charge 3 kg/mm²

tranchant de la valeur totale la biréfringence irréversible mesurée après relaxation complète.

Elle doit, si nos hypothèses sont correctes, être une mesure de la déformation réversible.

Pour cette dernière, j'ai montré [2] qu'elle suit une loi de la forme

$$\varepsilon = A - B_1 e^{-\alpha_1 t} - B_2 e^{-\alpha_2 t}$$

dans laquelle, A, B_1, B_2 dépendent du matériau et de la charge appliquée, tandis que α_1 et α_2 sont liés aux coefficients de viscosité.

Procédant comme dans le travail relatif aux déformations [2] on trouve que la biréfringence réversible est bien représentée par une équation de la forme

$$\varphi = 18564 - 2820\,e^{-0,4\,\cdot\,10^{-4}t} - 1920\,e^{-0,31\,\cdot\,10^{-5}t}$$

si φ est exprimé en minutes d'angle et t en secondes.

La fig. 12 donne la comparaison des valeurs calculées et des valeurs expérimentales. L'écart ne dépasse pas 1%.

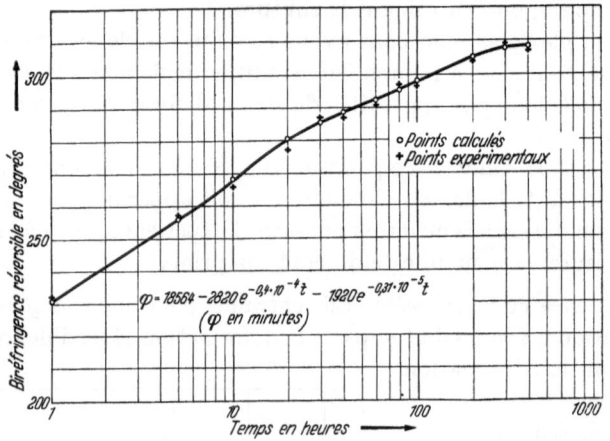

Fig. 12. Biréfringence réversible en fonction du temps. Charge 3 kg/mm²

Ceci confirme à la fois la dépendance entre biréfringence et déformation et cette forme générale de loi, déduite des équations de Lame complétées par les termes de viscosité [2].

Considérons maintenant la partie irréversible de la biréfringence relative à la même éprouvette et représentée par la seconde courbe de la fig. 11.

En appliquant le même procédé de calcul on constate qu'elle peut s'écrire avec les mêmes unités que ci-dessus.

$$\varphi = 1235 - 317\,e^{-0,088\,\cdot\,10^{-4}t} - 780\,\varepsilon^{-0,075\,\cdot\,10^{-5}t}.$$

La fig. 13 montre que les valeurs calculées par cette formule s'accordent très bien avec les valeurs expérimentales dans toute l'étendue des mesures.

On voit qu'une même forme de loi convient pour les déformations pseudo réversibles et pour les déformations irréversibles pour lesquelles les phénomènes ont la même origine (existence de la viscosité).

Cependant les coefficients des exponentielles sont nettement plus faibles dans le second cas.

Autrement dit la viscosité est beaucoup plus marquée pour les déformations plastiques que pour les déformations élastiques ce qui est tout-à-fait en accord avec les remarques faites par les auteurs déjà cités en ce qui concerne les métaux.

Dans ce dernier cas l'écart est tel que les déformations réversibles sont pratiquement instantanées alors que les déformations irréversibles présentent une réactivité notable.

Pour les corps comme le plexiglass, au contraire, les phénomènes affectent les deux domaines mais à des degrés très différents.

Il est remarquable que pour les deux coefficients α_1 et α_2 les rapports sont sensiblement du même ordre (4,5 pour α_1, et 4,13 pour α_2).

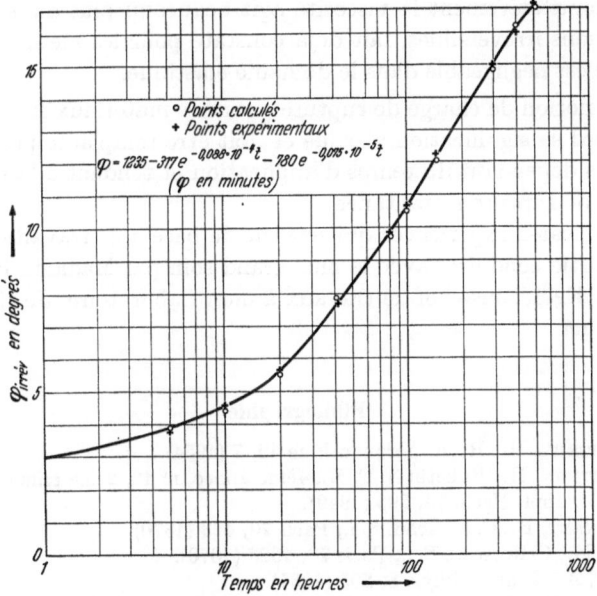

Fig. 13. Biréfringence irréversible en fonction du temps. Charge 3 kg/mm²

8. Conclusions. 1. Dans les matériaux fragiles pour lesquels la viscosité n'est pas négligeable il est possible de mettre en évidence des déformations irréversibles. Celles-ci ont pu être mesurées dans le plexiglass d'une part directement, d'autre part grâce à la biréfringence accidentelle par déformation.

Une très longue série d'expériences a permis d'étudier les déformations réversibles et irréversibles en fonction de la sollicitation de traction pure et du temps.

2. Le mécanisme d'apparition des phénomènes plastiques a été élucidé en considérant la représentation de Mohr et en utilisant les propriétés de la courbe intrinsèque.

3. Ces considérations permettent de préciser les notions de ductilité et de fragilité et conduisent à penser qu'au moment de la rupture proprement dite, il y a identité de comportement pour les deux catégories de substances.

4. On peut en déduire un critère précis de fragilité qui diffère d'ailleurs d'un type de sollicitation à l'autre.

5. La loi de variation avec le temps a la même forme pour les déformations irréversibles que pour la partie réversible et est conforme au résultat théorique trouvé antérieurement par l'auteur en s'appuyant sur la notion de viscosité.

6. Quantitativement la viscosité agit beaucoup plus dans le cas des déformations irréversibles, fait déjà constaté pour les métaux pour lesquels elle est négligeable dans le domaine élastique.

7. La notion de charge de rupture pour des matériaux fragiles et visqueux perd sa signification absolue et doit être remplacée par une série de valeurs dépendant du temps d'application et tendant à la limite vers la charge de rupture instantanée.

Les délicates expériences qui servent de base à ce travail ont été en grande partie conduites avec le plus grand soin par Madame Pauthier, Ingénieur Epci, sous-chef de travaux à mon Laboratoire. Je tiens à l'en remercier ici.

Bibliographie

[1] le Boiteux, H.: Rech. Aéronautique 42, 7 (1954).
[2] le Boiteux, H.: Bulletin S. F. M., 4ème année, n⁰ 12, 2ème trimestre 1954.
[3] Hill: J. appl. Mech. 16, 295 (1949).
[4] St Venant, B.: C.R. Acad. Sci., Paris 70, 473 (1870).
[5] Levy, M.: C.R. Acad. Sci., Paris 70, 1323 (1870).
[6] Nadai, A.: J. appl. Phys. 8, 205 (1937).
[7] Hencky, H.: Z. angew. Math. Mech. 4, 323 (1924).
[8] Reuss, A.: Z. angew. Math. Mech. 10, 266 (1930).
[9] Hoffman, O. et G. Sachs: Introduction to the theory of plasticity for Engineers. New York 1953.
[10] Resal: Résistance des Matériaux, 1898.

Viskosität und Zeitwirkungen im nichtlinearen Bereich

Von F. Schultz-Grunow, Aachen

Mit 28 Abbildungen

Viskosität und Zeitwirkungen treten sowohl in festen als flüssigen Stoffen auf. Deshalb sollen hier auch Flüssigkeiten in Betracht gezogen werden, um so mehr als in Flüssigkeiten, weil sie keine Fließgrenze besitzen, diese Eigenschaften klarer zutage treten. Bei den Zeitwirkungen wird auch der Einfluß großer Formänderungen auf die quantitative Beziehung zwischen Spannungs- und Formänderungstensor von Interesse sein. Wenn auch dem Folgenden der amorphe Körper zugrunde liegt, so wird sich doch ergeben, daß dieser Begriff nicht zu eng gefaßt zu werden braucht.

Es hatte nicht an Bemühungen gefehlt, den Begriff der NEWTONschen Viskosität η, definiert durch die Beziehung

$$\tau = \eta \, \dot{\gamma} \tag{1}$$

mit τ als Schubspannung, $\dot{\gamma}$ als Schergeschwindigkeit, auch im hochzähen und plastischen Bereich anzuwenden, indem man für η die Viskosität einer hypothetischen NEWTONschen Flüssigkeit einführte, die bei gleicher Schergeschwindigkeit die gleiche Schubspannung aufweist. Da aber diese Viskosität eine Abhängigkeit von den Größen zeigt, die sie in Beziehung setzt, darf sie hier nicht mehr als eine physikalische Konstante angesprochen werden.

Ferner wurde versucht, den Viskositätsbegriff dadurch zu erweitern, daß man zu der linearen Abhängigkeit der Dissipation von der zweiten Invarianten I_2 des Formänderungsgeschwindigkeitstensors, wie sie in der NEWTONschen Flüssigkeit besteht, eine lineare Abhängigkeit von der dritten, I_3 hinzunahm [1] in der Form

$$D = -4\,\eta\,I_2 + 6\,\eta'\,I_3 .$$

D bedeutet hier die sekundlich in Wärme verwandelte Reibungsarbeit und η' eine zweite Materialkonstante. Da I_3 vom dritten Grade in den Formveränderungsgeschwindigkeiten ist, kann das Zusatzglied positiv oder negativ sein. Es verletzt daher den 2. Hauptsatz.

Vielmehr muß man, um zu wirklichen physikalischen Konstanten zu kommen, eine nichtlineare Abhängigkeit der Schubspannung τ von der Schergeschwindigkeit $\dot\gamma$ zulassen, indem man zunächst

$$\dot\gamma = f(\tau) \qquad (2)$$

schreibt. Die Funktion $f(\tau)$ wird als Fließfunktion bezeichnet. Wegen der Forderung einer vom Maßsystem unabhängigen Aussage lassen sich zwei physikalische Konstanten A, C mit den gleichen Dimensionen wie τ, $\dot\gamma$ einführen [2]

$$\frac{\dot\gamma}{C} = f\left(\frac{\tau}{A}\right).$$

Nur im Fall der Proportionalität, die auch bei den hochzähen Flüssigkeiten im Grenzfall verschwindender Schubspannung besteht, vereinigen sich beide Konstante zur Newtonschen Zähigkeit

$$\eta = \frac{A}{C}.$$

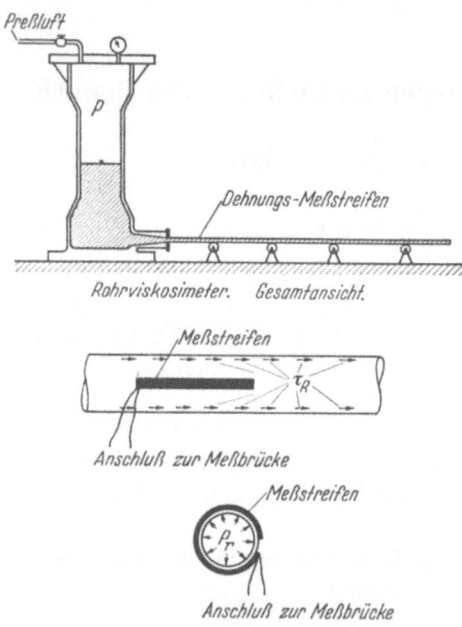

Abb. 1. Rohrviskosimeter. Messung des Druckabfalles mit Dehnmeßstreifen über die elastische Dehnung des Rohres. p Druck, τ Schubspannung

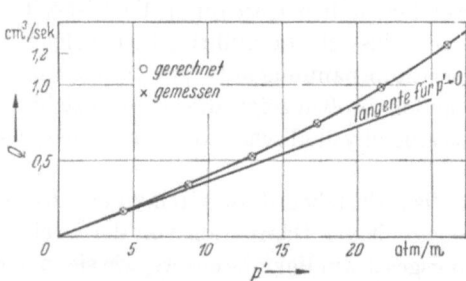

Abb. 2. Auswertung von Durchflußversuchen im Rohr mit Acrylsäurebutylester (Akronal 4 F) nach Gl. (3). Es ergeben sich die Konstanten $A = 4{,}61 \cdot 10^{-2}$ kg/cm², $C = 7{,}22$ s^{-1} und daraus die Viskosität $\eta = A/C = 0{,}64 \cdot 10^4 P$. Q Volumdurchsatz, p' Druckabfall über der Längeneinheit des Rohres

Die Fließfunktion kann man sowohl im Rohr- als auch im Couette-Viskosimeter unmittelbar aus dem Versuch ohne eine Hypothese über das Stoffverhalten ermitteln. Im Rohrviskosimeter läßt sich die Anlaufstrecke dadurch eliminieren [3], daß man den Druckabfall über die elastische Dehnung der Rohrwand mit elektrischen Dehnmeßstreifen mißt (Abb. 1). Ein anderer Vorteil dieser Meßmethode besteht darin, daß man diese Streifen axial und in Umfangsrichtung aufkleben und so den radialen und axialen Druck getrennt ermitteln kann. Tatsächlich hat sich bei fester Seife gezeigt, daß diese Drucke nicht gleich sind, sondern der radiale Druck 10% größer ist.

Auf diese Weise wurde Akronal und eine Polystyrol–Toluol-Mischung untersucht. Im Grenzfall kleiner Schubspannungen haben diese Stoffe eine rund 10^6mal so große Zähigkeit wie Wasser. Es ergab sich, daß diese Stoffe der PRANDTLschen Fließformel

$$\frac{\dot{\gamma}}{C} = \operatorname{Sinh} \frac{\tau}{A} \qquad (3)$$

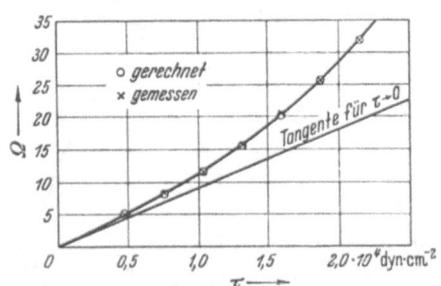

sehr genau genügen, wie Abb. 2 und 3 zeigen. Eine weitere Prüfung ergab sich aus dem Zusammenhang von A und C mit molekularkinetischen Daten [4]. Aus A konnte der Durchmesser des Polystyrolmoleküls ermittelt werden in einer Übereinstimmung mit anderen Messungen, die bemerkenswert ist, weil

Abb. 3. Auswertung von Versuchen im Rotationsviskosimeter mit einer Polystyrol–Toluol-Mischung nach Gl. (3). Es ergeben sich die Konstanten $A =$ $1{,}88 \cdot 10^{-2}$ kg/cm² und $C = 18{,}75$ s^{-1}. Ω Winkelgeschwindigkeit des Viskosimeters in 1/s, τ Schubspannung am äußeren Zylinder

A bei einer hohen Konzentration von rund 50% gemessen wurde, die anderen Messungen sich aber auf hochverdünnte Lösungen beziehen. Aus der Temperatur- und Konzentrationsabhängigkeit von C konnte die Aktivierungsenergie und die Anzahl der Fehlstellen in Übereinstimmung mit den physikochemischen Daten bestimmt werden.

Eine dritte Prüfung läßt der Vergleich der gemessenen Geschwindigkeitsprofile mit den von PRANDTL-VANDREY [5] theoretisch aus (3) errechneten zu. Im Versuch wurde das Profil aus Lichtbildern bestimmt, die man von einer anfangs ebenen Grenzfläche zwischen heller und schwarz gefärbter Flüssigkeit in bestimmten Zeitabständen auf-

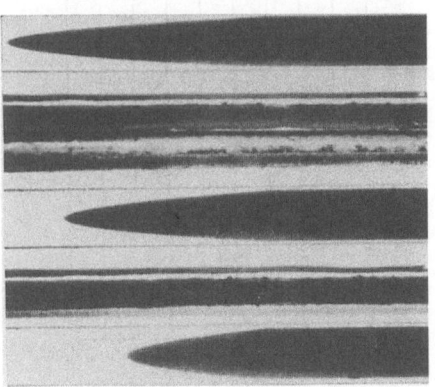

Abb. 4. Zeitliche Entwicklung der Grenzfläche zwischen heller und schwarz gefärbter Flüssigkeit in der Rohrströmung

genommen hatte. Abb. 4 zeigt solche Lichtbilder, Abb. 5 zeigt die volle Übereinstimmung zwischen Versuch und Theorie und die Abweichung von der POISEUILLE-Strömung.

Die übliche Annahme, daß bei Scherbewegungen der Stoff an den festen Begrenzungswänden haftet, hat sich bei einem plastischen Stoff, bei Ton von 19,7 Gew.-% Wassergehalt, als nicht zutreffend erwiesen. Wie Abb. 6 zeigt, fallen die beim Durchpressen durch zwei Rohre ver-

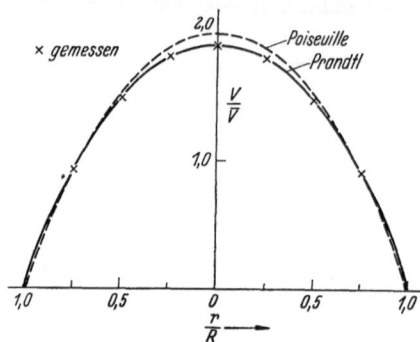

Abb. 5. Geschwindigkeitsprofil der ausgebildeten Rohrströmung. i mittlere Geschwindigkeit, v Geschwindigkeit am Halbmesser r, R Rohrhalbmesser

schiedenen Durchmessers ermittelten Wandschubspannungen τ_R bei gleicher mittlerer Durchflußgeschwindigkeit v_R zusammen. Das bedeutet, daß der Ton als starrer Pfropfen durch das Rohr gleitet. Diese Feststellung konnte durch rechnerische Auswertung der Fließfunktion bestätigt werden[1]. Hierzu muß die weitere Abhängigkeit

$$v_R = \varphi\,(\tau_R) \qquad (4)$$

eingeführt werden. Aus zwei Versuchen mit zwei verschiedenen Rohrhalbmessern R, aber gleichen Wandschubspannungen τ_R läßt sich die Fließfunktion als identisch Null und die Abhängigkeit (4) als eine lineare ermitteln.

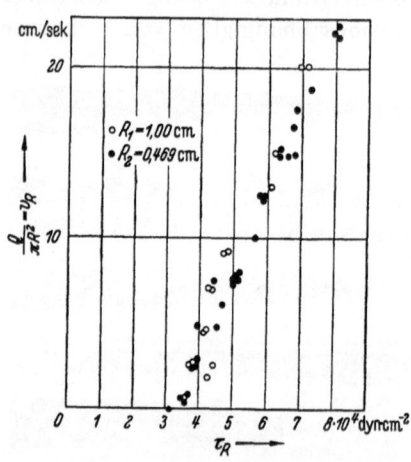

Abb. 6. Durchflußwiderstand von Ton mit 19,7 Gew.-% Wasser beim Durchpressen durch Rohre. Q Volumdurchsatz, R Rohrhalbmesser, τ_R Wandschubspannung, v_R Fließgeschwindigkeit

Eine Bestätigung dafür, daß A, C wirklich physikalische Konstanten sind, liefert die Dimensionsanalyse. Wenn in der Hydrodynamik die Widerstandszahl c_f, d. i. der mit dem Staudruck $\frac{\varrho}{2}\,v^2$ und einer charakteristischen Länge l dimensionslos gemachte Widerstand W von *einer* dimensionslosen Kombination der Einflußgrößen, der Reynoldsschen Zahl Re abhängt,

$$c_f = \frac{W}{\dfrac{\varrho}{2}\,v^2\,l^2} = f\,(Re)\,, \qquad (5)$$

so muß nun im Falle von zwei Materialkonstanten A, C der mit der Konstanten A dimensionslos ge-

---

[1] Aus einer noch nicht veröffentlichten Aachener Dissertation von J. Berghaus.

machte Widerstand von *zwei* dimensionslosen Kombinationen der Einflußgrößen abhängen:

$$\frac{W}{A\,l^2} = f\left(\frac{v\,l\,\varrho}{A/C}\,,\ \frac{v}{lC}\right).\tag{6}$$

Wegen der nur kleinen Fließgeschwindigkeiten v wird man von vornherein Trägheitseffekte und daher eine Abhängigkeit von $v\,l\varrho\,C/A$ ausschließen. Das Widerstandsgesetz lautet also

$$\frac{W}{A\,l^2} = f\left(\frac{v}{lC}\right).\tag{7}$$

Abb.7 zeigt den Widerstand von Akronal beim Durchfluß durch

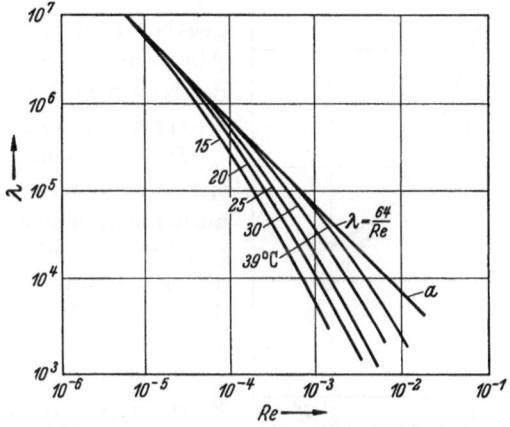

Abb. 7. Abhängigkeit der Rohrreibungszahl λ von der REYNOLDS-Zahl *Re* für Akronal 4 F bei verschiedenen Temperaturen. *a* POISEUILLEsches Widerstandsgesetz

Rohre in der hydrodynamischen Auftragung (6), also die Rohrwiderstandszahl in Abhängigkeit von der REYNOLDSschen Zahl. Die Kurven konnten einer Rechnung von PRANDTL-VANDREY [5] entnommen werden, in die die bei verschiedenen Temperaturen gemessenen Werte von A, C eingesetzt wurden. Die Temperaturabhängigkeit verschwindet in der Auftragung nach (7) (Abb. 8), die dortige Kurve stellt also tatsächlich das Widerstandsgesetz für alle möglichen A, C-Werte dar, wenn nur die PRANDTLsche Fließformel gilt.

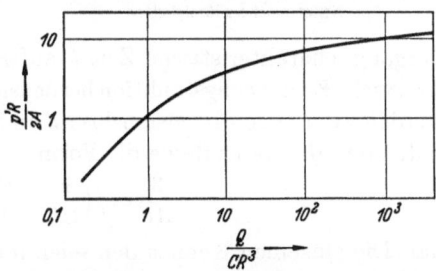

Abb. 8. Bestätigung des neuen Widerstandsgesetzes im Rohrversuch mit Akronal 4 F. p' Druckabfall über der Einheit der Rohrlänge, Q Volumdurchsatz, R Rohrhalbmesser, A, C Stoffkonstanten

Analoges ergab sich für den im COUETTE-Viskosimeter und in technischen Rührapparaten auftretenden Widerstand [6].

Das Widerstandsgesetz erlaubt auch die Beurteilung der Nützlichkeit viskosimetrischer Messungen. Offenbar sind nur solche Viskosimeter von Nutzen, die die Fließfunktion $f(\tau)$ und die Konstanten A, C zu ermitteln gestatten, also lediglich das Couette- oder Rohrviskosimeter. Liefert ein Viskosimeter wie etwa das Höpplersche Kugelfallviskosimeter eine Zähigkeit η gemäß der Definitionsgleichung (1), so ist sie wohl bei verschwindenden Schubspannungen physikalisch sinnvoll, aber der Grenzwert von τ, bis zu dem dies mit genügender Genauigkeit zutrifft, ist eine Funktion der Kenngröße v/lC, zu deren Anwendung C bekannt sein müßte. Die in Abb. 9 wiedergegebenen Messungen zeigen die Abweichung des Akronals von der Newtonschen Flüssigkeit im Höppler-Viskosimeter [6].

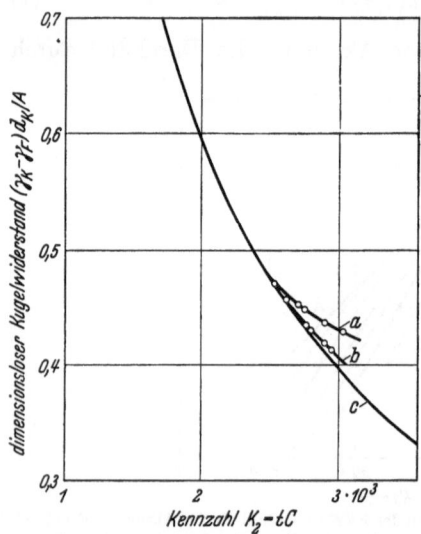

Abb. 9. Gültigkeitsbereich des Höppler-Viskosimeters für Akronal 4 F. γ_K, γ_F Wichte der Fallkugel vom Durchmesser d_K bzw. der Versuchsflüssigkeit, t Kugelfallzeit, A, C Stoffkonstanten, a Versuchsreihe mit steigender, b bei fallender Temperatur bei jedem nachfolgenden Versuch, c Verlauf für Newtonsche Flüssigkeiten

Das Widerstandsgesetz läßt sich auf nichtstationäres Fließen ausdehnen, ohne eine Hypothese annehmen zu müssen. Jetzt wird die Elastizität, charakterisiert durch den Schubmodul G, und die Nachwirkung, gekennzeichnet durch eine Zeitkonstante t^*, in Erscheinung treten. Die Fließformel wird die Form

$$\frac{\dot{\gamma}}{C} = f\left(\frac{\tau}{A}, \frac{\dot{\tau}}{GC}, \frac{t}{t^*}\right)$$

haben, wo nun die zeitliche Ableitung $\dot{\tau}$ der Schubspannung und eine für den vorliegenden Fließvorgang charakteristische Zeit t auftritt. Durch sie wird eine Boltzmannsche Erinnerungsfunktion berücksichtigt. Mit zwei neuen Konstanten werden zwei neue dimensionslose Kombinationen im Widerstandsgesetz auftreten, das demzufolge die Form

$$\frac{W}{Al^2} = f\left(\frac{v}{lC}, \frac{G}{A}, Ct^*\right) \tag{8}$$

hat. Die einzelnen Kennzahlen seien folgendermaßen bezeichnet

$$K_2 = \frac{v}{lC}, \quad K_3 = \frac{G}{A}, \quad K_4 = Ct^*. \tag{9}$$

Für v/l wird, wenn es sich um drehende Bewegungen handelt, vorteilhaft eine Winkelgeschwindigkeit ω eingeführt. Dann ist

$$K_2 = \frac{\omega}{C}. \tag{10}$$

Wenn auch z.B. die Konstanten G, A von gleicher Dimension sind, so brauchen sie doch nicht von gleicher Größe zu sein. Zwar ist in der MAXWELLschen Fließformel

$$\dot{\gamma} = \frac{t}{G} + \frac{\tau}{Gt^*} \tag{11}$$

$G = A$ gesetzt, hier wird sich aber zeigen, daß G, A sehr verschiedene Werte haben.

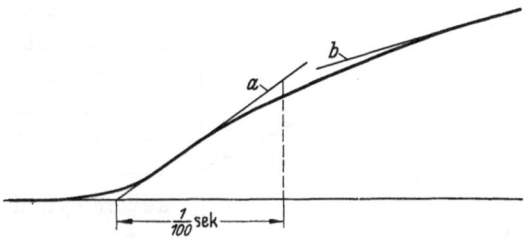

Abb. 10. Oszillogramm des Drehwinkels in Abhängigkeit von der Zeit beim Anfahren des COUETTE Viskosimeters mit konstantem Moment

Die Konstante t^* wurde im COUETTE-Viskosimeter durch einen Anfahrversuch bei konstantem Drehmoment bestimmt. Hier ist $t = 0$, so daß die Konstante G nicht zur Wirkung kommt und als einzige noch unbekannte Konstante in (8) t^* auftritt. Die in Abb. 10 dargestellte Messung des Drehwinkels φ in Abhängigkeit von der Zeit t zeigt zwei charakteristische Tangenten a, b und damit zwei charakteristische Winkelgeschwindigkeiten, ω_a im Augenblick des Anfahrens, ω_b für den endgültigen stationären Zustand. Der stetige Anstieg bis zur Tangente a ist durch die Apparateträgheit bedingt, wie eine einfache Rechnung zeigt. Trägt man die mit A dimensionslos gemachte Schubspannung über ω_a/C entsprechend dem Widerstandsgesetz (7) auf, so fallen nun nach Abb. 11 die bei verschiedenen Temperaturen aufgenommenen Meßreihen nicht zusammen. Hierin zeigt sich der durch

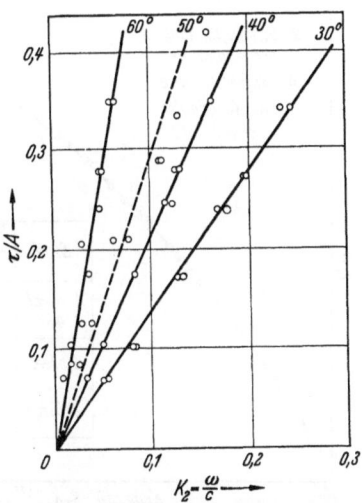

Abb. 11. Dimensionslose Auswertung von Anfahrversuchen im COUETTE-Viskosimeter mit Akronal 4 F bei konstantem Drehmoment. τ mittlere Schubspannung, ω Winkelgeschwindigkeit des inneren Zylinders im Augenblick des Anfahrens, A, C Konstante

(8) formulierte Einfluß der Kennzahl Ct^*. Setzt man in Übereinstimmung mit den weiter unten erwähnten theoretischen Vorstellungen

die Neigung der Versuchsgeraden in Abb. 11 dem Kehrwert von Ct^* gleich,

$$\frac{\tau/A}{\omega_a/C} = \frac{1}{Ct^*}, \qquad (12)$$

so ergibt sich die in Abb. 12 gezeigte Temperaturabhängigkeit der Zeitkonstanten t^*, die wie alle molekularkinetisch bedingten Zeiten mit der Temperatur abnimmt.

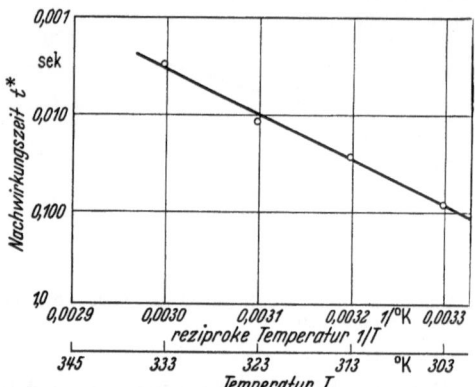

Abb. 12. Temperaturabhängigkeit der Zeitkonstanten t^*

Die Zeitkonstante t^* hat sich auch bei periodisch veränderlicher Scherbewegung als gültig erwiesen [7]. Im Couette-Viskosimeter war der äußere Zylinder periodisch angetrieben und der innere Zylinder elastisch gelagert. Weil auch die Spaltweite δ verändert wurde, lautet hier die dimensionslose Kennzahl K_2, wenn ω die Frequenz des periodischen Antriebs, a die relative Amplitude des einen Zylinders gegen den andern bedeutet

$$K_2 = \frac{v}{lC} = \frac{\omega a}{C\delta}.$$

Es wurde der in der Dimensionsanalysis übliche Potenzproduktansatz

$$\frac{W_{\max}}{A\,l^2} = K_2^{\alpha} K_4^{\beta} \qquad (13)$$

gemacht, wo $W_{\max}$ die Amplitude des periodisch sich ändernden Widerstandes bedeutet. Wie Abb. 13 zeigt, hat sich bei drei verschiedenen Temperaturen (13) bestätigt mit $\alpha = 1$,

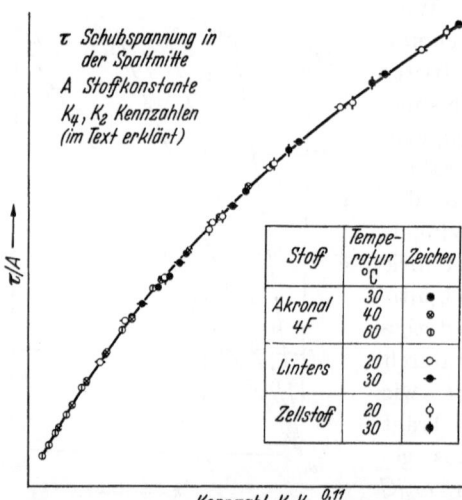

Abb. 13. Widerstandsgesetz für das Couette-Viskosimeter bei periodischer Scherbewegung und verschiedenen Spaltweiten. Meßwerte für Akronal 4 F, Linters (Polymerisationsgrad 900) und Zellstoff (Polymerisationsgrad 700 bis 800) bei verschiedenen Temperaturen

$\beta = 0,11$. Daß die Kennzahl G/A ohne Einfluß ist, wird sich noch erklären.

Die Konstante G als letzte der noch unbekannten Konstanten läßt sich aus einem Anfahrversuch mit konstanter Schergeschwindigkeit $\dot{\gamma}_0$ aus dem zeitlichen Anwachsen des Drehmoments im COUETTE-Viskosimeter ermitteln. Das Oszillogramm in Abb. 14 a, b zeigt dieses Anwachsen.

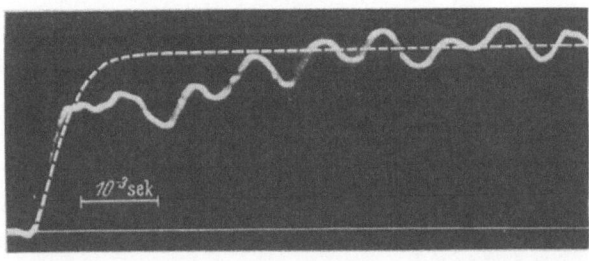

a

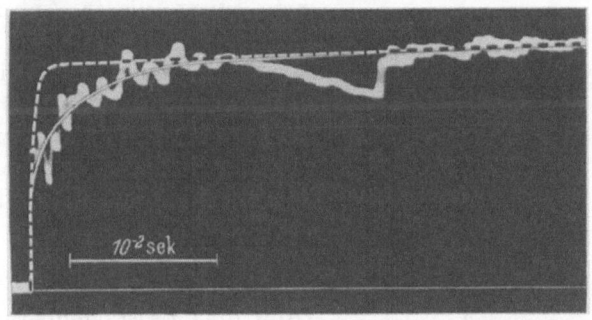

b

Abb. 14a, b. Oszillogramme des Drehmomentes in Abhängigkeit von der Zeit beim Anfahren des COUETTE-Viskosimeters mit konstanter Drehzahl, bei verschiedenen Zeitmaßstäben. a 200 U/min, b 240 U/min

Die Null-Linie in Abb. 14 b muß parallel zum gestrichelten Kurvenende liegen und nicht wie in der Abbildung geneigt.

Der geradlinige Anstieg lehrt, daß im Augenblick des Anfahrens allein die Elastizität wirksam ist und demzufolge

$$\frac{t}{G} = \dot{\gamma}_0 \qquad (14)$$

besteht. Aus der Neigung des geradlinigen Anstiegs wird für Akronal 4 F $G = 2000 \ \text{g/cm}^2$ bei $20°\text{C}$ ermittelt, während $A = 40 \ \text{g/cm}^2$ ist. Die Kennzahl $K_3 = G/A$ ergibt sich hiermit als wesentlich größer als die Kennzahlen K_2, K_4 in der oben erwähnten periodischen Scherströmung, was die Einflußlosigkeit von $K_3 = G/A$ in jenem Versuch erklärt. Es ist dort

$$K_2 = 0,5, \quad K_4 = 0,9, \quad K_3 = 50.$$

Die überlagerte Schwingung von 10^3 Hz in Abb.14a, b ist unvermeidbar, da ein Anfahren mit konstanter Geschwindigkeit einen Stoß bedeutet. Sie rührt von der immerhin geringen Elastizität her, über die die Kraftmessung erfolgte. Auf den ersten Augenblick des Anfahrens, der allein für die Auswertung wichtig ist, hat sie keinen Einfluß, da die sie verursachende Kraft nach (14) linear mit der Zeit wächst.

Unser Ansatz (14) steht nicht im Widerspruch zu der von Prandtl selbst [8] auf nichtstationäres Fließen erweiterten Formel

$$\frac{\dot{\gamma}}{C} = \frac{\tau}{GC} + \text{Sinh} \frac{\tau}{A} \tag{15}$$

genausowenig wie zu der Maxwellschen Formel (11), weil im ersten Augenblick des Anfahrens τ vorherrschend ist. Für spätere Zeiten ist aber (15) nicht in Übereinstimmung mit dem Versuch, wie die aus (15) errechnete Kurve, die in Abb.14a, b gestrichelt ist, zeigt.

Diese Abweichung läßt sich modellmäßig verstehen, denn (15) bedeutet nichts anderes als die Hintereinanderschaltung einer Feder und einer nichtlinearen Dämpfung in der Form

$$\frac{\tau}{A} = \text{Arsinh} \frac{\dot{\gamma}}{C}$$

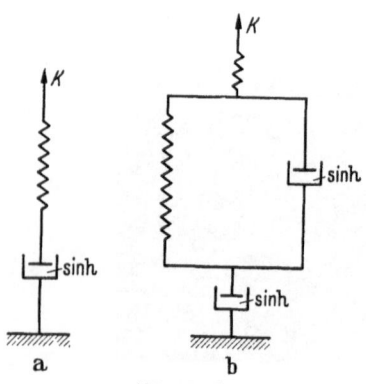

a   b

Abb. 15a, b.
Modelle mit nichtlinearer Dämpfung für nichtstationäres Fließen und Kriechen

(Abb. 15a). Es ist offenbar notwendig, noch eine Parallelschaltung hinzuzunehmen, wie dies Abb.15b andeutet. Diese Schaltung erklärt auch das Auftreten zweier Winkelgeschwindigkeiten ω_a, ω_b in Abb.10 bei dem Anfahrversuch mit konstantem Moment: Zuerst verlängern sich alle Elemente des Modells, was zu ω_a Anlaß gibt, im schließlichen stationären Fall verlängert sich aber nur die vorgeschaltete Dämpfung, die zu ω_b Anlaß gibt.

Jedoch ergibt sich mit dieser Modellvorstellung für den Beginn des Anfahrens die Beziehung (3), die sich in dem durch Abb.11 gegebenen Meßbereich zu

$$\frac{\dot{\gamma}}{C} = \frac{\tau}{A} \tag{16}$$

vereinfachen läßt. Das gleiche Resultat erhält man bei der herkömmlichen Annahme einer linearen Dämpfung. Abb.11 zeigt aber, daß nicht eine einzige unter 45° geneigte Gerade, wie sie (16) fordert, besteht. Das zeigt, daß das Modell nach Abb.15b nicht genügend Konstanten enthält, es fehlt die oben eingeführte Konstante t^*. Die Konstante läßt sich aber einführen, indem man das Modell nach Abb. 15b erweitert durch Hintereinanderschalten mehrerer Modelle, die je aus parallel geschalteter Dämpfung und Feder bestehen mit verschiedenen Konstanten C, wie dies PRANDTL [8] vorschlug und MUSSMANN [9] ausführte. Das PRANDTLsche Modell ist zwar ein anderes, es besteht aus zwei parallel geschalteten Stäben, die also gleiche Längenänderungen erleiden (Abb. 16). Der eine ist elastisch, der andere verhält sich plastisch nach (15), enthält also eine überlagerte Elastizität. Da aber die Querschnitte $\varkappa$, $(1 - \varkappa)$ der Stäbe in den Endformeln nur als Faktoren von Materialkonstanten auftreten, kann man an Stelle dieses Modells das in Abb. 15b gezeigte ohne die vorgeschaltete Dämpfung setzen, wenn man außerdem die Elastizität des plastischen Stabes in die

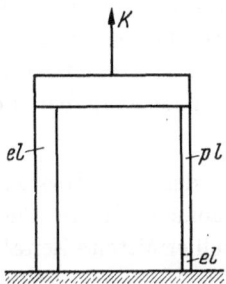

Abb. 16. PRANDTLsches Modell zur Erklärung des Kriechens. *pl-el* plastisch nach (15) fließender und *el* elastischer Stab parallel geschaltet

nachgeschaltete Feder verlegt. Die PRANDTLsche und MUSSMANNsche Rechnung läßt sich also auf das in Abb. 15b gezeigte Modell ohne vorgeschaltete Dämpfung übertragen. Die von PRANDTL hergeleitete Beziehung lautet

$$\gamma = \beta\, \tau \lg \frac{t + t'}{t'},\tag{17}$$

wo nun die gewünschte Zeitkonstante als t' auftritt. β ist eine weitere Konstante. Gleichung (17) ist bereits 1920 von BENNEWITZ [10] aus Versuchen an Glasfäden abgeleitet worden. Aus (17) ergibt sich

$$\dot{\gamma} = \beta\, \tau\, \frac{1}{t'}.\tag{18}$$

Der Vergleich mit (12) liefert die Beziehung der neuen Zeitkonstanten t' zur oben eingeführten t^*,

$$t' = \frac{1}{C^2\, t^*}.\tag{19}$$

Die für Akronal ermittelten Werte t' zeigt Tab. 1.

Tabelle 1. *Zeitkonstante t' für Akronal 4 F*

°C	20	40	50	60
t' [s]	0,18	0,13	0,11	0,09
t^* [s]	0,09	0,03	0,01	0,003

Für den Vergleich durfte $\omega = \dot{\gamma}$ gesetzt werden und der hierbei unterdrückte Faktor, der das Verhältnis vom Innen- zum Außendurchmesser des COUETTE-Viskosimeters enthält, der Konstanten β zugeschlagen werden. Durch diesen Faktor unterscheidet sich β von unserer Konstanten $1/A$.

Die Modellvorstellung mit nichtlinearer Dämpfung läßt sich auch auf die bei den festeren Stoffen auftretenden Kriecherscheinungen anwenden. Als Beispiel sei zunächst Ton gewählt, für den die Versuche von Geuze und Tjong Kie über die Torsion von Tonzylindern [11] vorliegen. Die Versuche ergaben die zeitliche Abhängigkeit

$$\theta = B\,(M - M_0)\,\Phi(t)\,, \qquad (20)$$

in der θ den Torsionswinkel je Längeneinheit, M das wirkende Torsionsmoment, M_0 die Fließgrenze, B eine Konstante und $\Phi(t)$ die experimentell ermittelte Zeitabhängigkeit bedeuten. Nimmt man für diese die logarithmische Zeitabhängigkeit von (17), so ergibt sich eine vollständige Übereinstimmung von Theorie und Versuch mit $t' = 0{,}56\,\mathrm{s}$, was Tab. 2 zeigt.

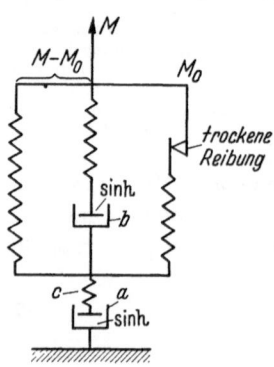

Abb. 17.
Modell, das die Nachwirkungs-erscheinungen von festem Ton wiedergibt mit nichtlinearer Dämpfung gemäß (3)

Man kann also feststellen, daß für Ton das in Abb. 17 dargestellte Elementarmodell zutrifft mit einer parallel geschalteten trockenen Reibung, die die Fließgrenze darstellt. Die vorgeschaltete Dämpfung a, die das stationäre Fließen bestreitet, blieb, weil sie offensichtlich wesentlich stärker als die parallel geschaltete b ist, in der Auswertung unberücksichtigt. Die vorgeschaltete Elastizität c konnte wegen ihres geringen Einflusses auf die Verformung nicht abgesondert werden.

Tabelle 2
Zeitliche Abhängigkeit $\Phi(t)$ der Verformung von Ton bei plötzlich aufgebrachter, konstant bleibender Last. $t' = 0{,}56$ s

t [h]	0,5	1	2,5	5	20	40
Φ experim.	1,68	2	2,5	2,9	3,1	3,2
Φ theoret.	1,68	2	2,5	2,9	3,1	3,2

Eine ähnlich befriedigende Übereinstimmung ergibt sich für polykristallines Aluminium mit den Versuchen von Ké [12], wenn man die beträchtliche elastische Verformung in Abzug bringt (Abb. 18), wie die Gegenüberstellung der theoretischen und experimentellen Werte in Tab. 3 zeigt.

Hier besteht also das in Abb. 19 gezeigte Modell. Die Tabellen entsprechen folgenden Zahlenwerten der Konstanten in (17) (vgl. Abb. 18):

$$t' = 50\ \mathrm{s}$$

$$e_{pl} = 0{,}68\,\lg\frac{t + t'}{t'}$$

Tabelle 3

Zeitliche Abhängigkeit der Verformung (Abb.18) bei plötzlich aufgebrachter und konstant bleibender Last für polykristallines 99,99% reines Aluminium bei 175° C. Vergleich des Experimentes mit Formel (17)

$t \cdot 10^{-2}$ [s]	2,5	.5	10	15	20	30
e_{pl} experim.	4,02	4,2	4,4	4,5	4,6	4,7
e_{pl} theoret.	4,03	4,21	4,4	4,51	4,6	4,71

Tabelle 4

Zeitliche Abhängigkeit der Verformung (Abb.18) nach plötzlicher Entlastung anschließend an den Lastfall von Tabelle 3. Zeitpunkt der Entlastung $T = 3250$ s nach Aufbringen der Last. Vergleich des Experiments mit Formel (21)

$t \cdot 10^{-3}$ [s]	3,6	3,9	4,55	5,2	5,8	6,5
Δe_{pl} experim.	0,45	0,26	0,10	0,10	0,05	0
Δe_{pl} theoret.	0,40	0,28	0,16	0,09	0,04	0

und entsprechend zu PRANDTL [8] für die Rückverformung nach der Entlastung (Tabelle 4)

$$\Delta e_{pl} = 0{,}58 \cdot 0{,}68 \lg \frac{t - T + t'}{T + t'}.$$

An die Stelle des theoretischen Faktors 0,5 tritt also hier der Wert 0,58. Eine weitere Abweichung der Theorie vom Experiment besteht in der Dauer der Rückverformung, die theoretisch dem Lastintervall gleich sein soll, nach BLEAKNEY [13] aber wesentlich länger als das Lastintervall ist.

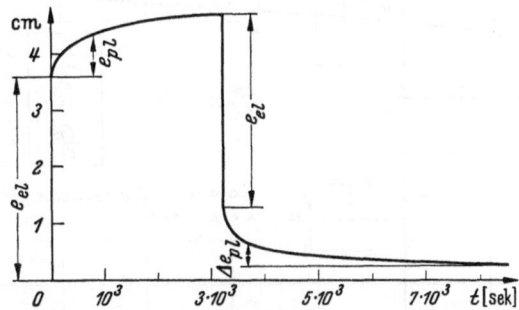

Abb. 18. Kriechen von 99,99% reinem, polykristallinem Aluminium bei 175°C nach Kê

Das stationäre Fließen des Tons durch Blenden fügt sich in den Rahmen des Widerstandsgesetzes (8), in dem als weiterer Parameter das Verhältnis r/R des Blenden- zum Rohrhalbmesser hinzutritt. Der Einfluß der Fließgrenze sei der Abhängigkeit von diesem Verhältnis zugeschlagen. Das Verhältnis G/A sei in Übereinstimmung mit den oben erwähnten Versuchsresultaten als einflußlos angesehen und auch die Kennzahl Ct^*, weil der Vorgang langsam vor sich geht. So ergibt sich für den Druckabfall Δp in der Blende, wenn man in die Kennzahl $K_2 = v/lc$ das sekundliche Durchflußvolumen Q und für l den Blendenhalbmesser einführt, die Abhängigkeit

$$\frac{\Delta p}{A'} = f\left(\frac{Q}{r^3 C'}, \ \frac{r}{R}\right),$$

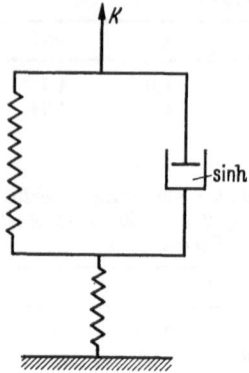

Abb. 19. Nichtlineares Modell, das die Kriecherscheinungen von 99,99% reinem, polykristallinem Aluminium bei 175°C wiedergibt

wo nun A', C' statt A, C geschrieben wird, weil die Verformung eine andere ist als die, auf die sich A, C beziehen.

Es wurde der gleiche Ton mit 17% und mit 18% Wassergehalt bei jeweils gleichen Verhältnissen r/R untersucht[1]. Relative A'-, C'-Werte konnte man dadurch ermitteln, daß zwei Punkte der für beide Wassergehalte bei gleichem Verhältnis r/R aufgenommenen Versuchskurven zur Deckung gebracht wurden (Abb. 20). Es ergaben sich, wie die Tabelle in Abb. 20 zeigt, gut übereinstimmende A'-, C'-Werte, obwohl durch den langgestreckten Kurvenverlauf die A'-, C'-Werte relativ genau festliegen. Der Kurvenverlauf läßt deutlich eine Zeitabhängigkeit erkennen. Der schwache Anstieg bei den größeren Abszissenwerten zeigt die innere Reibung. Die Plastizitätstheorie, die man

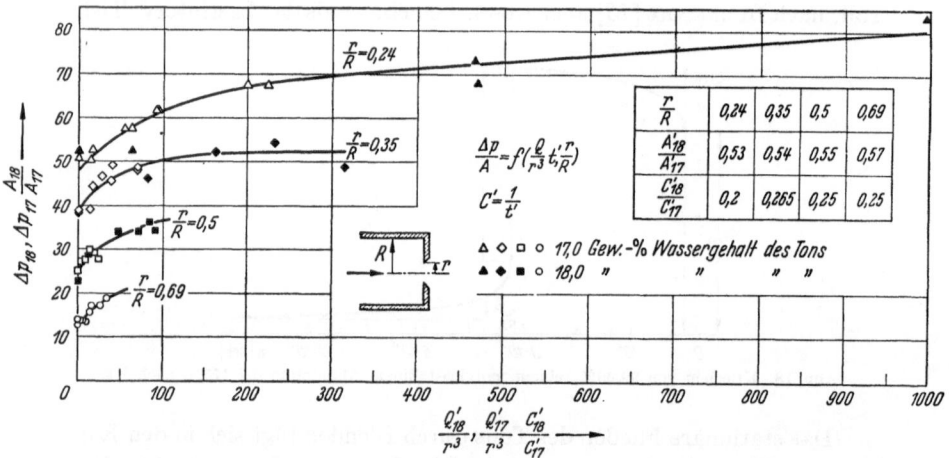

Abb. 20. Durchpressen von Ton mit 17 und 18% Wassergehalt durch Blenden. Q Volumdurchsatz, Δp Druckabfall in der Blende

als zutreffend halten würde, kennt dagegen keine innere Reibung und daher keinen Zeiteinfluß, sie liefert also in der Darstellung von Abb. 20 lediglich eine horizontale Gerade.

Einen weiteren Unterschied zur Plastizitätstheorie sieht man an der Gleitfläche, die das fließende Gebiet abgrenzt. Nach HILL [14] sollte diese Fläche an der Blendenkante ansetzen. Statt dessen setzt sie am Halbmesser r^* an (Abb. 21), und zwar ist das Verhältnis r^*/R unabhängig

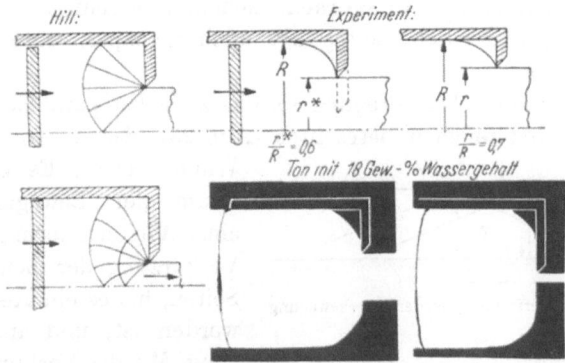

Abb. 21. Gleitfläche, die den fließenden vom ruhenden Ton abgrenzt im Vergleich mit den An-
nahmen der Plastizitätstheorie

von r/R. Nur bei Verhältnissen $r/R > r^*/R$ setzt die Gleitfläche an der Blendenkante an. Abb. 22 zeigt den nichtfließenden Teil, der beim Aus-bau an der Blende haftenbleibt, mit der scharf ausgebildeten Ansatz-kante der Gleitfläche und den fließenden Teil.

Abb. 22. Fließender und nichtfließender Ton nach Ausbau der Blende

Die bisherigen Ausführungen befassen sich vornehmlich mit der Schubspannung in der Gleitebene bei der Scherung, oder der Zugspan-nung beim einachsigen Zug. Von großer Bedeutung ist aber auch der Zu-sammenhang des Spannungs- mit dem Formänderungstensor. Die bisher aufgestellten Beziehungen gründen sich meist auf vereinfachte physi-kalische Vorstellungen [15] oder ermangeln einer physikalischen Begrün-dung [16, 17] und setzen zudem voraus, daß der Spannungszustand durch den augenblicklichen Formänderungszustand bestimmt sei. Das ist sicher unrichtig, denn das wesentliche Kennzeichen der im Zwischengebiet zwischen Elastisch und Fest liegenden Stoffe sind die erwähnten Zeitwirkungen. Daher wirkt die Vorgeschichte der Verformung über eine endliche Zeit nach und man muß, wenn man überhaupt zu einer physikalisch sinnvollen Beziehung zwischen den Tensoren kommen will, die endliche Formänderung in Betracht ziehen. Das gelingt für hochzähe Flüssigkeiten bei der Scherbewegung. Auch die experimentelle Prüfung ist hier möglich.

In der ebenen Scherbewegung sei das zunächst parallelogrammförmige Flüssigkeitsteilchen betrachtet (Abb. 23), das bis an die festen

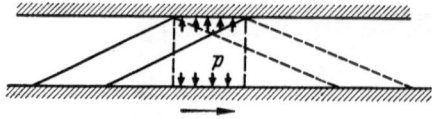

Abb. 23. Endliche Verformung bei der Scherbewegung

Wände reicht. Es erfährt im Laufe der Bewegung neben einer Winkeländerung auch eine Verkürzung der beiden freien Seiten, bis es ein Rechteck geworden ist, und anschließend eine Verlängerung. Mit der Verkürzung sind jedenfalls Druckspannungen verbunden, die noch nachwirken, wenn das Teilchen ein Rechteck geworden ist. Die endliche Verformung wirkt sich also in der Weise aus, daß die Platten, die die Flüssigkeitsschicht begrenzen, auch einen Normaldruck erfahren.

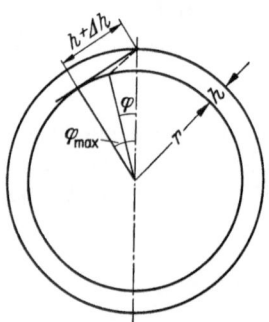

Abb. 24. Endliche Verformung im Ringspalt des Couette-Viskosimeters

Am Couette-Viskosimeter beschränkt sich diese Überlegung auf Nachwirkungszeiten, die kleiner oder höchstens gleich der Zeit sind, die zum Durchschreiten eines Drehwinkels φ_{max} benötigt wird, in dessen Bereich das Flüssigkeitsteilchen nicht auf dem inneren Zylinder aufgewickelt ist. In Abb. 24 liest man nun folgende geometrische Beziehung für die Verkürzung Δh bis zur radialen Lage des Teilchens ab, wenn h die Spaltweite, r der Radius des Innenzylinders und φ der Winkel ist, über den noch Nachwirkung merkbar ist:

$$(h + \Delta h)^2 = (r + h)^2 + r^2 - 2\,(r + h)\,r \cos \varphi,$$

woraus sich

$$\frac{\Delta h}{h} = \sqrt{1 + 2\left[\left(\frac{r}{h}\right)^2 + \frac{r}{h}\right](1 - \cos \varphi)}$$

ergibt. Hier ist also die Formänderung und nicht die Spannung gegeben.

Unter der Voraussetzung kleiner Zeiten, in der die Spannungen nachwirken, darf man elastische Effekte als vorherrschend ansehen, denn Abb. 14 lehrte, daß dann im Modell von Abb. 15b die nachgeschaltete

Feder vorherrschend ist. Deshalb kann die Normalspannung p der relativen Verkürzung Δh proportional gesetzt werden,

$$\frac{p}{p_0} = \sqrt{1 + 2\left[\left(\frac{r}{h}\right)^2 + \frac{r}{h}\right](1 - \cos\varphi)}$$

mit der Proportionalitätskonstanten p_0. Mit der Winkelgeschwindigkeit ω des Viskosimeters wird die Zeit t eingeführt,

$$\frac{p}{p_0} = \sqrt{1 + 2\left[\left(\frac{r}{h}\right)^2 + \frac{r}{h}\right](1 - \cos\omega t)}. \tag{22}$$

Aus den Bemerkungen über das dimensionslose Widerstandsgesetz (8), (9), (10) ist zu entnehmen, daß p/p_0 von ω/C abhängen muß. Die übrigen Kennzahlen sind hier einflußlos, weil ein und dieselbe Flüssigkeit betrachtet wird. Das bedeutet, daß für t eine Zeitkonstante einzuführen ist, die mit $1/C'$ bezeichnet sei, weil hier nicht die konstante Verformungsgeschwindigkeit besteht, auf die sich C bezieht. Damit ist die Konstante eingeführt, die in dem betrachteten Vorgang die Zeiteffekte charakterisiert. Aus dem gleichen Grund wurde als Proportionalitätskonstante p_0 und nicht A eingeführt. Die Größe von p_0 hängt von der Größe der Nachwirkungszeit ab. Aber C' müßte sich im Versuch von gleicher Größenordnung wie $1/t^*$, C ergeben.

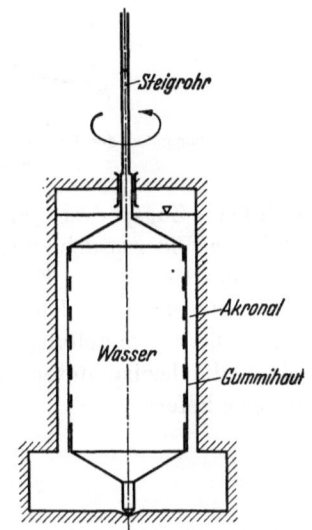

Abb. 25. Versuchsanordnung zur Messung des Normaldruckes bei reiner Scherbewegung im COUETTE-Viskosimeter

Die Versuchseinrichtung zeigt Abb. 25. Der innere, hohle und durchlochte Zylinder des COUETTE-Apparates ist mit einer Gummimembran überspannt. Wenn also ein Normaldruck auftritt, so steigt das im Zylinder befindliche Wasser in der Steigröhre (Abb. 24) hoch. Die Druckanzeige ist äußerst empfindlich wegen des winzigen Verhältnisses des Steigrohrquerschnittes zur Gesamtfläche der Löcher. Es ergab sich der in Abb. 26 in Abhängigkeit von der Winkelgeschwindigkeit gezeigte Normaldruck. Die Punkte sind die Meßergebnisse. Die Kurve stellt (22) dar mit den Konstanten

$$p_0 = 2{,}6\,\mathrm{g/cm^2}, \quad C' = 3{,}55\,\mathrm{s^{-1}} \text{ bzw. } \frac{1}{C'} = 0{,}284\,\mathrm{s}.$$

Die Zeitkonstante bzw. ihr reziproker Wert sind von der Größenordnung der Konstanten $C = 11\,\mathrm{s^{-1}}$, $t^* = 0{,}15\,\mathrm{s}$. Der Normaldruck beträgt rund 10% der Schubspannung.

Die Versuche beziehen sich auf eine Spaltweite von 4 und 8 mm bei 31,5 mm Halbmesser des Innenzylinders. Für beide Spaltweiten haben sich die angegebenen p_0, C'-Werte bestätigt. Hieraus sollte man schließen,

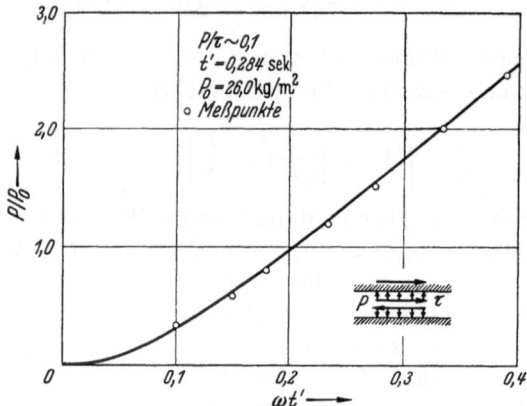

Abb. 26. Abhängigkeit des Normaldruckes im Couette-Viskosimeter von der Winkelgeschwindigkeit; Meßpunkte und gerechnete Kurve

daß keine Zugspannungen im Spaltquerschnitt wirken. Die Feststellung steht im Widerspruch zu einer Vermutung von Weissenberg [15], nach der man sich in der Flüssigkeit suspendierte elastische Fäden vorstellen soll. Diese Vorstellung kann aber nicht die beobachteten Druckspannungen liefern.

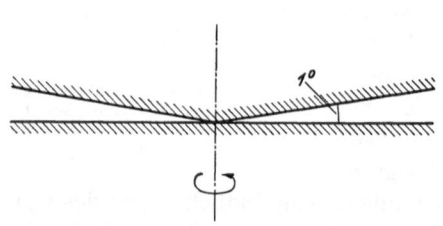

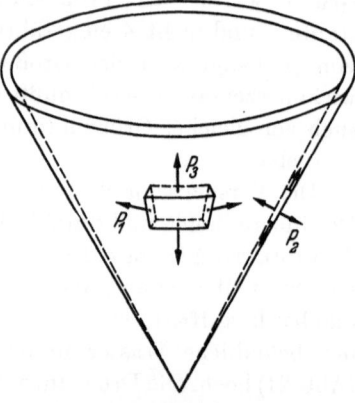

Abb. 27. Versuchsanordnung von Roberts   Abb. 28. Flüssigkeitsspalt zwischen koaxialen Kegeln

Unsere Feststellung steht auch im Widerspruch zu Roberts [18], der in einem von einer Planscheibe und einer dazu zentrischen, schwach konischen Scheibe gebildeten Flüssigkeitsspalt bei sich drehender Planscheibe den radialen und axialen Druck gemessen hat (Abb. 27). Seine Übereinstimmung mit Weissenberg besteht darin, daß er Gleichheit von radialem und axialem Druck feststellt. Es wird aber die Gleichgewichtsbedingung für einen konischen Spalt zwischen koaxialen Kegeln (Abb. 28) verwendet, welche mit den Bezeichnungen von Abb. 28 und dem Halbmesser r

$$\frac{d\,p_3}{d\ln r} = (p_1 - p_3) + (p_2 - p_3)$$

lautet. Jedoch ist der Spalt praktisch eben (Abb. 27), denn die eine Begrenzungswand ist eben und der Öffnungswinkel des Spaltes ist nur $1°$, d. h. die Spaltweite ändert sich nur um $1,7\%$ des Halbmessers, so daß in der Umgebung der Druckmeßanbohrungen praktisch der ebene Fall besteht, für den die Gleichgewichtsbedingung

$$\frac{d\,p_3}{d\ln r} = p_1 - p_3$$

lautet. Wenn also $p_2 = p_3$ festgestellt wird, so bestätigt das, daß im Versuch der ebene Fall vorlag.

Literatur

[1] RIVLIN, R. S.: Proc. roy. Soc., Lond., Ser. A 193, 260 (1948).
[2] SCHULTZ-GRUNOW, F.: Kolloid-Z. 138, 167 (1954).
[3] SCHULTZ-GRUNOW, F. u. H. WEYMANN: Kolloid-Z. 131, 61 (1953).
[4] WEYMANN, H.: Kolloid-Z. 138, 41 (1954).
[5] PRANDTL, L. u. FR. VANDREY: Z. angew. Math. Mech. 30, 169 (1950).
[6] SCHULTZ-GRUNOW, F.: Z. VDI 97, 409 (1955).
[7] LINDER, R.: Kolloid-Z. 143, 144 (1955).
[8] PRANDTL, L.: S. TIMOSHENKO 60th Ann. Vol., S. 184. New York 1939.
[9] MUSSMANN, H.: Ann. Phys. (5) 31, 121 (1938).
[10] BENNEWITZ, K.: Phys. Z. 21, 703 (1920).
[11] HARRISON, V. G. W.: Proc. 2. Intern. Congr. Rheology, London 1954, Beitrag von E. GEUZE u. TAN TJONG-KIE, S. 247.
[12] KÉ, T. S.: Phys. Rev. 71, 533 (1947).
[13] BLEAKNEY, H.: Canadian Journ. Chemistry 33, 55 (1955).
[14] HILL, R.: The Mathematical Theory of Plasticity, S. 185. Oxford 1950.
[15] WEISSENBERG, K.: Abh. preuß. Akad. Wiss. Nr. 2 (1931), und Arch. Sci. phys. nat. 17, 1 (1934).
[16] GARNER, F., H. NISSAN u. G. WOOD: Phil. Trans. roy. Soc., Lond., Ser. A 243, 37 (1950).
[17] MOONEY, M.: J. Colloid Sci. 6, 96 (1951).
[18] HARRISON, V. G. W.: Proc. 2. Intern. Congr. Rheology, London 1954, Beitrag von J. ROBERTS, S. 91.

Der Beschädigungsprozeß in Metallen beim Kriechen

Von **I. A. Oding** und **W. W. Burdukski**, Moskau

Die sinkende Festigkeit der Metalle beim Kriechen wird verschiedentlich erklärt [1]. In den Arbeiten von Greenwood [2], Crussard [3], I. A. Oding und W. S. Iwanowa [4] werden als die Hauptursache der Zerstörung des Metalls bei dessen dauernder Inanspruchnahme die Prozesse der Koagulation und der Auflagerung von Fehlstellen des Kristallgitters (Löcher) auf der Mikroporenoberfläche betrachtet.

Der gleiche Mechanismus ist von I. A. Oding [5] zur Erklärung der Ermüdungszerstörung der Metalle bei zyklischen Belastungen ausgenutzt.

In Entwicklung dieser Vorstellungen kann man eine Gesetzmäßigkeit der Ansammlung von Beschädigungen im Metall beim Kriechen feststellen, die schließlich zur Zerstörung führen.

Im Einklang mit diesen Arbeiten werden wir den Prozeß der Schwächung von Metallen unter Spannung ebenfalls als ein Ergebnis der Koagulation und der Auflagerung von Löchern auf der Mikroporenoberfläche betrachten. Die Mikroporen werden infolge dieser Auflagerung in Mikrorisse ausarten.

Die Geschwindigkeit der Auflagerung von Fehlstellen

$$V_{oc} = \frac{dN}{d\tau},$$

wo N die Zahl der sich auflagernden Löcher in einer Raumeinheit des Metalls ist, muß vor allem von der Konzentration der Löcher C (Anzahl der Löcher in einer Raumeinheit) abhängen. Je größer die Konzentration der Löcher ist, um so größer muß natürlich, bei sonst gleichen Bedingungen, die Geschwindigkeit der Auflagerung sein.

Aber die Konzentration von Löchern wird von der Kriechgeschwindigkeit V_p bestimmt werden, denn von ihrer Größe hängt die Intensität der *Bildung* von Löchern ab. Andererseits muß die Konzentration von Löchern auch noch von der Zeit der Inanspruchnahme des Metalls durch Spannung bei hohen Temperaturen abhängen, da der letztere Faktor, die Zeit, die Intensität der *Ansammlung* von Löchern bestimmt.

Die entstehenden Löcher erhöhen einerseits die Konzentration C und werden andererseits für die Auflagerung verausgabt. Aber ein Teil der Löcher wird außerdem bei einer Begegnung mit dislozierten Atomen vernichtet werden. Deswegen muß der Zeitfaktor bei der Geschwindigkeit der Auflagerung von Fehlstellen eine erhebliche Rolle spielen.

Also kann man annehmen, daß bei konstanter Spannung die Geschwindigkeit der Auflagerung von Fehlstellen zur Kriechgeschwindigkeit V_p und zur Zeit $\tau^{m'}$ direkt proportional sein wird. Das Potenzverhältnis zwischen V_{oc} und τ wird es ermöglichen, die Fähigkeit verschiedener Metalle zur Ansammlung von Fehlstellen wiederzugeben. Dann ist

$$V_{oc} = \frac{dN}{d\tau} = \varkappa' V_p \tau^{m'}$$

und

$$N = \varkappa V_p \tau^m + N_0,$$

wo N die Zahl der Löcher und $\varkappa$ der Koeffizient ist; m charakterisiert die Fähigkeit des Metalls zur Ansammlung von Löchern und N_0 offensichtlich die Zahl der Löcher vor dem Versuch.

Vorausgesetzt, daß die Beschädigung des Metalls zur Zahl der neugebildeten Löcher proportional ist, d.h. $\Delta = qN$, so kann man als erste Annäherung folgende Beschädigungsgleichung annehmen:

$$\Delta = M \tau^m, \quad M = q \varkappa V_p,$$

oder

$$\Delta = \left(\frac{\tau}{M*}\right)^m \ldots \tag{1}$$

Hier ist Δ die Spannung, die in einem bestimmten Zeitpunkt den Wert der Verminderung der ursprünglichen Grenze der Dauerstandfestigkeit anzeigt, und $M*$ ist ein Koeffizient, der den Einfluß anderer Faktoren berücksichtigt.

Der Charakter dieser Abhängigkeit kann für verschiedene Metalle sehr verschieden sein; die Größe m kann also einen Stoff bezüglich der Ansammlung von Beschädigungen darin charakterisieren.

Von großem Interesse ist auch *relative Beschädigung*, die vom Verhältnis der Größe Δ in einem bestimmten Zeitpunkt zur Größe $\Delta\tau_\sigma$ gekennzeichnet wird, d.h. zur Größe der Schwächung des Metalls vor seinem Bruch in der Probezeit τ_σ:

$$D = \frac{\Delta}{\Delta\tau_\sigma} = \frac{M_*^m \tau^m}{M_*^m \tau_\sigma^m} = \left(\frac{\tau}{\tau_\sigma}\right)^m \ldots \tag{2}$$

Die Größe der relativen Beschädigung kann folgendermaßen verwendet werden.

Nehmen wir an, daß die Grenzen der Dauerstandfestigkeit σ_τ und die Zeit bis zur Zerstörung τ_{σ_1} uns bekannt ist. Nehmen wir auch an (Abb. 1), daß wir beim Beginn des Experiments die Probe bis zur Spannung σ_1 belasten und das Experiment zur Zeit τ_1 unterbrechen, ohne daß die Zerstörung erfolgt ist, die zur Zeit τ_{σ_1} kommen sollte. Die Inanspruchnahme der Probe hat im Metall bestimmte Beschädigungen verursacht. Im Einklang mit der Gleichung (2) ist der Beschädigungsgrad folgendermaßen festzustellen:

$$D_1 = \left(\frac{\tau_1}{\tau_{\sigma_1}}\right)^m.$$

Wie man aus dieser Gleichung ersieht, kann sich die Größe D von 0 (bei $\tau_1 = 0$) bis 1 (bei $\tau_1 = \tau_{\sigma_1}$) bewegen.

Belastet man nun jetzt die gleiche Probe bis zur Spannung σ_2, und prüft man sie bei gleicher Temperatur bis zur Zerstörung, so erfolgt diese nicht zur Zeit τ_{σ_2}, sondern zur Zeit $\tau_2 \leqq \tau_{\sigma_2}$; denn die vorausgegangene Prüfung bei σ_1 hat sich auf das Metall in Form einer bestimmten Beschädigung ausgewirkt.

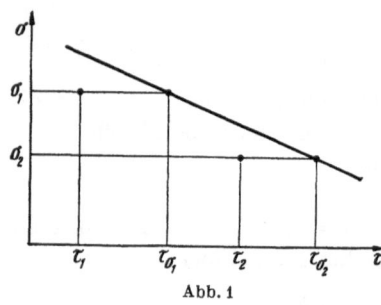

Abb. 1

Die Größe

$$D_2 = \left(\frac{\tau_2}{\tau_{\sigma_2}}\right)^m$$

stellt auch die relative Beschädigungsfähigkeit der Metalle auf dem zweiten Abschnitt dar. Wenn also das zweite Experiment das Metall zerstört hat, so muß die Summe von D_1 und D_2 offensichtlich 1 ausmachen:

$$D_1 + D_2 = a^m + b^m = 1 \ldots \tag{3}$$

bei

$$\frac{\tau_1}{\tau_{\sigma_1}} = a \quad \text{und} \quad \frac{\tau_2}{\tau_{\sigma_2}} = b.$$

Der gleiche Gedankengang ist auch in dem Fall annehmbar, daß nicht nur zwei verschiedene Belastungen, sondern mehrere erfolgen. In der vorliegenden Arbeit werden wir nur zweimalige Belastungen behandeln.

Wie bereits gesagt, ist m ein Kriterium, das die Fähigkeit des Metalls zur Ansammlung von Löchern charakterisiert. Wir wollen untersuchen, wie sich diese Größe für einige Metalle ändert, und wie die oben dargelegte Theorie den experimentellen Angaben überhaupt entspricht.

In der Tab. 1 sind die experimentellen Angaben über die Prüfung der Dauerfestigkeit bei $T = 600°$ C von drei Marken austenitischer Stähle angeführt: EI-432, EI-388 und EA-1 T im gehärteten und stabilisierten Zustand.

Der Stahl EI-432 wurde zuerst bei $\sigma_1 = 19{,}5$ kg/mm² erprobt, und nach 750 Stunden wurde die Spannung auf 20 kg/mm² erhöht; der Bruch erfolgte in 366 Stunden.

Der zweite Probestab wurde zuerst bei $\sigma_1 = 19{,}5$ kg/mm² im Laufe von 500 Stunden geprüft, dann wurde die Spannung auf 21 kg/mm² erhöht. Der Bruch erfolgte in 253 Stunden. Die Durchschnittsgröße m war 1,5 gleich.

Tabelle 1

Stahl-marke	Nr. des Probemusters	Die 1. Belastung				Die 2. Belastung				m
		Span-nung σ_1 kg/mm²	Zeit bis zur Zerstö-rung[1] τ_{σ_1} St.	Fak-tische Dauer-zeit der 1. Be-lastung τ_1 St.	$a = \dfrac{\tau_1}{\tau_{\sigma_1}}$	Span-nung σ_2 kg/mm²	Zeit bis zur Zer-störung τ_{σ_2} St.	Fak-tische Dauer-zeit bis zur Zer-störung τ_2 St.	$b = \dfrac{\tau_2}{\tau_{\sigma_2}}$	
EI-432	1	19,5	1050	750	0,72	20	690	366	0,53	1,4
	2	20,0	690	400	0,58	21	315	187	0,60	1,35
	3	19,5	1050	500	0,48	21	315	253	0,80	1,6
										1,5
EI-388	1	40	1175	350	0,27	41	900	338	0,38	0,65
	2	40	1175	350	0,27	42	675	202	0,30	0,55
	3	40	1175	350	0,27	39	1535	953	0,60	0,82
	4	40	1175	350	0,27	38	1900	1000	0,53	0,75
										0,70
EA-1 T	1	22	775	400	0,515	25	210	122	0,58	1,15
	2	23	500	250	0,500	25	210	93	0,442	0,95
	3	22	775	300	0,387	24	315	173	0,55	0,90
	4	25	210	150	0,715	24	315	83	0,28	1,00
										1,00

[1] Wird nach der Kurve der Dauerfestigkeit bestimmt.

Ebensolche Prüfungen wurden auch mit dem Probestab aus Stahl EI-388 und EA-1 T angestellt. Die Berechnung der Größe von m nach den experimentellen Angaben führte zu recht übereinstimmendem Ergebnis für jede Stahlmarke:

$$\text{EI-432} - m = 1,5\,,$$
$$\text{EA-1 T} - m = 1,0\,,$$
$$\text{EI-388} - m = 0,7\,.$$

Der Fall, daß m nahezu 1 ist, zeigt, daß die in der Spannung ausgedrückte Beschädigung zu der in Zeit ausgedrückten relativen Beschädigung proportional ist. In diesem Fall wird der Probestab, wenn bei der Spannung σ_1 ohne Zerstörung die relative Zeit $a = \tau_1/\tau_{\sigma_1}$ in Anspruch genommen wurde, beim Übergang auf die zweite Spannung σ_2 noch

$$\frac{\tau_2}{\tau_{\sigma_2}} = 1 - \frac{\tau_1}{\tau_{\sigma_1}}$$

dienen. So war es annähernd mit dem Stahl EA-1 T, für den $m = 1,0$ festgestellt wurde.

Wird 1 erheblich von m überstiegen, so bedeutet das, daß die erste Zeit der Prüfung sich auf den Probestab im Sinne der Beschädigung nur wenig ausgewirkt hat.

Also werden die Metalle, für die $m > 1$ ist, in den ersten Etappen der Inanspruchnahme durch Spannung und hohe Temperaturen relativ weniger beschädigt — die Intensität ihrer Beschädigung steigt in den folgenden Etappen der Probe.

Die Metalle hingegen, für die $m < 1$ ist, werden in den ersten Etappen der Inanspruchnahme erheblich beschädigt.

Also wird die vorgeschlagene Gesetzmäßigkeit „die Beschädigung des Metalls Δ ist proportional zur Kriechgeschwindigkeit V_p und zur Zeit der Probe in der Potenz m" von den experimentellen Angaben ausreichend gut bewiesen.

Zu einem interessanten Schluß führen weitere Erforschungen dieser Abhängigkeit.

Wie gesagt, gingen wir von der Voraussetzung aus, daß die Beschädigung des Metalls zur Zahl der aufgelagerten Fehlstellen proportional ist:

$$\Delta = q N$$

oder

$$\Delta = \varkappa_1 V_p \tau^m.$$

Wir wollen die Kriechgeschwindigkeit in dieser Gleichung durch die entsprechende Spannung ausdrücken.

Gegenwärtig haben sowohl analytische Forschungen als auch experimentelle Angaben gezeigt, daß das Verhältnis zwischen Spannung und Kriechgeschwindigkeit am besten durch den hyperbolischen Sinus ausgedrückt wird:

$$V_p = A_2 \sinh \alpha \sigma,$$

was für $\alpha \sigma > 1{,}5$ in

$$V_p = A_1 e^{\alpha \sigma}$$

übergeht.

Die letzte Gleichung, schon von Ludwick vorgeschlagen, wird gegenwärtig immer öfter als die den experimentellen Angaben am besten entsprechende betrachtet. Nehmen auch wir sie deswegen für unsere Zwecke an. So kommt

$$\Delta = \varkappa_1 A_1 e^{\alpha \sigma} \tau^m.$$

Wir bestimmen ferner die Zahl der Löcher N_i, die bei erforderlicher Spannung σ_i das Metall in τ_i Stunden zur Zerstörung führen wird, d.h. wir nehmen den äußersten Fall

$$N_i = A e^{\alpha \sigma_i} \tau_i^m.$$

Wir nehmen an, daß die erforderliche Zahl N_i um so größer sein muß, je niedriger die Größe der Spannung ist. Diese Abhängigkeit geben wir folgendermaßen wieder:

$$N_i = \frac{\text{const}}{e^{\beta \sigma_i}}$$

und dann

$$\text{const} = N_i e^{\beta \sigma_i} = A\, e^{(\alpha + \beta)\, \sigma_i} \tau_i^m.$$

Bezeichnen wir

$$\left(\frac{\text{const}}{A}\right)^{\frac{1}{m}} = B \quad \text{und} \quad \frac{\alpha + \beta}{m} = \gamma\,,$$

so wird endgültig

$$\tau_i = B\, e^{-\gamma \sigma_i}\,,$$

d. h. wir sind zu der Abhängigkeit zwischen der Spannung und der Zeit bis zur Zerstörung τ_i gekommen, die an Hand der experimentellen Angaben einer Reihe von Arbeiten aus der letzten Zeit empfohlen wird [6, 7, 8].

Aus diesen sehr einfachen Beispielen kann man sehen, daß die Verwendung der Vorstellung der Fehlstellen des Kristallgitters und die Rolle der Versetzungen im Problem der Warmfestigkeit mit großem Vorteil ausgenutzt werden können.

Literatur

[1] IWANOWA, W. S.: Metallowedenie i obrabotka Metallow 1, 19 (1955).
[2] GREENWOOD, J. N.: J. Iron Steel Inst. 171, 380 (1952).
[3] CRUSSARD, C. u. J. FRIEDEL: Proc. Symposium on Creep and Fracture of Metals at High Temperatures, 1954.
[4] ODING, I. A. u. W. S. IWANOWA: Doklady Akademii Nauk UdSSR 103, 77 (1955).
[5] ODING, I. A.: Doklady Akademii Nauk UdSSR 105, 1238 (1955).
[6] JURKOW, S. N. u. B. N. NARSULLAEW: Jurnal Technitscheskoi fisiki 23, 10 (1953).
[7] JURKOW, S. N. u. T. P. SANFIROWA: Doklady Akademii Nauk UdSSR 101, 237 (1955).
[8] ODING, I. A. u. W. S. IWANOWA: Teploenergetika 1, 24 (1955).

The Effect of Small Viscous Inclusions on the Mechanical Properties of an Elastic Solid

By **J. G. Oldroyd,** Swansea (Wales)

1. Introduction. As a guide to the interpretation of observed visco-elastic effects in solids that are almost ideally elastic, it is of interest to calculate what the effect would be, on mechanical properties, of local imperfections of different kinds distributed throughout an otherwise elastic solid. The presence of any flaws where work is done against internal friction when the material as a whole is deformed will give rise to delayed elastic response, and to energy losses when the material is subjected to varying deformations, for example during vibrations. To make the theoretical problem tractable, so that a quantitative answer can be obtained, it is necessary to consider a somewhat over-simplified picture of the type of imperfection which might arise in practice. In the present paper, what appears to be the simplest possible model of a visco-elastic solid is discussed.

A perfectly elastic solid S, which is considered to be homogeneous and continuous, and also HOOKEan and isotropic, is completely described by the two constants of LAMÉ $-\lambda$, μ (shear modulus $= \mu$, bulk modulus $\varkappa = \lambda + {}^2/_3\,\mu$). We shall consider an imperfectly elastic solid whose main constituent is S, but in which a number of small, widely dispersed, spherical cavities are present, each filled with a viscous liquid S'. The proportion c of the whole volume which is occupied by S' is assumed to be very small, so that c^2 is negligible compared with unity. To fix ideas, S' is taken to be ideally viscous with a constant viscosity η when deformed at constant volume, and ideally elastic with a constant bulk modulus $\varkappa'$ when subjected to varying isotropic stresses.

The combined elastic and viscous properties of some other disperse systems have been determined by a perturbation method, first used by FRÖHLICH and SACK [1], and later by the present author in a paper subsequently referred to as I [2]. Previous calculations of this kind have dealt with liquid suspensions and emulsions, whose components have been regarded as incompressible. The extension of the method to the essentially solid system now being discussed, when the bulk and shear moduli of S are of comparable magnitudes, is not difficult; and the reader is referred to I for a fuller account of the method. Where a mathematical argument follows closely the same sequence of steps as are given in detail in I, though with different algebra, it is not repeated in the present paper. It has been shown in I that the perturbation method cannot be relied upon to give results more accurate than to the first order in c.

It is our purpose to find general stress-strain relations for the imperfectly elastic solid system comprising S and S', which will describe the behaviour of a typical macroscopic element of the system when subjected to arbitrary (varying) stresses. For small strains the relationships may be assumed to involve the stress and strain components linearly, but must be expected to involve time derivatives of these quantities as well as the quantities themselves.

2. Method. To fix ideas, the liquid inclusions in the S–S' system are supposed of constant small radius a, and it is assumed that there exists an unstressed equilibrium state of the composite material which we shall take as the unstrained reference configuration. It is useful to think of a fictitious, homogeneous, isotropic solid S^* with the same macroscopic rheological properties as the composite material, and it is, in effect, the properties of S^* that we seek to determine. These must be such that if a small part of a macroscopic element of S^* were replaced by the actual components S and S', no difference in rheological behaviour could be detected by macroscopic observations. The simplest possible perturbation of the macroscopic element forms the basis of the FRÖHLICH and SACK method.

A macroscopic element of the material must be taken to mean an element whose linear dimensions are large compared with the radius a and the mean distance between neighbouring inclusions. Two systems are now considered: in the first, the whole of a comparatively large region of space is filled with S^*; in the second, a single cavity full of S' ($r < a$ referred to spherical polar coordinates r, θ, Φ) is surrounded by S in the space $a < r < b$, where $b^3 = a^3/c$, the remaining space $r > b$ being filled with S^*. The two systems are required to be indistinguishable, with the greatest possible accuracy, if observations are made at any distance $r = R$, sufficiently large compared with b. The homogeneous or composite sphere $r \leqq R$ is to be regarded as a single macroscopic element subjected to an arbitrary variable state of stress. It is assumed that distances and displacements throughout the region considered are so small by ordinary standards that inertia and body-force terms in the equations of motion may be neglected and it is not necessary to distinguish material and partial time derivatives.

The stress-strain relations for the solid S are, in the usual cartesian-tensor notation, referred to coordinates x_i $(i = 1, 2, 3)$,

$$p_{ik} = 2\mu\, e_{ik} + \lambda\, e\, \delta_{ik}, \tag{1}$$

where p_{ik} denotes the stress tensor, the components of strain e_{ik} are
defined in terms of cartesian components of displacement u_i by

$$e_{ik} = \frac{1}{2}\left(\frac{\partial u_k}{\partial x_i} + \frac{\partial u_i}{\partial x_k}\right), \tag{2}$$

$e(= e_{11} + e_{22} + e_{33})$ is the dilatation, and δ_{ik} is the KRONECKER delta
$(= 1$ if $i = k$, otherwise $= 0)$. In terms of the bulk and shear moduli,
the equations of state of S take the form

$$p_{ik} = 2\mu\left(e_{ik} - \frac{1}{3}e\,\delta_{ik}\right) + \varkappa e\,\delta_{ik}. \tag{3}$$

The equations of state of the liquid S' can be written for our present
purpose as

$$p_{ik} = 2\eta\,\varDelta\left(e_{ik} - \frac{1}{3}e\,\delta_{ik}\right) + \varkappa' e\,\delta_{ik}, \tag{4}$$

where $\varDelta \equiv d/dt$, i.e. they are of the same form as (3) with an operator
$\eta\,\varDelta$ instead of a constant shear modulus. The properties of S^* will also
be described by a set of linear relations which can be written in an identical form, namely

$$p_{ik} = 2\mu^*\left(e_{ik} - \frac{1}{3}e\,\delta_{ik}\right) + \varkappa^* e\,\delta_{ik}, \tag{5}$$

where μ^* (and similarly $\varkappa^*$) is not necessarily a constant but may be an
operator of the general form

$$\mu^* = \mu_0 \frac{1 + \lambda_2\varDelta + \nu_2\varDelta^2 + \xi_2\varDelta^3 + \cdots}{1 + \lambda_1\varDelta + \nu_1\varDelta^2 + \xi_1\varDelta^3 + \cdots}, \tag{6}$$

$\mu_0, \lambda_1, \lambda_2, \nu_1, \nu_2$, etc., being constant coefficients. This operator notation
can be used without confusion (cf. [2]) if it is understood that any
equation relating two functions of time which is written as

$$F(t) = \mu^* f(t)$$

means precisely

$$(1 + \lambda_1\varDelta + \nu_1\varDelta^2 + \cdots)F(t) = \mu_0(1 + \lambda_2\varDelta + \nu_2\varDelta^2 + \cdots)f(t).$$

The existence of equations of state of identical form (3), for all the
materials with which we are concerned, means that the analysis of
deformation is *formally* the same as if all the materials were elastic solids,
the coefficients having in general an operational significance. To make
the parallel complete we define

$$\mu' = \eta\,\varDelta, \quad \lambda' = \varkappa' - \frac{2}{3}\mu', \quad \lambda^* = \varkappa^* - \frac{2}{3}\mu^*,$$

so that equation (1) characterizes also S' or S^* when the coefficients are
distinguished by primes or asterisks. The deformation of a composite
elastic body with spherical symmetry can be determined in terms of
spherical harmonics by well-known methods [3].

Any variable state of stress acting on a macroscopic material element can be obtained by superposing an isotropic variable pressure

$$p_{ik} = \begin{pmatrix} -P & 0 & 0 \\ 0 & -P & 0 \\ 0 & 0 & -P \end{pmatrix} \tag{7}$$

and not more than six axially symmetric deviatoric stress systems, each of the form

$$p_{ik} = \begin{pmatrix} 2T & 0 & 0 \\ 0 & -T & 0 \\ 0 & 0 & -T \end{pmatrix} \tag{8}$$

when referred to suitably chosen *fixed* axes (see [2]); P and T are functions of the time t. Since the equations of state are linear, the principle of superposition applies and it suffices to consider a macroscopic element of S^* subjected to arbitrarily varying states of stress (7) and (8) (separately). It is found (in §§ 3—4) that a study of the applied stresses (7) determines $\varkappa^*$, and a study of (8) determines μ^*, independently.

In the sequel the form of the expressions for the displacement and stress as functions of position and time are stated without proof, in the interests of conciseness. It may be readily verified that the quoted expressions satisfy the appropriate stress-strain relations and the equations of motion. Working in terms of displacements $(u, v, 0)$ in spherical polar directions, the relevant formulae for the non-vanishing physical components of strain are

$$e_{rr} = \frac{\partial u}{\partial r}, \quad e_{\theta\theta} = \frac{u}{r} + \frac{1}{r}\frac{\partial v}{\partial \theta}, \quad e_{\Phi\Phi} = \frac{u}{r} + \frac{v}{r}\cot\theta, \left.\begin{matrix}\\\\\\\\\end{matrix}\right\} \tag{9}$$
$$e_{r\theta} = \frac{1}{2}\left[\frac{1}{r}\frac{\partial u}{\partial \theta} + r\frac{\partial}{\partial r}\left(\frac{v}{r}\right)\right].$$

In S the equations of motion, when expressed in terms of the displacements, reduce to

$$\frac{1}{r^2}\frac{\partial}{\partial r}\left(r^2\frac{\partial u}{\partial r}\right) + \frac{1}{r^2\sin\theta}\frac{\partial}{\partial \theta}\left[\sin\theta\left(\frac{\partial u}{\partial \theta} - 2v\right)\right] - \frac{2u}{r^2} = -\frac{\lambda+\mu}{\mu}\frac{\partial e}{\partial r}, \tag{10}$$

$$\frac{1}{r^2}\frac{\partial}{\partial r}\left(r^2\frac{\partial v}{\partial r}\right) + \frac{1}{r^2\sin\theta}\frac{\partial}{\partial \theta}\left(\sin\theta\cdot\frac{\partial v}{\partial \theta}\right) + \frac{2}{r^2}\frac{\partial u}{\partial \theta} - \frac{v}{r^2\sin^2\theta} = -\frac{\lambda+\mu}{\mu r}\frac{\partial e}{\partial \theta}, \tag{11}$$

where

$$e = \frac{1}{r^2}\frac{\partial(r^2 u)}{\partial r} + \frac{1}{r\sin\theta}\frac{\partial(v\sin\theta)}{\partial \theta}; \tag{12}$$

the corresponding equations in S' or S^* have λ and μ distinguished by primes or asterisks.

3. Determination of $\varkappa^*$. The conditions in the homogeneous and composite macroscopic elements are now considered separately, in the case when the state of stress observable macroscopically is given by (7), i. e., in the usual notation for physical components of stress in spherical polar directions,

$$\widehat{rr} = \widehat{\theta\theta} = \widehat{\Phi\Phi} = -P, \quad \widehat{\theta\Phi} = \widehat{\Phi r} = \widehat{r\theta} = 0 \tag{13}$$

on $r = R$, for sufficiently large R, where P is an arbitrary function of the time.

a) *Homogeneous element.* It is clear that the displacement corresponding to (13) throughout the material S^*, is given as a function of position and time by

$$u = -\frac{1}{3} P r / \varkappa^*, \quad v = 0. \tag{14}$$

b) *Composite element.* A solution of the following form gives a finite displacement at the origin and satisfies the equations of motion:

$$u = \alpha' r, \qquad\qquad v = 0 \quad \text{in } S'\,(r < a), \tag{15}$$

$$u = \alpha r + \beta r^{-2}, \qquad v = 0 \quad \text{in } S\,(a < r < b), \tag{16}$$

$$u = \alpha^* r + \beta^* r^{-2}, \quad v = 0 \quad \text{in } S^*\,(b < r), \tag{17}$$

where α', α, α^*, β and β^* are functions of the time to be determined. (In each region, e is independent of position.) Continuity of v and $\widehat{r\theta}$ across $r = a$ and across $r = b$ is already ensured by equations (15)–(17), and that of u and $\widehat{rr}$ requires

$$\left.\begin{aligned}
\alpha' = \alpha + a^{-3}\beta, \quad 3\varkappa'\alpha' = 3\varkappa\alpha - 4\mu a^{-3}\beta, \\
\alpha^* + b^{-3}\beta^* = \alpha + b^{-3}\beta, \quad 3\varkappa^*\alpha^* - 4\mu^* b^{-3}\beta^* = 3\varkappa\alpha - 4\mu b^{-3}\beta.
\end{aligned}\right\} \tag{18}$$

The requirement that the composite element shall be macroscopically indistinguishable from the homogeneous element, no matter what is the function $P(t)$, imposes further conditions on the unknown functions of the time, which must be satisfied with the greatest possible accuracy when R is large. The displacement and stress at $r = R$ are the same in the two elements, if

$$\left.\begin{aligned}
\alpha^* + R^{-3}\beta^* = -\frac{1}{3} P / \varkappa^*, \quad 3\varkappa^*\alpha^* - 4\mu^* R^{-3}\beta^* = -P, \\
3\varkappa^*\alpha^* + 2\mu^* R^{-3}\beta^* = -P.
\end{aligned}\right\} \tag{19}$$

There is in this case an *exact* solution of the equations (18)–(19) (in which β^* vanishes), provided that

$$\varkappa^* = \frac{(3\varkappa' + 4\mu)\varkappa + 4\mu(\varkappa' - \varkappa)c}{3\varkappa' + 4\mu - 3(\varkappa' - \varkappa)c}. \tag{20}$$

Since the expression for $\varkappa^*$ does not involve Δ, the solid is ideally elastic under isotropic pressure (to the order of accuracy attained here), with a constant bulk modulus $\varkappa^*$ defined by (20).

4. Determination of μ^*. Although the algebra is somewhat more complicated, the logical sequence of steps which determines the operator μ^* is the same as that in § 3. The homogeneous and composite macroscopic elements are now considered to be subjected to a state of stress (8), i.e., by choice of polar axis,

$$\widehat{rr} = T\,(3\cos^2\theta - 1), \quad \widehat{\theta\theta} = T\,(2 - 3\cos^2\theta), \quad \widehat{\Phi\Phi} = -T, \atop \widehat{\theta\Phi} = 0, \quad \widehat{\Phi r} = 0, \quad \widehat{r\theta} = -3\,T\cos\theta\sin\theta \right\} \tag{21}$$

on $r = R$, for sufficiently large R, where T is an arbitrary function of the time.

a) *Homogeneous element.* The components of displacement defined by

$$u = \frac{1}{2}\,T\,r\,(3\cos^2\theta - 1)/\mu^*, \quad v = -\frac{3}{2}\,T\,r\cos\theta\sin\theta/\mu^* \tag{22}$$

correspond to stresses (21) throughout the material; the dilatation vanishes everywhere in this element.

b) *Composite element.* A form of solution of the equations of motion in which the dilatation is the sum of axially symmetric spherical harmonic functions of degrees 2 and -3 can be constructed (cf. [2]). If, in S, we write

$$e = -\,(\mu/\lambda)\,(A\,r^2 + B\,r^{-3})\left(\frac{3}{2}\cos^2\theta - \frac{1}{2}\right), \tag{23}$$

the associated components of displacement are

$$u = \left[\frac{A\,r^3}{14} + \frac{(3\,\lambda + 5\,\mu)\,B}{12\,\lambda\,r^2} + C\,r - \frac{3\,D}{2\,r^4}\right](3\cos^2\theta - 1), \tag{24}$$

$$v = -\left[\frac{(5\,\lambda + 7\,\mu)\,A\,r^3}{14\,\lambda} + \frac{\mu\,B}{2\,\lambda\,r^2} + 3\,C\,r + \frac{3\,D}{r^4}\right]\cos\theta\sin\theta, \tag{25}$$

where A, B, C and D are functions of the time only, to be determined. The corresponding non-vanishing stresses in S ($a < r < b$) are

$$\widehat{rr} = \mu\left[-\frac{A\,r^2}{14} - \frac{(9\,\lambda + 10\,\mu)\,B}{6\,\lambda\,r^3} + 2\,C + \frac{12\,D}{r^5}\right](3\cos^2\theta - 1), \tag{26}$$

$$\widehat{\theta\theta} = \mu\left[-\frac{5\,A\,r^2}{14} + \frac{5\,\mu\,B}{6\,\lambda\,r^3} + 2\,C - \frac{3\,D}{r^5}\right](3\cos^2\theta - 1) - {} \atop {} - \mu\left[\frac{(5\,\lambda + 7\,\mu)\,A\,r^2}{7\,\lambda} + \frac{\mu\,B}{\lambda\,r^3} + 6\,C + \frac{6\,D}{r^5}\right](2\cos^2\theta - 1), \right\} \tag{27}$$

$$\widehat{\Phi\Phi} = \mu\left[-\frac{5\,A\,r^2}{14} + \frac{5\,\mu\,B}{6\,\lambda\,r^3} + 2\,C - \frac{3\,D}{r^5}\right](3\cos^2\theta - 1) - {} \atop {} - \mu\left[\frac{(5\,\lambda + 7\,\mu)\,A\,r^2}{7\,\lambda} + \frac{\mu\,B}{\lambda\,r^3} + 6\,C + \frac{6\,D}{r^5}\right]\cos^2\theta, \right\} \tag{28}$$

$$\widehat{r\theta} = -\mu\left[\frac{(8\,\lambda + 7\,\mu)\,A\,r^2}{7\,\lambda} + \frac{(3\,\lambda + 2\,\mu)\,B}{2\,\lambda\,r^3} + 6\,C - \frac{24\,D}{r^5}\right]\cos\theta\sin\theta. \tag{29}$$

Exactly similar expressions for the displacements and stresses, in which the symbols A, B, C, D, λ and μ are distinguished by primes or asterisks, are valid in $S'(r < a)$ and $S^*(r > b)$. There are twelve unknown functions of the time A, A', A^*, B, B', B^*, C, C', C^*, D, D' and D^*, to be determined in terms of $T(t)$ from boundary and continuity conditions; and [since $\lambda^* + \frac{2}{3}\mu^*$ is known from (20)] there is in effect one unknown operator to be determined by the condition that a solution is possible.

Some simplications in the continuity conditions can be made by considering the origin and external conditions first. The displacement at the origin is finite only if

$$B' = 0, \quad D' = 0. \tag{30}$$

The components of stress and displacement at $r = R$ can be made to agree with (21) and (22), with a proportional error of the order of $D^*/(C^* R^5)$ — which can be made as small as we please by taking R sufficiently large — provided that

$$A^* = 0, \quad B^* = 0, \quad C^* = \frac{1}{2} T/\mu^*. \tag{31}$$

The equations expressing continuity of u, v, $\widehat{rr}$ and $\widehat{r\theta}$ across $r = a$ and $r = b$ can then be written as

$$\frac{a^2 A}{7} + \frac{(3\lambda + 5\mu)B}{6\lambda a^3} + 2C - \frac{3D}{a^5} = \frac{a^2 A'}{7} + 2C', \tag{32}$$

$$\frac{(5\lambda + 7\mu)a^2 A}{14\lambda} + \frac{\mu B}{2\lambda a^3} + 3C + \frac{3D}{a^5} = \frac{(5\lambda' + 7\mu')a^2 A'}{14\lambda'} + 3C', \tag{33}$$

$$\mu\left[-\frac{a^2 A}{7} - \frac{(9\lambda + 10\mu)B}{3\lambda a^3} + 4C + \frac{24D}{a^5}\right] = \mu'\left[-\frac{a^2 A'}{7} + 4C'\right], \tag{34}$$

$$\mu\left[\frac{(8\lambda + 7\mu)a^2 A}{7\lambda} + \frac{(3\lambda + 2\mu)B}{2\lambda a^3} + 6C - \frac{24D}{a^5}\right] = \mu'\left[\frac{(8\lambda' + 7\mu')a^2 A'}{7\lambda'} + 6C'\right], \tag{35}$$

$$\frac{b^2 A}{7} + \frac{(3\lambda + 5\mu)B}{6\lambda b^3} + 2C - \frac{3D}{b^5} = 2C^* - \frac{3D^*}{b^5}, \tag{36}$$

$$\frac{(5\lambda + 7\mu)b^2 A}{14\lambda} + \frac{\mu B}{2\lambda b^3} + 3C + \frac{3D}{b^5} = 3C^* + \frac{3D^*}{b^5}, \tag{37}$$

$$\mu\left[-\frac{b^2 A}{7} - \frac{(9\lambda + 10\mu)B}{3\lambda b^3} + 4C + \frac{24D}{b^5}\right] = \mu^*\left[4C^* + \frac{24D^*}{b^5}\right], \tag{38}$$

$$\mu\left[\frac{(8\lambda + 7\mu)b^2 A}{7\lambda} + \frac{(3\lambda + 2\mu)B}{2\lambda b^3} + 6C - \frac{24D}{b^5}\right] = \mu^*\left[6C^* - \frac{24D^*}{b^5}\right]. \tag{39}$$

It is of interest that λ^* is absent from these equations, and the solution for μ^* is therefore obtained without making use of the result (20).

A set of equations which is formally the same as (32)—(39) with $\varkappa$ and $\varkappa'$ put infinitely large was solved in I, and the same sequence of steps now determines the operator μ^* that permits a general solution for the eight remaining unknown functions of the time: it is found that

$$\mu^* = \mu \frac{(9\varkappa + 8\mu)\mu + 6(\varkappa + 2\mu)\mu' - (9\varkappa + 8\mu)(\mu - \mu')c}{(9\varkappa + 8\mu)\mu + 6(\varkappa + 2\mu)\mu' + 6(\varkappa + 2\mu)(\mu - \mu')c}. \tag{40}$$

Since $\mu' = \eta \Delta$, we have

$$\mu^* = \mu_0 \frac{1 + \lambda_2 \Delta}{1 + \lambda_1 \Delta}, \tag{41}$$

where the constants μ_0, λ_1 and λ_2 are given by

$$\mu_0 = \mu \left[1 - \frac{5(3\varkappa + 4\mu)c}{9\varkappa + 8\mu} \right], \tag{42}$$

$$\lambda_1 = \frac{6(\varkappa + 2\mu)\eta}{(9\varkappa + 8\mu)\mu} \left[1 - \frac{5(3\varkappa + 4\mu)c}{9\varkappa + 8\mu} \right], \tag{43}$$

$$\lambda_2 = \frac{6(\varkappa + 2\mu)\eta}{(9\varkappa + 8\mu)\mu} \left[1 + \frac{5(3\varkappa + 4\mu)c}{6(\varkappa + 2\mu)} \right]. \tag{44}$$

5. Discussion. The significance of the results (20) and (41) is most clearly seen if we consider separately the two terms of the equations of state (5). The stress p_{ik} is the sum of two parts:

$$p_{ik} = p'_{ik} + p'' \delta_{ik}, \tag{45}$$

where p'_{ik} is the partial stress associated with distortion without change of volume, defined by the deviatoric strain tensor $e'_{ik} = e_{ik} - \frac{1}{3}e\,\delta_{ik}$, and $p''\delta_{ik}$ is the isotropic stress related directly to volume changes. The relationships between the partial stresses and partial strains can be written as

$$\left(1 + \lambda_1 \frac{d}{dt} \right) p'_{ik} = 2\mu_0 \left(1 + \lambda_2 \frac{d}{dt} \right) e'_{ik} \tag{46}$$

and

$$p'' = \varkappa^* e, \tag{47}$$

where μ_0, λ_1, λ_2 and $\varkappa^*$ are constants defined by (42)—(44) and (20). Equation (46) implies, for example, that a varying simple shear stress τ is related to the angle of shear γ which it produces by the differential equation

$$\left(1 + \lambda_1 \frac{d}{dt} \right) \tau = \mu_0 \left(1 + \lambda_2 \frac{d}{dt} \right) \gamma \tag{48}$$

in the solid with local imperfections, to be compared with the algebraic relation $\tau = \mu\gamma$ in the perfectly elastic solid S.

All the constants in the stress-strain relations are seen to be independent of the actual size of the viscous inclusions, and depend only on the overall volume concentration. Their physical significance can be summarized as follows: $\varkappa^*$ is the bulk modulus, μ_0 the shear modulus for very slowly applied stresses, λ_1 is a relaxation time in the sense that shear stresses will decay as $\exp\left(-t/\lambda_1\right)$ when the strain is reduced to zero, and λ_2 is a retardation time in the sense that shear strains decay as $\exp\left(-t/\lambda_2\right)$ when stresses are removed. It is observed that $\lambda_1 < \lambda_2$ (in contrast to what has been found in liquid disperse systems [1, 2], where viscosity and rate of strain replace rigidity and strain in the equation corresponding to (46), and λ_1 exceeds λ_2). The inequality in this sense is not unexpected if we note that an equation of the type (46), if valid for all continuous changes of stress, implies

$$\lambda_1 p'_{ik} - 2\lambda_2\mu_0 e'_{ik} = \text{continuous} \qquad (49)$$

in the limit when the change of stress is a sudden jump [1]. Hence $\lambda_2\mu_0/\lambda_1$ is the effective modulus for very rapid stressing, and we should expect this to exceed the modulus for slow stressing when there is some internal friction present.

In conclusion it is observed that the formulae (20) and (40) are of much wider application than to the particular case discussed in detail in this paper. The change in properties of any uniform visco-elastic solid, due to the presence of small inclusions of another material, whether solid or liquid, with different elastic and viscous properties, is readily deducible from these formulae. The whole of the above analysis remains valid if the continuous component S and the disperse component S' have each viscous and elastic properties combined. The equations of state of S and S' for small strains can still be written in the form (3) provided the symbols $\mu, \varkappa, \mu'$ and $\varkappa'$ denote the appropriate *operators* of the form (6); and the properties of the composite solid are obtained by substituting these operators in (20) and (40). As an example, if S is perfectly elastic as before but S' is now taken to be a MAXWELLian elastico-viscous liquid, so that

$$\mu' = \frac{\eta\,\Delta}{(1 + \lambda_0\,\Delta)}, \qquad (50)$$

where η and λ_0 are constants, it is seen from (40) that μ^* can still be written in the form (41), with the same value of μ_0 and values of λ_1 and λ_2 each increased by λ_0.

The results obtained by MACKENZIE [4] for the elastic constants of a solid containing spherical holes follow from the above formulae (20) and (40) in the special case when μ' and $\varkappa'$ both vanish. Similar results obtained by HASHIN [5] for the moduli of an elastic solid containing rigid spherical particles are also included as a limiting case in which μ' and $\varkappa'$ are infinitely large constants. From the formula (20), applied to the particular case of a liquid system in which $\varkappa$ and $\varkappa'$ are constants, with $\varkappa$ infinitely large, and $\mu = \eta\,\Delta$, we obtain the approximate expression

$$\varkappa^* = \frac{\varkappa'}{c} + \frac{4\eta}{3c}\,\Delta,$$

in agreement with SIR GEOFFREY TAYLOR's result for the volume viscosity $\left(\dfrac{4\,\eta}{3\,c}\right)$ of an incompressible liquid containing air bubbles [6]. In this application, the interfacial tension will influence the effective value of $\varkappa'$.

References

[1] FRÖHLICH, H. and R. SACK: Proc. Roy. Soc., London A 185, 415 (1946).
[2] OLDROYD, J. G.: Proc. Roy. Soc., London A 218, 122 (1953) (referred to as I).
[3] LOVE, A. E. H.: A treatise on the Mathematical Theory of Elasticity, Chap. XI, 4th ed. Cambridge 1927.
[4] MACKENZIE, J. K.: Proc. Phys. Soc., London B 63, 2 (1950).
[5] HASHIN, Z. Bull. Res. Council Israel 5 C, 46 (1955).
[6] TAYLOR, SIR GEOFFREY: Proc. Roy. Soc., London A 226, 34 (1954).

Dislocation Relaxations in Metals and Single Crystal Quartz

By **W. P. Mason**, Murray Hill (N. J.)

With 8 figures

1. Introduction. During the last several years, a number of ultrasonic attenuation measurements have been made in face centered metals which show attenuation maxima at temperatures related to the metal and the measuring frequencies. The first of these measurements were made by BORDONI [1], and other measurements were made by BÖMMEL [2] and WELBER [2]. For lead at a frequency of 10.1 Kc, BORDONI's measurements are shown by fig. 1. All measurements reported are for strains of 10^{-8} or less. The data of fig. 1, which is typical of other metals, show an attenuation peak followed by an attenuation which increases exponentially with the temperature. The activation energy for lead is 4350 calories/mole. Fig. 2 shows the details of the low temperature peaks found by BORDONI for four face centered metals. Mild cold work increases the height of the peaks.

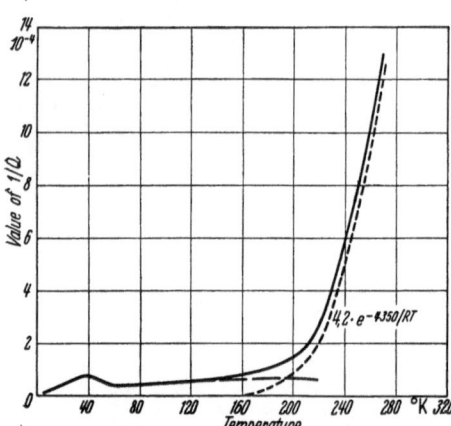

Fig. 1. Internal friction Q^{-1} of polycrystalline lead measured at 10.1 Kc (after BORDONI)

2. Experimental evidence for relaxations in face centered metals. To prove that the low temperature peaks are relaxations one must measure the attenuation over a wide frequency range. Dr. H. BÖMMEL [2] at Bell Telephone Laboratories has made measurements in the megacycle range, and these measurements taken together with BORDONI's [1] and WELBER's [2] show that the relaxation frequencies define a relaxation curve with an activation energy of 1000 calories per mole for lead. The low temperature experimental arrangement has been described previously [3], and fig. 3 shows measurements of 10.1 megacycle shear waves and 26.5 Mc longitudinal waves

along the [100] direction for path lengths shown on the figure. The
limiting attenuation shown by the dashed lines is mainly due to misorien-

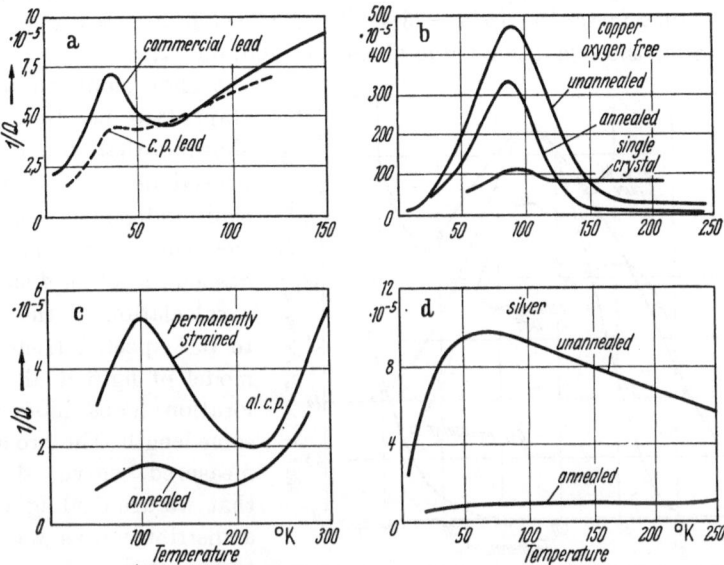

Fig. 2. Internal friction peaks at low temperatures for face centered metals. Frequencies of measurement from 10.1 to 40 Kc (after BORDONI)

tation caused by the softness of the lead. This loss can be evaluated by
standard water bath techniques and the dashed lines show the evalua-

tion. It appears that the loss
in the superconducting state
at the lowest temperature
measured — 1.5° K — is nearly
zero. If one keeps the crystal
normal conducting by applying a magnetic field, a high
loss occurs at low temperatures due to the viscous electron
lattice interaction as discussed
previously [4]. This loss disappears when the electrons
are in the superconducting
state showing that the lattice
cannot transfer momentun to
the electrons.

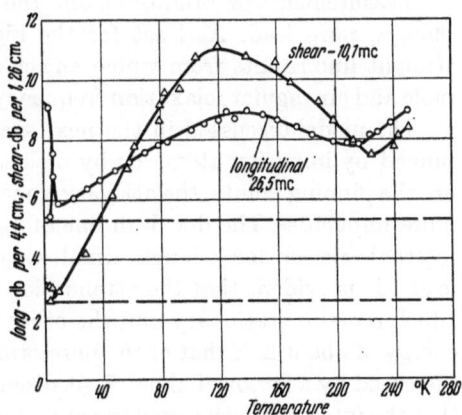

Fig. 3. Ultrasonic attenuation for shear and longitudinal
waves for 99.99% pure lead (after BÖMMEL)

From the attenuation and velocity measurements [5], the Q^{-1} factor
(the decrement $\delta = \pi Q^{-1}$) can be determined. Fig. 4 shows this value for

26.5 Mc longitudinal waves plotted against the temperature. The exponentially rising term, shown by the dashed line, occurs again with the same activation energy as for fig. 1. If one takes account of the proportionality factor [5] $\Delta\mu/\mu = (3/4)[(\lambda + 2\mu)/\mu] (\Delta c_{11}/c_{11}) = 5.2 \, \Delta c_{11}/c_{11}$ to transfer the relaxation value along the [100] direction to that along

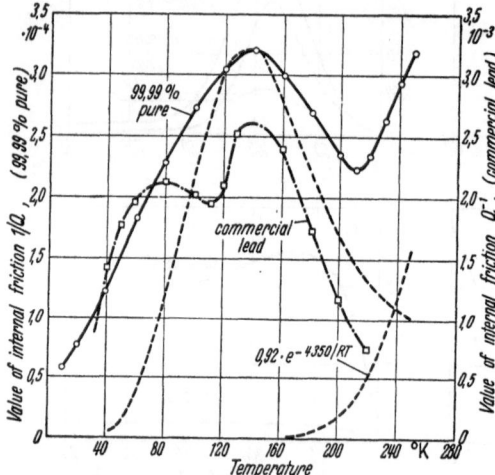

the glide plane, the constants are nearly equal showing that the exponential term is substantially independent of the frequency. The second dashed line of fig. 4 shows a calculation of the loss to be expected from the model of fig. 6 if all dislocation loops have the same length. The broader measured curve shows that we are dealing with a distribution as was to be expected.

Fig. 4. Internal friction for single crystal lead. Solid line for 99.99% pure lead for longitudinal waves measured at 26.5 Mc. Dot-dash line for commercial lead for longitudinal waves of 21.5 Mc. Unlabelled dash line is theoretical calculation of relaxation for single loop length. Dashed line labelled $0.92 \cdot e^{-4350/RT}$ is exponentially increasing loss

If we plot the log of the measured relaxation frequencies against $1/T$, fig. 5 shows the results of all measurements of BORDONI's and the other results which were all for 99.99% pure lead. At least for the higher frequency points, a good straight line results from which an activation energy of 975 calories per mole and an angular relaxation frequency constant of $5 \cdot 10^9$ are obtained.

The model discussed in the next section considers dislocation loops pinned by impurity atoms or by dislocation nodes. If impurity atoms are the pinning points, the activation energy should be lowered if we have more impurities. The dot-dash line of fig. 4 shows a measurement of the internal friction for a longitudinal wave of 21.5 Mc for a "commercial lead". It is evident that the attenuation is broken into two parts, one of which remains stationary and the other of which has a lower activation energy of about half that of the pure sample. The internal friction is also increased by a factor of about 8. As discussed below, this result indicates that the internal friction consists of two parts, one of which is determined by the impurity content and the other determined by the dislocation node structure which does not vary with impurity content.

3. Theoretical dislocation model for low temperature relaxation. The model considered for this effect is shown by fig. 6. It consists of edge dislocations bound at such points as A and B either by impurity atoms or by dislocation nodes. The experimental evidence presented above indicates that both are present. At low temperatures, the edge dislocations are assumed to be along minimum energy positions. Thermal agitation causes the dislocation loops to vibrate about their position of minimum potential energy, but due to the PEIERL's restoring force, there is not enough thermal energy to surmount the energy barrier. Thermal energy will impart an energy equal to kT to all the normal modes of the dislocation considered as a vibrating string. Ocassionally, these will combine to cause small segments of the dislocation loop to cross the energy barrier. However, the potential well model for this effect will be one for which the height H of the potential well model of fig. 6 c is small and scarcely larger than the bottom of the second well whose value is A. Under these conditions, the

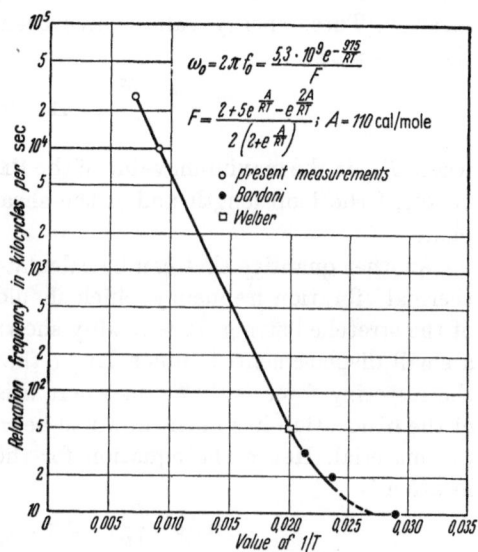

$$\omega_0 = 2\pi f_0 = \frac{5.3 \cdot 10^9 e^{-\frac{975}{RT}}}{F}$$

$$F = \frac{2 + 5e^{\frac{A}{RT}} - e^{\frac{2A}{RT}}}{2\left(2 + e^{\frac{A}{RT}}\right)}; \; A = 110 \, cal/mole$$

○ *present measurements*
● *Bordoni*
□ *Welber*

Fig. 5. Plot of logarithm of the relaxation frequency versus $1/T$. Open circle and square are for 99.99% pure lead. Solid points after BORDONI

mean life of the dislocation in such a well will be very small and hence a short loop will have little effect on the internal friction. The most pro-

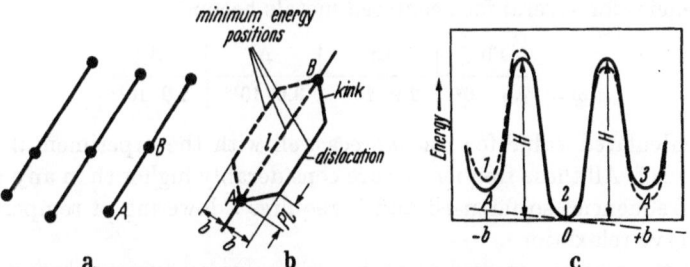

Fig. 6. Proposed dislocation model for relaxation effect. a) dislocations pinned by impurity atoms or dislocation modes, b) shape of dislocation displacements, c) equivalent potential well model

bable type of loop will be one which coincides in shape with the first normal mode of the vibration and hence the loop shape assumed is shown by fig. 6.1 b. This form is assumed since it represents the form

most nearly consistent with the first mode of motion of the loop that
has a minimum potential energy in the first displaced position.

Taking the model shown by fig. 6b, the height of the barrier H of
fig. 6c can be evaluated as the sum of the energy required for the dislocation and its two kinks to surmount the energy barrier determined by
the PEIERL's restoring stress and the energy required to stretch the dislocation. These energy values have been calculated previously [5] and
the values are

$$H = \frac{T_{13_0} b^2 \mathfrak{l}}{\pi}, \quad A = 2 b^3 \sqrt{\frac{T_{13_0} \mu}{\pi}} \tag{1}$$

where T_{13_0} is the maximum value of the PEIERLS stress, b is the BURGER's
vector, $\mathfrak{l}$ the loop length and μ the shearing constant along the glide
plane.

Another quantity that can be calculated from this same model is the
thermal vibration frequency which should correspond to the first mode
of the stretched string. It is readily shown that the PEIERL's stress, for
a small displacement, is much larger than the stretching force so that
the restoring force is $2\pi T_{13_0} \mathfrak{l} d$ where d is the displacement. The mass
of the dislocation has been shown to be $\pi \varrho b^2 \mathfrak{l}$ where ϱ is the density of
the material. Hence the equation for the angular frequency of a dis-
location is

$$\pi \varrho b^2 \mathfrak{l} \frac{d^2 d}{d t^2} + 2 \pi T_{13_0} \mathfrak{l} d = 0. \tag{2}$$

Solving for the angular frequency $\omega = 2\pi f$, we find

$$\omega = \frac{1}{b} \sqrt{\frac{2 T_{13_0}}{\varrho}}. \tag{3}$$

For ratios of $T_{13_0}/\mu = 6 \cdot 10^{-6}$ as will be found presently, the angular
frequencies for several face centered metals become

Pb	Cu	Al	Ag
$\omega = \quad 7.8 \cdot 10^9$	$2.9 \cdot 10^{10}$	$3.6 \cdot 10^{10}$	$1.9 \cdot 10^{10}$.	$\tag{4}$

This calculated value for lead agrees well with the experimental value
of $5.3 \cdot 10^9$. All these frequencies are considerably higher than any ultrasonic frequencies so far used and it requires a lowering of temperature
to observe relaxations.

All the stresses applied in these measurements are much too small
to cause the loops to cross the energy barrier H. The stress can, however,
bias the potential wells as shown by fig. 6c and cause more dislocation
loops to be in position 1 than in position 3 thus causing a plastic strain.
When the sign of the stress reverses, position (3) will have more dislocations than position (1), but the displacement lags the stress causing

a hysteresis loop and hence an energy loss. This loss will be a maximum when the relaxation time for the dislocation jump equals the frequency of the applied stress. By a straightforward application of rate theory, the loss can be calculated and it has been shown to equal [5]

$$\frac{1}{Q} = \frac{2\,e^{-A/RT}}{1+2\,e^{-A/RT}} \left[\frac{N_0\,l^2\,(1-p)^2\,b^4\,\mu^E}{k\,T} \right] \frac{\omega/\omega_0}{1+(\omega/\omega_0)^2} \quad \text{where} \quad \omega_0 = \gamma_0\,e^{-(H-A)/RT}. \quad (5)$$

In this equation N_0 is the number of dislocation loops, l the loop length assumed constant for this calculation, μ^E is the elastic shear constant along the glide plane, k is BOLTZMANN's constant, T the absolute temperature, and γ_0 the angular frequency given by equation (4).

From the measured Q^{-1} curve for 99.99% lead of fig. 4 some quantitative results can be obtained. Equation (5) is for a single loop length and as can be seen the measured value is broader than the single loop value showing that we are dealing with a distribution of loop lengths. To take account of the broadening, we double the height of the maximum in order that the area of the theoretical curve shall equal the value of the measured curve. One also multiplies by 5.2 in order that the measured value shall correspond to the energy loss along the glide plane. Hence

$$3.2 \cdot 10^{-4} \cdot 2 \cdot 5.2 = \frac{e^{-A/RT}}{1+2\,e^{-A/RT}} \left[\frac{N_0\,l^2\,(1-p)^2\,b^4\,\mu^E}{k\,T} \right]. \quad (6)$$

The first factor is 2/3 since A is small compared to RT at 140° K. $p \doteq .05$, $b = 3.5 \cdot 10^{-8}$ for lead, $\mu^E = 7 \cdot 10^{10}$ dynes/cm^2, $k = 1.38 \cdot 10^{-16}$, $T = 140°$ K and hence we have

$$N_0\,l^2 \doteq 1.3 \cdot 10^3. \quad (7)$$

The exponentially rising loss gives a value of $N_0 l = \bar{N}$ the number of dislocations per sq. cm. of $3 \cdot 10^6$. Hence the average loop length is about $4 \cdot 10^{-4}$ cm. This value together with the measured activation energy of 1000 cal/mole gives a method for calculating the ratio of the PEIERLS stress to the shear modulus. From equations (1) we have

$$H - A = \left[\frac{T_{13_0}}{\mu} \left(\frac{\mu\,b^2\,l}{\pi} \right) - \frac{2\,b^2\,\mu}{\sqrt{\pi}} \sqrt{\frac{T_{13_0}}{\mu}} \right] \frac{6.025 \cdot 10^{23}}{4.182 \cdot 10^7} = 1000. \quad (8)$$

Employing the values already determined, we find $T_{13_0}/\mu = 6.8 \cdot 10^{-6}$.

The measurements for the "commercial lead" show that the average loop length for the impurity component is about half this value or $2 \cdot 10^{-4}$ cm. The height of this peak shows that the number of impurity pinned loops is about 30 times as large. In addition there is a skelital structure of node pinned loops of length about $4 \cdot 10^{-4}$ cm. The loops are about 8 times as numerous as for the pure lead. BORDONI's peak mesurements shown by fig. 2 show that ratios of PEIERLS stress to μ run from 4.8 to $7 \cdot 10^{-6}$ giving a round figure of $6 \cdot 10^{-6}$.

4. Internal friction increasing exponentially with the temperature. The exponentially increasing loss shown by figs. 1 and 4 appears to be independent of the frequency. This same type of loss is always found for metals but the temperature is usually higher than for lead. The activation energies are of the right order to agree with the binding energies of impurity atoms as calculated from COTTRELL's formula [5].

$$U = \frac{1}{3}\left(\frac{1+\sigma}{1-\sigma}\right)\mu\, b^3\, \varepsilon\,, \qquad (9)$$

where σ is POISSON's ratio and ε is the difference between the impurity radius and the metal radius divided by the metal radius. Hence it seems probable that the loss is connected with the breakaway or pinning by impurity atoms of dislocation loops.

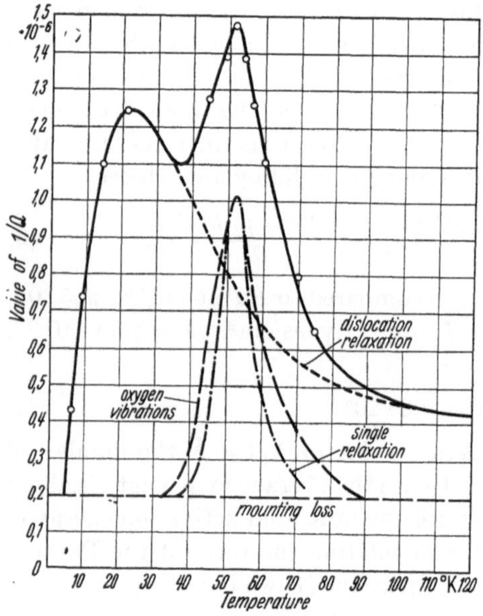

Fig. 7. Internal friction for single quartz crystal at 5 Mc. Loss divided into dislocation relaxation, oxygen vibrations, and mounting loss

Such a process can cause a loss to the ultrasonic vibration since the vibration can add energy to the thermal vibration equal to on the average

$$\frac{1}{2}\,\dot{u}^2\, m\,, \qquad (10)$$

where $m = \pi \varrho\, b^2 \mathfrak{l}$ is the effective mass of a dislocation loop and $\dot u$ is the particle velocity. If the loop becomes unpinned or pinned by another atom during the vibration, the amount of energy lost to the vibration will be $^1/_4\, m\dot u_0^2$ per loop where $\dot u_0$ is the maximum velocity of the loop.

The frequency with which the dislocation attempts to pull away or approach a repinning atom is twice the frequency $f = \omega/2\pi$ of equation (3). The probability of success is $e^{-U/RT}$. We have to multiply these factors by the number of dislocation loops in the volume under consideration which for a time $dt = V_s dt$, where V_s is the shear wave velocity. The result of this multiplication is

$$\Delta E = \frac{\dot u_0^2 \sqrt{\varrho\,\mu}\sqrt{\dfrac{2\,T_{13_0}}{\mu}}\; N_0\, \mathfrak{l} b V_s\, e^{-U/RT}}{4}\; dt\,. \qquad (11)$$

The input energy is $\frac{1}{2}\sqrt{\varrho\mu}\,\dot{u}_0^2$, dence the e nergy loss in time dt can be regarded as the first term of the expansion

$$E = \frac{\dot{u}_0^2}{2}\sqrt{\varrho\mu}\,e^{-\delta t} = \frac{\dot{u}_0^2}{2}\sqrt{\varrho\mu}\left(1 - \frac{\sqrt{\frac{2\,T_{13_0}}{\mu}}\,V_s N_0\,l\,b\,e^{-U/RT}}{2}\,dt\right). \qquad (12)$$

Since $Q^{-1} = \delta/\pi$, we have finally that the internal dissipation introduced by this source of loss is

$$Q^{-1} = \sqrt{\frac{2\,T_{13_0}}{\mu}}\,V_s N_0\,l\,b\,e^{-U/RT}. \qquad (13)$$

This loss is independent of the frequency and from the measured results of figs. 1 and 4 gives a value of $N_0 l = \bar{N}$, the number of disloca- tions per sq. cm.

$$N_0 l = \bar{N} = 3 \cdot 10^6. \qquad (14)$$

5. Dislocation relaxations in crystal quartz. A similar relaxa- tion in crystal quartz has recently been observed [6]. This work was started since previous results for well mounted quartz had shown that the internal friction was pro- portional to the frequency up to 100 Mc. Fig. 7 shows a measure- ment of the internal friction of a 5 Mc AT cut crystal as a function of the temperature. There are a broad relaxation peak labelled dislocation relaxation, a nearly single relaxation peak labelled oxygen vibrations, and a cons- tant loss labelled mounting loss which is the total loss below 6° K.

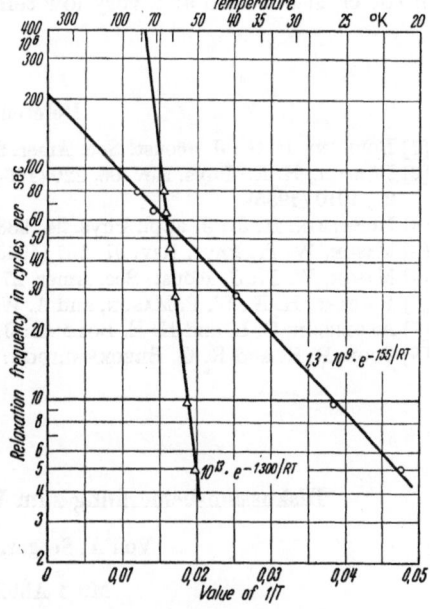

Fig. 8. Plot of logarithm of two relaxation frequen- cies versus $1/T$. Curve with triangles due to oxyge vibrations. Curve with circles due to dislocation relaxations

Further contouring of crystals has reduced the mounting loss to $5 \cdot 10^{-8}$.

The internal friction was measured from 5 to 80 Mc and fig. 8 shows a plot of the two relaxation frequencies against $1/T$. From the constants of the sharp relaxation, it is of atomic nature and in fact is similar to one measured previously in fused silica [7]. It appears that the sharp peak with an activation energy of 1300 calories per mole is due to oxygen vacancies in the crystal quartz which distorts the lattice structure and allows the sidewise vibrations of the oxygen atoms that are responsible for the internal friction of fused silica.

An exponentially increasing internal friction has been observed for crystal quartz by Cook and Breckenridge [8] and from their values and the measurements given here, one can evaluate the parameters as

$$N_0 l = \bar{N} \doteq 10^8, \quad \bar{l} \doteq 2 \cdot 10^{-3} \, cm, \quad \frac{T_{13_0}}{\mu} \doteq 5 \cdot 10^{-8}. \tag{15}$$

The low value of Peierls stress to shearing modulus indicates that the dislocations in quartz are wide dislocations.

The small residual aging of the elastic properties of quartz, noted after processing the crystal, is probably due to the stabilization of dislocation positions as a function of time. This aging should be eliminated if the crystal is held at a very low temperature.

References

[1] BORDONI, P. G.: J. acoust. Soc. Amer. 26, 495 (1954).
[2] BÖMMEL, H. E.: Phys. Rev. 96, 220 (1954). — B. WELBER: J. acoust. Soc. Amer. 27, 1010 (1955).
[3] McSKIMIN, H. J.: J. appl. Phys. 24, 988 (1953).
[4] MASON, W. P.: Phys. Rev. 97, 557 (1955).
[5] MASON, W. P.: J. acoust. Soc. Amer. 27, 643 (1955).
[6] BÖMMEL, H. E., W. P. MASON, and A. W. WARNER: Phys. Rev. 99, 1894 (1955).
[7] ANDERSON, O. L. and H. E. BÖMMEL: J. Amer. ceram. Soc. 38 (1955).
[8] COOK, R. K. and R. G. BRECKENRIDGE: Phys. Rev. 92, 1419 (1951); see fig. 2.

Diskussionsbemerkung zum Vortrag von W. P. Mason

Von A. Seeger, Stuttgart

Mit 1 Abbildung

Die folgenden Bemerkungen beziehen sich auf das zuerst von BORDONI [1] beobachtete Dämpfungsmaximum in Metallen bei tiefen Temperaturen. Hinsichtlich der Deutung dieses Maximums stimme ich mit Dr. MASON darin überein, daß es sich dabei um eine mit der PEIERLS-Spannung von Versetzungen verknüpfte Erscheinung handelt. Meine eigene Deutung [2] weicht jedoch von derjenigen von Dr. MASON in wesentlichen Einzelheiten ab. Da sowohl theoretische Überlegungen als auch experimentelle Ergebnisse gegen die Vorstellungen von Dr. MASON sprechen, seien die beiden Auffassungen einander kritisch gegenübergestellt.

1. In der MASONschen Theorie hängt die Aktivierungsenergie des Relaxationsvorganges linear vom Abstand l zwischen den Verunreinigungsatomen ab, welche die Versetzungen festhalten. Es sollten deswegen große Schwankungen der Aktivierungsenergie von Probe zu Probe und eine starke Abhängigkeit vom Verfor-

mungsgrad auftreten. Beides ist nicht der Fall, wie aus den von Dr. MASON zitierten Messungen an Blei sowie aus den ausführlichen Messungen von NIBLETT und WILKS [3] an Kupfer hervorgeht. Bei den von Dr. MASON in Abb. 4 wiedergegebenen Messungen an Blei kommerzieller Reinheit tritt zwar ein Nebenmaximum auf, das (im Gegensatz zum Hauptmaximum) die von Dr. MASON vertretene Ursache haben könnte. Es erscheint jedoch in Anbetracht der Ergebnisse an Kupfer [3], wo bei allen Verunreinigungsgraden und Vorverformungen stets ein Nebenmaximum mit etwa der halben Aktivierungsenergie des Hauptmaximums gefunden wurde, sehr nahe-liegend, beide Maxima in gleicher Weise zu deuten. Die unter 3. zu gebende Erklärung verlangt die Existenz zweier Maxima. In diesem Fall muß das Fehlen des Nebenmaximums bei reinem Blei besonders erklärt werden.

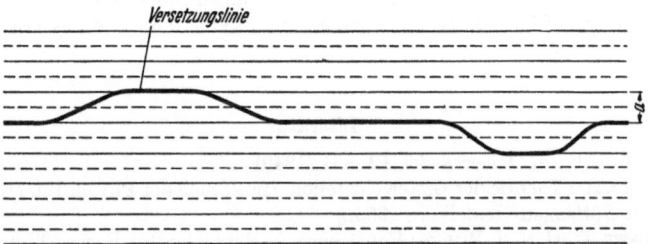

Abb. 1. Wellen in einer Versetzungslinie. Die ausgezogenen Geraden geben die Versetzungslagen mini-maler potentieller Energie, die gestrichelten Geraden diejenigen maximaler potentieller Energie an.

2. Von Dr. MASON wird die Annahme gemacht, daß eine meßbare Anzahl von Versetzungen als gerade Linien in den Potentialmulden der PEIERLS-Spannung liegt. Es läßt sich schwer beurteilen, ob dies am absoluten Nullpunkt der Fall ist oder nicht; man kann jedoch sicher sagen, daß bei der zur Messung verwendeten Tem-peratur die Versetzungen in starkem Maße Wellen von der in der Abbildung dar-gestellten Art aufweisen (von SHOCKLEY [4] „kinks" genannt). Dies rührt davon her, daß eine Versetzungslinie flexibel ist und bei endlichen Temperaturen durch die Bildung von Wellen ihre freie Energie vermindert. Die Existenz von Wellen macht aber den von Dr. MASON diskutierten Prozeß unmöglich.

3. Die Existenz von Wellen im thermischen Gleichgewicht ist die Grundlage des in [2] vorgeschlagenen Prozesses. Zu ihrer Bildung muß eine Aktivierungsenergie U aufgebracht werden. Die vermehrte Bildung von Wellen unter der Wirkung von Wechselschubspannungen führt zu einem Relaxationsprozeß mit der Aktivierungs-energie U. Diese kann mit Hilfe eines in anderem Zusammenhang ausführlich dis-kutierten Modells [5] berechnet werden und ergibt sich im einfachsten Falle eines sinusförmigen Verlaufs der potentiellen Energie einer Versetzung als Funktion ihres Ortes näherungsweise zu

$$U = \frac{4a}{\pi} \sqrt{\frac{2 E_0 a b \tau_p^0}{\pi}}. \tag{1}$$

Hierbei ist a die Periode des Potentials, b die Versetzungsstärke, E_0 die Linien-energie der Versetzung und τ_p^0 die PEIERLS-Spannung am absoluten Nullpunkt (bei Vernachlässigung quantenmechanischer Effekte).

In diesem Modell ist angenommen, daß die Versetzungslinie im Mittel näherungs-weise parallel zu einer dichtest gepackten Richtung verläuft. Da sie dies auf zwei Arten tun kann, nämlich als Schraubenversetzung oder mit einem Winkel von 60° zwischen BURGERS-Vektor und Versetzungslinie, erhält man zwei verschiedene U-Werte, den beiden unter **1.** erwähnten Maxima entsprechend.

Für quantitative Zwecke ist Gleichung (1) wohl noch zu ungenau, da man ziemlich sicher noch ein zweites Sinusglied in der Fourierentwicklung der potentiellen Energie berücksichtigen muß. Die Formel für U wird dann aber zu umfangreich, um hier wiedergegeben zu werden. Eine ausführliche Darstellung der Rechnungen und ein Vergleich mit den Experimenten wird an anderer Stelle erfolgen [6]. Es sei jedoch schon hier erwähnt, daß eine weitere Ausdehnung und Verfeinerung der experimentellen Ergebnisse sehr erwünscht ist, da gegenüber der einfachen, hier skizzierten Theorie zwei zusätzliche Einflüsse auftreten können:

a) Bei hohen Frequenzen und entsprechend niedrigen Temperaturen können sich quantenmechanische Effekte bemerkbar machen [7].

b) Da die Versetzungen im vorliegenden Fall in Halbversetzungen aufgespalten sind, kann selbst eine geringe Temperaturvariation der Stapelfehlerenergie die Aktivierungsenergie U beträchtlich ändern und damit eine falsche Aktivierungsenergie vortäuschen.

Literatur

[1] Bordoni, P. G.: Ricerca Sci. 19, 851 (1949).
[2] Seeger, A.: Theorie der Gitterfehlstellen, Handbuch der Physik Bd. 7/I, S. 602 ff. Berlin/Göttingen/Heidelberg 1955.
[3] Niblett, D. H. u. J. Wilks: Conférence de Physique des Basses Temperatures, Paris, September 1955, S. 484 ff.
 — Phil. Mag., demnächst.
[4] Shockley, W.: Trans. Amer. Inst. Mining metallurg. Engr. 194, 829 (1952).
[5] Frank, F. C. u. J. H. van der Merwe: Proc. roy. Soc., Lond., Ser. A 198, 205, 217 (1949); 200, 125 (1950); 201, 261 (1950).— A. Kochendörfer u. A. Seeger: Z. Phys. 127, 533 (1950). — A. Seeger u. A. Kochendörfer: Z. Phys. 130, 321 (1951).— A. Seeger, H. Donth u. A. Kochendörfer: Z. Phys. 134, 173 (1953).— A. Seeger: Z. Naturforsch. 8a, 246 (1953).
[6] Seeger, A.: Phil. Mag., demnächst.
[7] Siehe den Beitrag von G. Lelbfried in diesem Band.

Berichtigung — Erratum

Auf Seite V, Zeile 11 (Vorwort) ist der Name G. MILLÁN hinzuzufügen.
Please add on page VII line 10 (Preface) the name G. MILLÁN.

Grammel, Madrid-Kolloquium